This book compares resistance to technology across time, nations and technologies. Three post-war technologies – nuclear power, information technology and biotechnology – are used in the analysis. The focus is on post-1945 Europe, with comparisons made with the USA, Japan and Australia. Instead of assuming that resistance contributes to the failure of a technology, the main thesis of this book is that resistance is a constructive force in technological development, giving technology its particular shape in a particular context. Whilst many people still believe in science and technology, many have become more sceptical of the allied 'progress'. By exploring the idea that modernity creates effects that undermine its own foundations, forms and effects of resistance are explored in various contexts.

The book presents a unique interdisciplinary study, including contributions from historians, sociologists, psychologists and political scientists.

Resistance to
new technology

Resistance to new technology

nuclear power
information
technology
and
biotechnology

edited by
MARTIN BAUER

The National Museum of Science & Industry
Science Museum

CAMBRIDGE
UNIVERSITY PRESS

Published by the Press Syndicate of the University of Cambridge
The Pitt Building, Trumpington Street, Cambridge, CB2 1RP
40 West 20th Street, New York, NY 10011-4211, USA
10 Stamford Road, Oakleigh, Melbourne 3166, Australia

First published 1995

Printed in Great Britain at the University Press, Cambridge

A catalogue record for this book is available from the British Library

Library of Congress cataloguing in publication data

Resistance to new technology: nuclear power, information technology and
biotechnology/edited by Martin Bauer.
p.   cm.
Papers from a three-day conference held at the Science Museum, London,
5–7 April 1993.
Includes index.
ISBN 0 521 45518 9 (hardback)
1. Technology assessment – Congresses. 2. Technology assessment –
Europe – Congresses. 3. Nuclear energy – Social aspects – Congresses.
4. Information technology – Social aspects – Congresses.
5. Biotechnology – Social aspects – Congresses. I. Bauer, Martin.
T174.5.R48 1995
303.48′3–dc20   94-26745   CIP

ISBN 0 521 45518 9 hardback

Contents

Contributors

MARTIN BAUER
Department of Social Psychology
London School of Economics
Houghton Street
London WC2A 2AE

ANTONIO J J BOTELHO
830 Montgomery Avenue, apt #202
Bryn Mawr, PA 19010
USA

KRISTINE BRULAND
Department of History
University of Oslo
p. b. 1008 Blindern
0315 Oslo
Norway

MARLIS BUCHMANN
Department of Sociology
ETH Zentrum
Universitätsstrasse 13
CH-8092 Zürich
Switzerland

ROBERT BUD
The Science Museum
Exhibition Road
London SW7 2DD

DANCKER D L DAAMEN
Department of Social and Organizational Psychology
Leiden University
Wassenaarseweg 52
2333 AK Leiden
The Netherlands

SHEILA S JASANOFF
Department of Science and Technology Studies
Cornell University
Ithaca, NY 14853
USA

HANS MATHIAS KEPPLINGER
Institut für Publizistik
Johannes Gutenberg Universität
Jakob-Welder-Weg 20
6500 Mainz
Germany

ROY M MACLEOD
Department of History
University of Sydney
Sydney, NSW 2006
Australia

RODERICK MARTIN
Department of Management Studies
Glasgow University Business School
53–59 Southpark Avenue
Glasgow G12 8LF

IAN MILES
PREST
Mathematics Tower
The University of Manchester
Oxford Road
Manchester M13 9PL

DOROTHY NELKIN
Department of Sociology
University of New York
269, Mercer Street, 4th floor
New York, NY 10003
USA

JOACHIM RADKAU
Fakultät für Geschichtswissenschaft und Philosophie
Universität Bielefeld
Bultkamp 16
4800 Bielefeld
Germany

ADRIAN J. RANDALL
School of Social Sciences
University of Birmingham
Edgbaston
Birmingham B15 2TT

DIETER RUCHT
Wissenschaftszentrum Berlin für Sozialforschung
Reichpietschufer 50
W-1000 Berlin 30
Germany

JOHN M STAUDENMAIER S J
MIT, Dibner Institute
38 Memorial Drive
Cambridge, MA 02139
USA

GRAHAM THOMAS
Department of Innovation Studies
University of East London
Romford Road
London E15 4LZ

ALAIN TOURAINE
CADIS
Ecole des Hautes Etudes
54, Boulevard Raspail
75006 Paris
France

IVO A VAN DER LANS
Department of Social and Organizational Psychology
Leiden University
Wassenaarseweg 52
2333 AK Leiden
The Netherlands

Preface

This book combines contributions from a conference on 'Resistance to New Technology – Past and Present', held at the Science Museum, London, on 5–7 April 1993. The event brought together historians, sociologists, political scientists, media researchers and psychologists to reflect on the problem of 'resistance' in relation to technological developments. Around 150 people from twenty countries gathered in the unusual setting of steam engines, automobiles and space rockets provided by the Science Museum.

The aims of the meeting were (a) to take stock of the forms and effects of resistance in the recent past; (b) to compare different technologies in this context; and (c) to think about, and work towards, a functional analysis of resistance in the process of technological development. The meeting provided material to **overcome the technocratic bias** according to which resistance is nothing but a nuisance in the technological process.

When comparing the forms and effects of resistance the book focuses on three major developments since 1945: nuclear power, computing and information technology, and biotechnology. The story told is mainly, but not exclusively, European. In making comparisons the contributions reach out historically to the origins of the idea of 'progress' and the Luddite revolt of the early nineteenth century, and geographically to Australia, North America and Japan. The scope of the book prohibits the inclusion of several dimensions of the problem of resistance worth mentioning. First, the book excludes the problems of resistance or non-resistance to new technology in authoritarian and totalitarian systems such as Eastern Europe between 1945 and the collapse of communism, the USSR, or China. Secondly, it excludes the problem of resistance to new technology in developing regions such as South America, India and South East Asia: these topics provide the scope for another conference.

I would like to thank all contributors and reviewers for their cooperation in reworking the papers making the publication of the book in its present form possible. I would like to thank Kathy Angeli, who handled much of the organization of the conference; Jane Gregory and Victoria Smith, for help during various stages of the editorial process; and Fiona Thomson, at Cambridge University Press, for her patience and genuine interest. Finally, my special thanks go to John Durant, whose projects I have shared for the last three years and whose friendly encouragement made this book possible.

Martin Bauer, London, February 1994

Resistance to new technology and its effects on nuclear power, information technology and biotechnology

MARTIN BAUER

Basic questions

The word 'resistance' has become unsuitable for use in the context of new technology. The allegation is that it serves mainly to blame those who resist; talking about resistance implies a managerial and technocratic bias. However, in developing the idea for this conference, I was confident that 'resistance' would prove ambiguous in meaning and rich in connotations, particularly in the European context.[1]

Historians of technology recently rediscovered 'resistance' as a 'force' that shapes technology which requires an adequate analysis (Mokyr 1990, 1992). For the economist resistance is basically the vested interests of old capital in ideas, skills and machinery. In addition, in the light of the critique of the 'Whiggish' historiography of technology (Staudenmaier 1985), it seems reasonable to lift 'resistance' from the dustbin of history.

Artefacts such as machines, power stations, computers, telephones, broadcasts and genetically engineered tomatoes, and the practice of their production, handling, marketing and use – in other words, technological innovations – are not the only factors of historical change. Technological determinism seems an inadequate account of our history. Various social activities give form to processes and products, facilitate their diffusion and mitigate their consequences. However, technology is not neutral. It creates opportunities and simultaneously constrains human activity. We experience the latter as being paced by 'machines' rather than controlling them. The selection of options is not neutral; it is likely to be contested and in need of legitimation. The control of technology by those affected by it remains a desirable agenda. From here the disagreement begins: who should be accountable to whom, and which procedures ensure this accountability?

In this context it constitutes a challenge to reflect on the concepts of resistance, and to review historical events, in order to rehabilitate the notion. In my view this may be achieved by analysing resistance in terms of its various consequences. This means starting with a functional hypothesis, instead of assuming *a priori* dysfunctional consequences of resistance for the 'progress' of society. Here we are less interested in the causes of resistance than in its effects. Resistance as the 'enemy of progress' is only a part of the story, as we will see, both historically and for more recent events.

Methodologically resistance to new technology is as much an independent as it is a dependent variable: the public reaction to technology influences new technology in a circular process. The study of resistance is located halfway between traditional impact analysis and the recent focus on socio-technical networking of humans and artefacts (Latour 1988; in German, *Technikgenese*: e.g. Joerges 1989).

The term 'resistance' elicits contradictory connotations. It seems necessary to justify an academic contribution on 'resistance' because the term is loaded with a managerial and modernisation bias and is not suitable for an impartial analysis of social events.[2] A participant at the conference put it bluntly: resistance implies a 'whitewash of big industry' in technological controversies. The term 'resistance' elicits quite different connotations. Traditions of philosophical thinking give resistance a moral dignity; and events of the twentieth century impute 'resistance' (*Widerstand, resistence, resistenza*) with an aura of 'heroism' in the struggle against totalitarianism. In South America 500 years of colonialism give it the 'heroic' meaning of fighting a lost cause that may finally prevail. We may deal with a real cultural difference in semantics.

The managerial and technocratic discourse stipulates resistance as a structural or a personal deficit. Resistance is irrational, morally bad, or at best understandable but futile. In contrast, the German discussion of new technology since the 1970s has been conducted under the term 'acceptance crisis'; the debate carefully avoids the term 'resistance', which is reserved for the respectable part of the national identity for which post-war historians, not only in Germany, have fought (Schmädeke & Steinbach 1985).

Because of these semantic confusions authors have argued that the term be dropped. I disagree, and suggest that we keep the term and stress the intended meaning. The ambiguity of the term 'resistance' allows us to ask different questions about an old problem of social change. This volume provides a review of the notions of resistance in technological controversies. The nineteen contributions compare forms and effects of resistance across time, space and technology: from machine breaking and technology transfer in the nineteenth century, Fordism in the early twentieth century, to three base technologies after 1945: civil nuclear power, information technology, and new biotechnology.

The comparison is done, naturally, using different approaches. The scope of material is limited geographically. The technological processes in countries of the Far East, in China, Taiwan and Japan are excluded; equally excluded are countries of the Third World and of Eastern Europe, where the problem of resistance to new technology may have a different angle altogether. These chapters mark the ambitious beginnings of an attempt to map intellectual territory: to study the contributions of resistance to the 'progress' of technology.

In the following I elaborate the rationale that brought contributors together to address the following key questions:

What are the forms of resistance?
What is being resisted?
Who are the resistant actors?
What are the effects of resistance?
What are the (dis)analogies between technologies?
The concept of 'resistance': towards a functional analysis?

My own theoretical inclination is towards a functional analysis of resistance by its consequences in a wider context. We do better to study functions of a process first, and to study dysfunctions afterwards as a dynamic aberration of normal processes. To date we have assumed the dysfunctions and neglected the functions of resistance to new technology. My framework draws upon recent developments in the theory of autonomous systems and elaborates a functional analogy between resistance and acute pain with reference to processes of self-monitoring. Metaphorically speaking, resistance is the 'acute pain' of the innovation process. I cannot assume that contributors subscribe to this framework, so I put it at the end of the book in an attempt to summarize. I develop this framework which both stimulated the idea of the book, and, I dare to hope, embraces many of the issues in a coherent manner. The core thesis states:

> Resistance affects socio-technical activity like acute pain affects individual processes: it is a signal that something is going wrong; it reallocates attention and enhances self-awareness; it evaluates ongoing activity; and it alters this activity in various ways to secure a sustainable future.

Three base technologies after 1945

Someone interested in contemporary history might ask: why compare nuclear power, information technology, and biotechnology? The automobile and space technology are equally major technological innovations of the twentieth century.[3] There are several reasons for the focus on these three technologies. The choice depends on the frame of comparison or '*tertium comparationis*'.

Table 1.1 *Invention, innovations and Organisation for Economic Co-operation and Development (OECD) attention*

Technology	Invention	Innovation	OECD attention
Nuclear	1942 first nuclear chain reaction	1955–6 first nuclear power stations (USSR, UK)	**1956 first report**
Information technology	1943 ENIAC 1947 transistor	1954 commercial	**1960 micro** 1965 office automat
	1959 integrated circuit 1959 micro processor	1961 commercial 1965 micro computer 1975 home computer 1981 IBM PC	1971 IT policy series 1979 new IT series
Biotechnology	1944 DNA 1947 double helix 1973 rDNA	1975 CETUS US 1977 BIOGEN (Europe) companies founded	**1982 first report**

Sources: OECD publications catalogue; Wright, 1986; Rüdig, 1990.

Similarities

Five similarities of nuclear power, information technology, and biotechnology suggest a viable comparison. I am using these technologies generically as clusters of innovations that are distinct objects of R & D policy, planning and public perceptions.

First, economic historians suggest a periodization of time since about 1780 in cycles of roughly 50 years, commonly known as 'long waves' or 'Kondratieff cycles' of the world economy. Each upswing is based on the scientific and technical ideas developed during the previous downswing for which capital becomes available in the new upswing. Evidence indicates that the fourth wave turned into the downswing in the early 1970s, and a fifth long cycle may have taken off since.[4] The technologies commonly associated with this hypothetical fifth upswing are **civil nuclear power**, the new source of energy; **microelectronics**, with its ramifications into computer and communication technology, the new form of communication (Freeman 1985; DeGreene 1988; Ayers 1990); and **biotechnology and genetic engineering**, the new forms of food production, animal breeding and medical care. Resistance to new technology can take the form of social movements. Hobsbawm (1976) suggested that the size and intensity of social movements, in his case nineteenth century labour movements, relate to long waves of economic development. Screpanti (1984) showed that strike activity intensified during long economic upswings between 1860 and 1970 to reach peaks at the upper turning points of the long cycles; long periods of depressions showed the lowest level of strike activity. More recent observers see the decoupling of social protest movements from the economic system as characteristic of the post-war period (Pakulski 1993), suggesting that the link between economic cycles and popular unrest is historically contingent.

Secondly, nuclear power, information technology and biotechnology reach the attention of planners, forecasters and policy makers in a time series. Table 1.1 compares basic inventions, first-time commercial innovations, and the first related OECD policy reports for the three technologies.[5] This indicator shows that civil nuclear power gained policy focus as early as 1956; information technology after 1971, when the OECD starting a series of reports on information technology policies;[6] biotechnology came into the international policy focus not earlier than 1982. In terms of government policy nuclear power is older than information technology; and information technology is older than biotechnology.

Thirdly, each of these technologies gave rise to sociological imaginings of the 'coming new era', sometimes optimistic and enthusiastic in terms of revolutions, pessimistic in the light of doom, but often ambiguous. We find an abundance of books and articles with titles ranging from the 'atomic age', 'nuclear state' (Jungk 1979), the 'micro-electronic revolution', the 'computer age', the 'information society', the 'electronic society' (Dertouzos & Moses 1979; Lyon 1988), to the dawning of 'biosociety', the 'age of biology' or 'biotechnics' (see Bud 1993).[7]

Fourthly, all three technologies gained considerable media attention at different times, although in a controversial manner mainly after the 'Oil Crisis' of 1973. The cycle of press coverage for nuclear power, information technology and biotechnology confirms the time series in which these three technologies have entered the public arena. Figure 1.1 shows the shifting of media attention between these technologies, based on several European sources.[8] The peak of the coverage is indexed with the value 100 in each case. In Germany nuclear power reached a peak of press attention in 1979 (Kepplinger 1988, p. 665) coinciding with large scale anti-nuclear demonstrations.[9] The coverage of information technology peaks in 1984 (Sensales 1990, p. 66), the year ominously associated

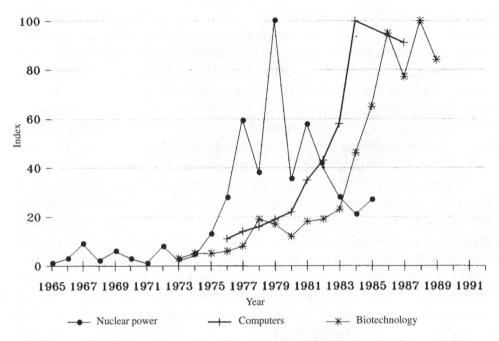

Figure 1.1. Press coverage of nuclear power, information technology and biotechnology in Europe 1965–90. The graph shows the development of press coverage
on three technologies. The coverage is indexed with the peak = 100. Data on nuclear
power and biotechnology from Germany; data on information technology from Italy.

with George Orwell's dystopian novel. This peak was probably reached earlier in
Germany, during the public debate on the Population Census in 1983 (Mathes &
Pfetsch 1991). The coverage of biotechnology, genome analysis and new genetics
peaks in Germany in 1989 (Kepplinger, Ehmig & Ahlheim 1991; Brodde 1992;
Ruhrmann 1992), coinciding with the publication of the parliamentary enquiry,
the Catenhusen report, into the dangers and benefits of biotechnology (see Bud,
Chapter 14, or Radkau, Chapter 16). In Britain, to date, press coverage on
biotechnology peaked in 1992, and it is likely to rise in the coming years as the
public debate intensifies (Durant, Hansen & Bauer 1995). The evidence shows
that the Asilomar conference in February 1975 (Wright, 1986), where a group
of US geneticists put a moratorium on their own research in order to explore the
social consequences first, was an event with limited impact on the European
public. Nuclear energy, information technology, and biotechnology form a series
of policy and media attention. Media attention lags behind policy attention but
the gap is decreasing (compare Table 1.1), an indication that public sensitivity to
technological issues is increasing. The 'reaction time' from first innovations to
public controversy in Europe is about 20 years for nuclear energy, 12–15 years
for information technology, and about 8–10 years for biotechnology. Public

opinion expressed in media coverage seems to function as feedback information for a process that is already under way.

Fifthly, all three technologies have been the subject of public controversy and social mobilization. Nuclear power has been contested by mass movements since the 1960s; local protests merged into anti-nuclear movements all over Europe and gained significant influence in energy policy (see Rucht, Chapter 13). The debate on information technology has been more of a concern for intellectuals; it rarely mobilized large resistance beyond local actions (Kling & Iacono 1988; Martin, Chapter 9, or Miles & Thomas, Chapter 12).[10] Issues of unemployment, social control, privacy, and security of information resulted in a cluster of opinion surveys in the mid-1980s (Jaufmann & Kistler 1986; Bauer 1993).[11] Biotechnology is the current issue: public opinion is forming on genome analysis, gene therapy, and genetically engineered plants and animals, to date without having mobilized mass actions. On the whole one could say that nuclear power has a long history of debate and has mobilized large-scale resistance which moved from local to trans-national activity. Information technology did not result in large collective resistance; resistance is local on the level of particular industries, work-place actions or consumer behaviour. Biotechnology is the great unknown. Local resistance has occurred against field testing of genetically altered plants, and public opinion has been gauged in recent years (Marlier 1992; European Commission 1993). Jeremy Rifkin and his associates campaign worldwide against new biotechnology and its ramifications.

In summary, nuclear energy, information technology and biotechnology have in turn been viewed as leading technologies with long-term impacts; each of them stimulated dreams of a 'new era', from the atomic age, to the information society, to the dawn of the biosociety. Policy concern and public debate about these technologies came in a sequence. One can assume that the resistance against nuclear power conditioned the resistance and its effects for later developments (see Radkau, Chapter 16). It is likely that we are dealing with an example of institutional learning, the building of new procedures and fora of debates at the interfaces of science, technology and society.

Together with similarities, we need to identify differences between these technologies. Such differences may be candidate variables for explaining the variance in resistance that accompanied these developments.

Dissimilarities

Analogies are useful, but may also be misleading (see Radkau, Chapter 16). In substance the three technologies contribute to different functions of human life. Nuclear power is a source of energy; information technology deals with ways of storing and processing information; biotechnology deals with enhanced production of food (fermentation, plant and animal biotechnology), and with new forms of medical care (pharmaceutical and genetic screening and therapy). These developments build on each other. Computers are to a large extent involved in the development of new biotechnology for process automation, modelling and information storage; and computers and biotechnology rely on available energy.

The three technologies differ in the choices they offer to the public. Buying a computer is different from buying a nuclear power plant. Choices offer points of resistance. Many different choices diffuse user resistance, and may prevent it from becoming a social movement. Nuclear energy is mainly a question of binary choice – yes or no – with some leeway on how large the percentage of the total energy production of a country should be in the form of atomic energy. In contrast, information technology and biotechnology offer wider choices. People have some choice over the extent of computerization they want for their lives: they may restrict the computer to assisting in a few tasks at home or at work, or conduct all their lives with 'computer assistance', from driving a car, to cooking a meal, to making coffee in the morning, to finding a partner. These choices are not free from constraints, as Dorothy Nelkin (Chapter 18) will argue on the issue of penetration of privacy. Biotechnology is similar to information technology in terms of choice among various food products; however, when dealing with the problem of production processes, the experimental release of genetically altered organisms, and the issue of patenting life forms, we may be confronted with binary choices of yes or no (see Jasanoff, Chapter 15, or Bud, Chapter 14).

Risk is both a unifying and a distinctive feature of these three technologies. The capital limits of insuring the potential damage of a large technological project marks a crisis point of modern societies and the transition into 'risk societies'. According to this view it is the success of institutions, not their failure, that undermines their basis and creates space for new political processes to emerge. The quest for technological control leads paradoxically to increased uncertainty, and undermines the capability for action. The distribution of unintended consequences of technological progress becomes a conflict area that cuts across traditional party political lines (Beck 1993).

Nuclear power is 'technically' a low risk area, with large dangers but small probability. Incidences such as Three Mile Island or disasters such as Chernobyl have low probabilities, but, as we all know, have happened – in 1979 and 1986, respectively. In contrast information technology poses small dangers with high probability. Computer addiction (Shotton 1989), exposure to VDU (visual display unit) radiation, and posture pains such as repetitive strain injury (RSI) are widespread, but are not regarded as alarming.[12] The empirical risk of suffering from RSI is probably larger than being contaminated by nuclear fallout. However, risk comparison is notoriously controversial (Covello 1991). Jasanoff (Chapter

15) will show that German, British and US expert proceedings attribute different kinds and sizes of risks to biotechnology. Comparisons are furthermore complicated by the fact that public risk perception is not cumulative; it does not add up many small risks to give one large risk. Risk perception feeds on the size and controllability of danger, and less on its probability. The magnitude of potential damage differs for nuclear power, information technology, and biotechnology. Nuclear power and biotechnology share the possibility, in the former case real, in the latter hypothetical, of large scale, unlimited in time and space, social and physical damage – a problem that is to a lesser extent associated with information technology. This difference may explain the presence or absence of large-scale organized resistance.

The type of risk makes a difference for public perception. The health risks of the three technologies vary. Radiation touches the problem of physical well-being of individuals and society. Leukaemia, cancer and malformations at birth are issues central to people's life concerns. By contrast the problems of information technology are more abstract: it seems to alter the way we think and make decisions; 'artificial intelligence' seems to bother mainly people with expertise in the field of 'thinking', such as intellectuals and academics. Information technology has not lent itself to mass mobilization of resistance that goes beyond local issues except in Germany in 1983 and 1987 where a computer readable identity card and the population census became a major issue (Mathes & Pfetsch 1991). Radiation fallout from accidents like Chernobyl poses **risks without limits**; the event had consequences in Northern Scandinavia within days. Power stations may pose geographically concentrated risks of leukaemia and cancer in the vicinity. The problematic impacts of information technology are more widely scattered and mostly transitory; and impacts such as enhanced social control and penetration of privacy are more symbolic and difficult to recognize (see Nelkin, Chapter 18). In new areas of biotechnology and genetic engineering health risks are an open question, and more of a diffuse but widespread concern than well defined.

The three technologies differ in capital intensity and geographical concentration. Nuclear power is the most expensive technology per investment unit. It may have once been part of the expert imagination that each household would have a nuclear power generator in the back garden, in the basement or in the automobile;[13] but nuclear energy production became concentrated in large plant sites – 423 sites in 24 countries by 1990. Information technology is radically different. Computing units have become increasingly smaller and cheaper and are widely distributed in business and in households as mainframes, micros, PCs (personal computers), laptops or notebooks, or as integrated parts of an increasing number of artefacts. Similarly, biotechnology does not require the large-scale investments of nuclear energy with all that it implies for the disposal or re-processing of nuclear waste. The human genome project is a small investment compared to the nuclear programmes over the last 40 years. The biotechnology industry started up in small businesses linked to university departments (Wright 1986), and turned into a venture of established chemical industries in recent years. Within one country nuclear power is centralized technology, while information technology and biotechnology tend to be distributed technologies.

Whether the enterprise is private or public makes a difference. Nuclear power and computers were initially a state enterprise from research and development to production, a matter of national security linked to the capacity to produce 'the Bomb' and to make large-scale calculations in missile technology. Private companies take over later: nuclear power remains a state enterprise in many countries until the present time while the diffusion of information technology and biotechnology is predominantly a matter of private business. Public control is limited to setting legal boundaries and incentives for investment. The involvement of the state makes nuclear power a direct political issue. Nuclear power is public technology; information technology and biotechnology are primarily private technologies. This leads to a different culture of industry. The nuclear power industry inherited a tradition of secrecy from its military roots. Research and development was conducted behind closed doors, dictated by national security during the Cold War years. The chemical, computer and communications industries are more open. Their processes and products are more visible in everyday use. Public enterprise does not mean an open culture, and private developments do not imply a culture of secrecy.

The public discourse of technology varies across and within technology over time. The geographical concentration of nuclear power makes it visible in the landscape as an icon of 'progress', 'doom', or 'a devil's bargain' (Gamson & Modigliani 1989). The Cold War favoured images of secrecy and national security in relation to nuclear power, while the 1980s favoured the imperative of international economic competition. Neither information technology nor biotechnology is a major issue of national security, but does impact on national competitiveness. Confronted with the threat of terrorism nuclear power implied intensified social control, perhaps even a police state within the 'nuclear state' (Jungk 1979): it has similarities with information technology in terms of political risk. Another variable of discourse is the 'newness' of technology. Jasanoff (Chapter 15) shows that for biotechnology the context dictates the rhetoric: for purposes of fund raising and innovation policy 'novelty' and 'revolution' are appealing arguments. However, to prevent legislation and judicial activity, 'novelty' is an undesirable argument as existing regulations are supposed to suffice; new biotechnologies 'become' forms of brewing and breeding, or 'old wine in new bottles'. The effects of 'newness' on the mobilization of resistance may be seen by comparison with 'old' technologies, e.g. the automobile or hydroelectric energy. According to all of these criteria – choices offered, risks involved, capital intensity, public or private, the culture of industry, and public image – nuclear power stands apart from information technology and biotechnology.

Larger contexts may account for similarities

Beyond these differences and the similarities of high expectations, innovation policy focus, media coverage and public controversies, all three technologies share the historical context of the post-war period. Four features of that period may explain the similarities: the end of Utopian visions, the Cold War period, the changing role of expertise, and the shift in cultural values.

The last waves of Utopianism motivated the 1968 student protests and have calmed down since; Utopias seem no longer suitable for conceiving the future. The criteria for criticizing our civilization towards the end of the twentieth century are diffuse. The malaise concerns the question: where are we going? The notion of 'Progress', writ large, is in crisis again (see Touraine, Chapter 2). The equation of progress and new technology is no longer taken for granted. Bold expectations of the future give way to concern for the present; optimism transforms into caution; democratic expectations become immediate and more widespread. In the search for sustainable development from a global perspective we need to co-opt dissent, rather than ignoring it, blinded by a vision for the future. We are interested in those who resisted without a vision of the future and trace the precursors of modern technology assessments: the actions of the latter-day guilds and Luddites to new technology. We are busy reinventing institutions that traditionally bridged the gap between the place of action and the place of consequences; institutions that got lost in the way of 'Progress' (Schot 1991).

With the fall of the Berlin Wall in 1989, the Cold War came to an end. The mobilization of science and technology to ensure a balance of military forces framed the public debates in all Western countries. Decontextualizing the three technologies from this common context may lead to false conclusions. The Cold War affected public discourse at all levels. National security was a whitewash for much mismanagement, many disasters and cover ups. Whole research programmes were conducted secretly. The discourse of new technology was framed by the East–West divide. Secrecy throws expertise into disrepute (Knorre 1992). The present situation may bring a different pattern of public response to new technology. The lack of democratic decision making on grounds of security is no longer justified; the end of the Cold War strengthens the aspirations of various groups of people to have a say.

We have witnessed over the last 30 years the disenchantment with expertise. Technocracy is not a socially acceptable form of government. The idea of technocracy is a modern idea, and equally the struggle against technocracy is part of modernity; resistance is part of this parallel process to secure freedom of choice. Anti-big business and anti-technocracy unites all three movements of resistance (see Touraine, Chapter 2). Bureaucracy is the target, rather than engineering or science itself. We find some similarity of motives with the Luddites (Randall, Chapter 3), who opposed a particular way of putting machines to work to avert the dissolution of their life world. Current technological resistance equally fights the process, not the product, of technological development. The issues are often public deceit and lies; manipulation and exclusion; pollution and exploitation; expert conspiracy; and the unequal distribution of risks.

The argument of a changing culture ties in with Inglehart's (1990, p. 143) diagnosis of a value shift in advanced industrial society from a material to a post-material orientation. Post-materialists believe in ideas and celebrate personal relationships, and want a say in their jobs, freedom of speech, and democracy. In contrast the materialists favour above all law and order, economic growth, and low inflation. The end of the Cold War, the loss of Utopia, the disenchantment with expertise and the culture shift point to a common context for nuclear power, information technology and biotechnology in the development of science, technology and industry. This opens up the possibility of analysing the functions of resistance for these three post-1945 developments: similar functions achieved with changing structures.

What are the forms of resistance?

The problem of definition

The first problem is to define resistance. In the following pages I take stock of ways that 'resistance' has been defined in various contexts. Mokyr (1990) lists resistance among the major variables that explain variance in technological development such as life expectancy, nutrition, willingness to bear risks, geography, path dependency, labour costs, science, religion, values, institutions and property rights, politics and the state, war, openness to new information, and demography.

According to the Oxford English Dictionary, 'resist' means 'to withstand, to prevent, to repel, to stand against, to stop' in four different contexts: to describe human actions, to describe the power or capacity of such actions, to describe the opposing relation of forces, and to describe specifically the non-conductivity of electricity, magnetism or heat. 'Resistance' as human action means the '**refusal to comply with some demand**'; and historically the **clandestine insurgence** against an occupying, illegitimate power during the Second World War of which the French '*Resistance*', the German '*Widerstand*' or the Italian '*Resistenza*' are well-known examples. I see three distinctions constitutive of what counts as resistance in the present context: rationality/irrationality, resistance/opposition, and the problem of self-reference.

Resistance as diversion from the 'one best way'

In the context of (social) engineering, resistance is traditionally the deviation from the Rational writ large, or F. W. Taylor's 'one best way'. Planners, engineers, managers and designers encounter the world as a set of ill-defined problems (see Staudenmaier, Chapter 7). To structure this uncertainty increases predictability and control over matter and people. Resistance is the uncertainty of the designer struggling with matter and people that needs to be reduced. Rationality is claimed by the designer, and actions that challenge his or her proposal are 'resistance'. However, a design is necessarily 'bounded' within constraints of space and time (Simon 1981) and the notion of a single best solution – Rationality writ large – is hardly justifiable. Under any constraint different assumptions suggest different solutions and legitimate the 'resistance' to a single best design.

Any design reduces some kind of uncertainty while creating others both for the designer and for the users. Risk analysis quantifies this as the product of damage and its probability. To take expert calculations and deviations from this baseline as 'bias' in risk perception is a common way of defining resistance (see Bauer critically on 'technophobia', Chapter 5). Bias is the outcome of either missing information or 'inadequate' information processing. 'Debiasing' popular perceptions with additional information and special training makes good business, but seems naïve as it assumes that additional information leads to the same conclusions for everyone. People are affected differently by the same information. Different assumptions suggest different conclusions on the same information.

Staudenmaier (Chapter 7) points out that historical narratives of great technological projects may pay legitimate tribute to 'great human achievements'; but they are 'Whiggish' in outlook and selective with facts. Complementary stories need to be told to understand the dynamics of technological developments: the stories of losers, sidetrackings, failures and obstacles on the way. Even the story of 'great men' is incomplete. Human achievements are ambiguous. Human actions encounter durability, material and cultural, which a universal notion of Rationality cannot displace without violence. Alain Touraine (Chapter 2) discusses this as the crisis of Progress leaving us in a conflict between the Scylla and Charybdis of narcissistic particularism and the universalism of large projects, both equally undesirable in the light of historical experience. Resistance re-emerges as the dignity of bearing the tensions in between two illusory seductions in the late twentieth century without giving in to either of them (Touraine 1992).

Resistance and opposition

Both activities, resistance and opposition, challenge a given project, although in different relations to social institutions. Opposition is challenging activity within institutional boundaries. 'Institutional' activities are those that can be expected to happen. Resistance is best understood as **activity** which is **unexpected** in both content and form by the innovators. This does not preclude an observer predicting such activities; the surprise of the innovator and designer is crucial. This distinction allows us to describe the transition from resistance to opposition, whereby activities are institutionalized and become predictable; when institutions are dismantled, opposition activities can become resistance.

Furthermore, resistance is a form of **risky behaviour**. Someone who violates norms and rules runs the risk of exclusion, punishment or material sanctions. The legal discussion of the 'right to resistance' tackles the paradox of justifying actions outside the law without undermining the rule of law. This is done with a view to law as a process (*Rechtsfortbildung*); the legal system is always imperfect and in need of improvement along the lines of the constitution (Rhinow 1985). If resistance breaks a norm and institutions play a role in defining that norm, resistance as action is the **product of normative communication**. The more restrictive the norms, the larger the range of activities that count as resistance. Resistance is defined by a system that is in and of this conflict.[14]

Self-reference and resistance

Several questions point to the problem of self-reference. Does resistance always have clear motives and purpose, or can activities challenge a technological project without such motives and purpose? Do the resistant persons see themselves as being resistant or not; or, in other words, is 'resistance' an observer category or a category of self-description, or both? What activity counts as an act of resistance? We are dealing with the logical hierarchy of actions and behaviour: various behaviours can qualify as resistance actions depending on the context of motivation.

A clear **motive** and a **conscious purpose** to challenge a project in unexpected ways define an act of resistance. 'Purposeful' resistance means that the action is willed and planned, and that some consequences, positive or negative, likely or unlikely, are taken into account. It is risky action. Resistance may follow a moral imperative in justifying the motive. An ethical and moral discourse may provide the grounds for resisting a certain project.

Purpose and motive make the difference between resistance and mere

avoidance behaviour, such as not buying something, or not consuming or using it once a device has been bought. If avoidance meant resistance, this would imply that there is an obligation to consume in the first place. Some people would strongly argue the need to distinguish between deliberate boycott and the reluctance to consume or to change from typewriter to word processing out of ignorance or lack of opportunity. 'Not buying' becomes resistance only in purposeful consumer boycott. Conscious purpose and motive transform mere behaviour into dignified resistance. Defining resistance as the continuum from 'not consuming' to 'risking one's life' seems inadequate. Risking one's life for a moral reason has a dignity that does not seem comparable to mere non-consumption for non-specific reasons.

However, the problem of consciousness complicates the issue. To define resistance by its purpose may do justice to the moral claim, but it may be that a variety of events, often silent, unspectacular, everyday actions are equally functional in challenging a technological project, or setting limits to its power. It may be a way of life not to go along with a project, not out of high moral purpose, but out of tradition or parochial concerns (Scott 1985). Wynne (1993) shows how forms of so-called 'public ignorance' of science and technology are unorganized ways of negotiating an acceptable relationship with experts by limiting their cognitive control. Avoidance behaviour is an observer category; people buy goods or not, use them or not, are indifferent and frustrate one's expectations. Our actions involve variable degrees of awareness. We may be fully conscious, partially conscious, temporarily unconscious, or entirely unaware of what we are achieving with our actions. The attitude often follows the action, and we come to intend what we are already doing. In raising consciousness about what we are doing, we may adjust our attitudes to our behaviours. Behaviours without a conscious motive cannot be ruled out as forms of resistance by their consequences. In that sense the high moral purpose seems not to be a sufficient criterion for resistance actions.

Actions may differ in their ethical and moral dignity, but functionally they may well have similar effects.[15] For the present purpose we shall explore effects of resistance with or without motives; intentions serve to classify effects as anticipated or not. The discourse on the morality of resistance is a variable that is itself an indicator of the intensity of resistance. The more people reflect on the dignity of resistance the more pressing is the problem of resistance. Miles and Thomas (Chapter 12) see resistance as a narrow range of actions on one end of the continuum of acceptance of information technology. We should perhaps talk about resistance in the functional sense with a small 'r' and in the moral sense with a capital 'R'.

Towards a taxonomy

A taxonomy of resistance tries to distinguish different qualities and intensities of resistance. Kinds and strength of resistance are important for assessing differential effects on new technologies in various contexts. Identifying forms of resistance is also a precondition for the analysis of the internal dynamics of resistance, as it transforms with contexts and time.

Active/passive and individual/collective

Resistance is frequently described as active or passive. Active refers to purposeful actions that may involve violence against persons or objects: direct attacks against machines, the destruction of machines, the intrusion or occupation of building sites. This may include forms of violence that are often confined to a small circle of determined, specially trained and motivated activists. By contrast, passive resistance comprises forms of wilful inactivity, where bodily movements are restricted: non-compliance, blocking, sit-ins, not doing things one is expected to do by law. Passive resistance is more easily turned into mass action, as people do not need special training and it may be physically less demanding.

Another distinction draws a line between resistance as an action of one or many individuals: individual, cumulative or collective resistance. Isolated individuals may resist a project; large numbers of people can resist with cumulative effect without coordinating their actions.[16] An observer can identify the effects of these actions. People themselves may not see it. The cumulative effect of uncoordinated actions can be functionally equivalent to organized actions in a repressive context. Formal structures are both enabling and constraining, and some observers take them as a necessary criterion for resistance (see Nelkin, Chapter 18). The degree of coordination of resistance is a variable. Individuals may coalesce into movements, new movements may co-opt existing structures. Mazur (1975) has shown how the US anti-nuclear movement rode on the back of the older environmental movement. Jeremy Rifkin explicitly tries to co-opt animal rights groups, vegetarians, environmentalists, and Third World awareness groups into a 'rainbow' coalition for his apocalyptic struggle against biotechnology incorporated.

International activities must be distinguished by the mode of aggregation: multinational resistance operates from a hierarchical centre that coordinates and plans analogously to a multinational business (e.g. Greenpeace; Jeremy Rifkin); in contrast there is trans-national resistance which works as a forum for discussion more like the United Nations: heterarchical with various autonomous centres of activity (e.g. the international anti-nuclear movement).

Classes of actions and levels of analysis

In a stocktaking exercise a variety of behaviour patterns can be identified as 'resistance', of which civil disobedience is the most salient one. Sharp (1973) distinguishes symbolic actions, such as collecting signatures, distributing flyers and pamphlets and rallying, from non-cooperation and non-violent interventions. Among the latter are site occupations; setting up anti-nuclear camps; mass demonstrations or sit-ins that intervene with the normal course of the daily activities. Many of these forms are ritualistic, and so inhibit violence in similar ways to the rituals of the Luddites (see Randall, Chapter 3).

Economic behaviours include consumer boycotts, avoiding the consumption of certain products, or refusing to give services to certain institutions. We find campaigning by consumer organizations on behalf of users and consumers. Technological innovations may precipitate industrial action.

Cultural and symbolic activities are the public display of the ambiguity towards new technology (see Staudenmaier, Chapter 7). Cultural activities include film making, writing popular books, giving talks, ordering arguments and organizing conferences on controversial issues. Ambiguity is displayed in variations on the image of the 'mad or irresponsible scientists' or of 'Dr Frankenstein and his monster' as the motive force in science fiction writing and film. Resistance relies on alternative value patterns, an ethos of dissent, or religious values. Sociological evidence suggests that absolute religious values have become more salient in recent years and with them has come an uneasiness with compromise (see Nelkin, Chapter 18). The media's coverage of science and technology is an indicator of cultural resistance which Kepplinger (Chapter 17) analyses and critically addresses. Anxieties and fears ('technophobia', 'nuclearphobia', and 'cyberphobia') may result in a retreat from public life into the confines of privacy and into a controllable small world of non-engagement and alternative lifestyles.

Legal actions involve a call upon the state to regulate, in some countries by calling for a plebiscite. Thus, political lobbying and voting behaviour may indicate technological resistance (Buchmann, Chapter 10). In the USA activists force judicial rulings to change the law in litigations against individuals, corporations, or institutions. Court cases of this kind are called in the name of civil rights that are violated in the context of new technology.

We may distinguish levels of analysis. Our daily activity resists physical influences; we resist the weather, the sun, the heat and the cold in various ways; our immune system resists noxious agents. At the level of cognition we resist the 'external' influence on our experience that is mediated by film, radio or print media. At the organizational level the work-place is an arena of resistance against new technology, abject working conditions and exploitations. At the cultural level we may resist changes that are suggested or even imposed by exponents of other cultures on our way of life; where cultures meet there will be resistance. Symbolic meaning is both a resource and an object of resistance as shown by Staudenmaier (Chapter 7) in the case of popular images of Henry Ford in the USA.

Measuring the intensity of resistance

Measuring the intensity of resistance is essential to compare its effects across time, technology and national contexts. Measures of resistance allow us to associate resistance and effects in a systematic manner. Several measures of resistance will be put forward in this book.

I myself shall critically discuss how resistance to new technology has been measured psychometrically, and has been defined as a clinical problem. A test score defines 'cyberphobia', a supposedly 'irrational' form of anxiety towards information technology, that is claimed to be widespread. The measurement consists of two steps: first, to define 'cyberphobia' on a test distribution; secondly, to screen a population using this test as a criterion.

Daamen and van der Lans (Chapter 4) show how resistance to nuclear power, information technology and biotechnology is measured by negative attitudes on a multi-item scale of survey responses. The validity of survey measures is critically discussed in the light of the assimilation effect. Attitude measures are sensitive to the immediate context of data collection. Context effects throw light on artefacts occurring in survey measurements.

Rucht (Chapter 13) counts and analyses press coverage of protest events across countries. The type of event, the number of participants, and the degree of violence provide an indicator of resistance to nuclear power in different countries over an extended period of time.

Botelho (Chapter 11) demonstrates how diffusion studies measure resistance in a market system by the differentials in the diffusion rates of new products or processes. The later and the slower the process, the larger is the resistance of the system. Geographical concentration of an idea, process or product is another indicator for resistance in diffusion systems (Hagerstrand 1967). Some regions adopt new ideas and products, others not, so that over time clusters of diffusion appear in the landscape. Such processes are modelled by degrees of resistance in regions.

Miles and Thomas (Chapter 12) arrange forms of resistance to information technologies on an ordinal classification from non-acceptance at one extreme to resistance at the other. Degrees of non-acceptance of consumer electronics are: reluctant purchase; infrequent use, once purchased; partial use of functions; bewilderment over choice; and the moratorium 'no'. The criterion of acceptance is the purchase of consumer goods.

Comparing the intensity of resistance across the three technologies leads to the following conclusion: resistance is strongest against nuclear power, weakest in the case of information technology, and, to date, medium for biotechnology.[17] Information technology imposes small damage with high probability: no large-scale resistance has formed, yet. Nuclear power is characterized by high damage potential with relatively low probability. The risks of biotechnology are still largely unknown. The resistance against information technology is mostly local and a matter of 'intellectuals'; it is mainly informal, individual, and passive, such as refusal to work with computers. Organized resistance, as in the British printing industry (see Martin, Chapter 9), seems to be infrequent.

What is being resisted and why?

The title of this book suggests that the object of resistance **is** technology. This common-sense assumption is, as we shall see, problematic. What is being resisted is normally complex and requires empirical analysis. The study of resistance needs to face the possibility that resistance is more easily characterized by the process and its effects than by its antecedents; the causes may be more numerous than the effects.

Resistance may be directed against the machinery and the technical devices involved, as they become the symbolic focus of what infringes people's livelihood; this is traditionally the meaning of the term 'Luddism' or 'machine breaking' (see critically Randall, Chapter 3). It may be useful to distinguish resisting hardware from resisting its consequences. The latter case may target a parameter of the design, rather than the design as a whole. Effects have different ranges: personal, local, national, international. Local consequences differ from consequences in distant places by their significance.

Often it is neither design nor consequences that are resisted, but the process by which the technology is put to work that is found wanting. Manipulation, being patronizing and breaking informal contracts are the problems.[18] However, it makes only limited sense to separate the design idea from its implementation. The implementation **is** the design in time and specific contexts.

Big business and state power appear to be the principle object of resistance (Touraine, Chapter 2, or Evers & Nowotny 1987). New technologies are financed, developed and implemented by large corporate sectors or state bureaucracies under expert guidance, often in the context of defence and warfare. 'Technocracy' seeks to abolish politics and, by implication, to exclude the citizen and the public from the decision-making process; technocrats regard politics as an interference in the rational techno-logic. In such a context, neither science nor technology themselves are resisted, but their exclusive control by experts who are not held accountable in order to enhance control over non-experts. Intertwined as it became historically with a technocratic world project, not least in its Soviet version, 'scientific universalism' lost more and more of its appeal.

Another distinction is that between product innovation and process innovation. Process innovations increase the efficiency of production with new procedures and capital goods. Product innovations are new consumer goods. People resist new processes at the work-place and as investors and producers (see Bruland, Chapter 6); they resist new products as consumer (see Miles & Thomas, Chapter 12). Process innovations will be a focus for labour organizations, and product innovations for consumer organizations.

It is the new technology that is resisted, rather than the old one. 'Newness' makes a difference, not least because risks are difficult to assess, and the resulting uncertainty stifles action. 'Newness' is relative to place and time. Whether a technology is new is a matter of debate. Newness is claimed and highly desirable in the context of patenting, but it is not necessarily desirable in public debates on regulations. The features of public resistance are themselves the outcome of public debates, not least in the form of social scientific inputs into that debate.

Historically resistance is found in the context of technology transfer of fishing nets and techniques, the steam saw and automobiles in Scandinavia (Bruland, Chapter 6); and against the introduction of textile machinery in the early nineteenth century in Britain (Randall, Chapter 3). Its targets are specific implementations, a particular mix of techniques, or a technology at large. Furthermore, technology is resisted because it signifies the hegemony of a foreign power (Touraine, Chapter 2).

With regard to nuclear technology, various events are the object of resistance: nuclear power plants *per se*, government nuclear policy, the siting of particular plants, the disposal of nuclear waste in the sea or under ground, its transport over long distances, increased levels of radiation in the vicinity of installations, and nuclear power as foreign domination (see Rucht, Chapter 13, or MacLeod, Chapter 8).

With information technology, consequences are the object of resistance: the intrusion of privacy with sophisticated marketing control methods (Nelkin, Chapter 18); the threat posed to freedom and democracy by centralized social control over and misuse of information. VDU radiation and repetitive strains pose some health risks. Obfuscation of decision making with artificial intelligence is a major concern for people with intellectual commitments (e.g. Weizenbaum 1976). Loss in quality of working life, de-skilling, changing job structures, bad quality of user interfaces, and job redundancies are resisted. Electronic publishing revolutionized the printing process, making various intermediary tasks redundant. The conflict at Fleet Street, London, tells this story (see Martin,

Chapter 9). Products such as minitel, audiotext and videotext diffuse at different rates in different countries (Miles & Thomas, Chapter 12). Current concerns about new media focus upon the spreading of pornographic material and of computer addiction, a protective concern on behalf of youth, with a legal angle when 'hackers' intrude and interfere with sensitive data bases.

Resistance to biotechnology targets the siting of research facilities, as, for example, in Switzerland (see Buchmann, Chapter 10), the field testing of genetically altered plants, the breeding of transgenic animals, the patenting of life forms, and experimentation with human embryos and new reproductive technologies for humans. The control of the genetic code is open to abuse. For Jeremy Rifkin this constitutes the latest in a series of secular 'enclosures' for purposes of exploitation: from land, to sea, to air, and now to genes. Processes, products, and their implications are resisted, where the sanctity of life and nature, biodiversity, and the sustainable development of the world are at stake.

The concept of resistance motivation summarizes those processes which start and end the resistance activity, and which maintain the momentum of activity in the face of obstacles, and maintain vigilance in the absence of a real concern (Haltiner 1986). It covers all processes of individual and collective mobilization. Motives describe particular processes of that kind. Owing to its ambiguous nature the technological process is accompanied by anticipations of known and unknown dangers and risks. Fear and anxiety are primarily evolutionary achievements that prepare the organism for adequate action in situations of danger. Emotions such as fears and anxieties redirect energies and prepare for necessary actions. Resistance often reveals competing 'Weltanschauungen'. Images and visions of the future are motives both for technological design and for resistance to technology (see Staudenmaier, Chapter 7). Anticipations of doom may motivate actions to avert the catastrophe scenario. The diagnosed shift in cultural values (Inglehart 1990) influences people's preferences in private consumption and political behaviour. Finally there is motivation resulting from the recursive effects of resistance on itself; initial success mobilizes individual and collective energies for future action and rallies support and new resources; failure may lead to both increased determination or to loss of confidence; resources are depleted and resistance may come to an end.

Who is resisting?

Any study of resistance needs to identify the actors. Who are the people that resist a particular technology or new technology in general; how do they differ from other social groups; how large is this group, and where are they located within the structures of society? Generally there are two ways of defining social groups, statistically in terms of socio-demographic variables ('statistical groups'), or according to self-reported membership ('natural groups').

Natural groups

Natural groups provide a social identity and are often formally organized with a legal status as a club or society that carries a name. Specific groups that have been associated with resistance to new technologies are groups with a cultural mission. C. P. Snow's distinction of a world of arts and a world of science (Snow 1959; Hultberg 1991) has recently been called upon. Arts and the social sciences are associated with ignorance of and resistance to science and technology. Kepplinger (Chapter 17) recognizes in particular the post-1968 generation as political editors and agenda setters within the German press, the '**reflective elite**' that through its position influences public opinion about technology negatively.

New **radical social movements** such as feminists, environmentalists and animal rights movements are associated with resistance to new technology. Religious groups rally for moral concerns and stricter regulation of biotechnology. Activists such as Jeremy Rifkin in the USA rally rainbow coalitions around campaigns, which recently included chefs signing up to refuse to use genetically altered vegetables in their cooking. Much organized resistance is community and culture based, rather than based on economic categories such as labour or consumers. Local networks of people galvanize into large-scale activity under favourable conditions such as economic growth, persistent leadership and coalitions.

The **anti-nuclear movement** is made up of numerous groups with an international network and constitutes a well-defined topic of social research (Rüdig 1990). The resistance to computer technology is less organized. At times, but not generally, trade unions are the organizational resources for resisting changes in production technology (see Martin, Chapter 9). Kling & Iacono (1988) identify three core actors of the '**counter-computer movement**': libertarians concerned with civil rights and privacy issues; consumerists concerned with new forms of consumer credit; trade unionists concerned with impacts on the number and quality of jobs; and pacifists concerned with computerized warfare and the likelihood of wars. All these groups are formed around wider issues, and take up the resistance to certain developments in computing and information technology as part of their activity.

'**Green' parties** are trans-national and core activists like Greenpeace or Jeremy Rifkin are international actors of resistance to nuclear power and biotechnology, less so to information technology. Touraine (Chapter 2) points to the resistance of **governments**, for example in Islamic countries, who for reasons of nationalist politics may resist new technologies to block foreign hegemony. Botelho (Chapter

11) shows how a **civil service elite** with a particular vision of research is a factor of resistance, in this case unintended, in the development of semi-conductor technology in post-war France.

Statistical groups

Various classes of actors can be distinguished on purely conceptual criteria. Such criteria are not normally used by people to define their social identities. Sociological analysis suggests that the social basis of new social movements, some of which take among other preoccupations an anti-technological position, is diverse, but to a large extent **middle class** (Pakulski 1993). However, the correlation between an economic position and social mobilization weakens in the post-war period. According to Touraine (Chapter 2) cultural conflicts will increasingly supersede socio-economic ones. Value orientations increasingly explain political mobilization better than income or property based social stratifications (Buchmann, Chapter 10). Beck (1993) predicts a realignment of the political sphere over the distribution of large technological risks which is superimposed on the traditional conflicts over income distribution.

General public opinion is continuously gauged in social surveys and opinion polls. It caused a shock wave in 1984 when a German Allensbach poll found that 10% of the German population regarded technology as evil, and 54% partly so. The percentage of those who regarded technology as 'a good thing' dropped from 72% in 1966 to 32% eighteen years later (Jaufmann & Kistler 1986). Opinion polls and attitude research located negative attitudes towards information technology and computing in the mid-1980s among the less skilled, low earning, female, rural, older and left-orientated parts of the population (Bauer 1993). Studies on 'cyberphobia' (see Bauer, Chapter 5) report its prevalence among women, the less educated and the marginalized.

Vested interests in past investment is a frequently mentioned category of actors; Bruland (Chapter 6) shows how in Scandinavia holders of old capital obstructed the building of new capital in the form of new machines and new methods of production. Historically labour elites such as cotton spinners or wool combers, or groups such as small farmers, labourers, brick makers and paper makers, were resisting new technology during the industrial revolution (see Randall, Chapter 3).

Buchmann (Chapter 10) shows how in Switzerland the combination of **traditional values, social marginality and habitual nay-saying** forms a cluster of motives that explains negative votes in a national referendum on the regulation of biotechnology and reproductive medicine.

A rising number of people choose 'soft medical treatment', and practise **astrology and parasciences** which coexist with beliefs in, activities in and support for science proper. Observers often tend to construe such practices as expressions of anti-scientific attitudes; however, these practices serve different needs and are therefore not necessarily contradictory (Touraine, Chapter 2).

In organizational research it was a traditional claim that resistance is a matter of **corporate hierarchy**: the lower in the hierarchy the more likely is the resistance to innovations (Lawrence 1954; Johns 1973). The arrival of information technology and computers since the 1960s has altered this empirical finding. For some time resistance to the use of computers polarized on the top **and** on the bottom of the corporate hierarchy, with middle management being the innovators. In the context of new manufacturing methods (e.g. CIM, production islands), the middle strata are likely to resist the innovations (Klein 1984; Littek & Heisig 1986; Carloppio 1988). The position of the middle management is often at risk in the flattened hierarchical structures that computers may bring. In the corporate context resistance is no longer related to hierarchy; depending on the context, resistance is found at all levels of the corporate hierarchy.

Non-users, purchasers of electronic equipment who make only partial use of it, and family members who do not use the electronic devices that other members have bought (Miles & Thomas, Chapter 12) are **categories of consumers** that resist technology by underuse. Non-using may be deliberate resistance, or it may be an indicator of uselessness, functionally equivalent to pushing a product out of the market. Similarly the biotechnology industry anxiously anticipates consumers who will hesitate to buy genetically altered foodstuffs, such as the 'flavr savr' tomatoes specially engineered for a longer shelf life. The controversy on what information product labels should contain speaks clearly of the nervousness in the biotechnology industry.

What are the effects of resistance?

The literature on resistance is rich in descriptions of the various forms of resistance, the analysis of its actors, and of the causes, motives and conditions. Most studies focus on resistance as the dependent variable. The analysis of resistance by consequences, with resistance as an explanatory and independent variable of technological change, is rare; corresponding chapters in textbooks tend to be thin. Findings on consequences of resistance are best developed for the anti-nuclear protest movements (Rüdig 1990; see Rucht, Chapter 13).

Principally four dimensions of effects may be distinguished:

* foreseen and unforeseen effects,
* direct and indirect effects,

* impact and recursive effects,
* functional and dysfunctional effects.

These analytical dimensions define in combination a theoretical space of sixteen types of effects: from foreseen–direct–functional impact to unforeseen–indirect–recursive and dysfunctional effects.

Direct effects are the immediate effects on the innovation process, while indirect effects impinge on the technological process via mediating processes. Indirect effects are often instrumental in reaching the objectives of actors. The distinction of foreseen and unforeseen refers to the point of view of resistance actors. Unforeseen effects cannot be intended, while foreseen effects may be intended but need not. It may be helpful to distinguish the effects of resistance on the technological processes, and recursive effects on the resistance itself in terms of structural changes: growth, motivation, disintegration and decline. Functional or dysfunctional refers to the long-term effects of resistance on the technological process. Functional effects increase the viability of a project, dysfunctional effects decrease it, depending on the level of analysis; effects that are dysfunctional on one level may be functional on another level and vice versa. Events are not *a priori* functional or dysfunctional; it depends on the process dynamics. Processes assumed to be functional may turn dysfunctional under conditions that have to be clarified empirically; events may be both functional and dysfunctional simultaneously but for different processes. Commonly, functional and dysfunctional correspond with the positive or negative value. Actors try to reduce dysfunctional effects and to increase functional ones. However, from an observer's point of view we are able to appreciate the positive consequences of seemingly dysfunctional events.

Finally, resistance does not determine its own consequences; its effects are contingent. Similar forms and intensities of resistance result in different nuclear power programmes in different countries. Traditions of legal regulation channel the voicing of resistance and constrain its effects on technological developments (see Rucht, Chapter 13, Jasanoff, Chapter 15, or Martin, Chapter 9). In analysing effects we need to take into account the reactions of innovators and regulatory bodies. These contingencies make it difficult to attribute 'causality' between resistance and its effects. More realistic are circular influences where past effects condition the future relationship between resistance and its effects. This relationship is itself a variable.

Direct and foreseen effects

Direct effects are often reactions of actors such as governments, international agencies or corporate management challenged by resistance whose actions alter the trajectory of technology. On the one hand, actors may try to calm a conflict situation with the revision of a project or a policy reform: a response that may be in timing and content the effect of resistance activities. On the other hand, forms of repression and police actions need to be expected as resistance often transgresses the boundaries of legality. The state is called to enforce the rule of the law. Repression may be successful in temporarily suppressing resistance activity; however, in many cases this is only to strengthen its resilience, and to widen its mobilization basis, and media coverage, and to lead to silent conversion among the wider public (Raschke 1988, p. 355; Mugny & Perez 1991, p. 154).

Preventing an event from happening is the strongest direct effect. A project may be abandoned in response to resistance as in the case of the Austrian nuclear power plant 'Zwentendorf' near Vienna, mothballed before it started to produce energy. Other effects are: delaying the resisted events; postponing a final decision; abandoning a project for good, or for only a limited period of time. The Luddite struggle against machinery is an example of the latter. The introduction of textile machinery was postponed in a particular area of Northern England (Randall, Chapter 3).

Resistance lowers the level of expectations of innovators. The exuberant expectations of many a project face a 'reality test' on resistance, and get adjusted. A striking example is shown by the international projections for nuclear energy. The exuberant predictions of the 1950s and 1960s were massively reduced in the 1970s. Figure 1.2 shows the projections for 1977, 1980 and 1990 together with the actual development of world nuclear energy production. The actual production in 1990 is about half the projection of 1977. The projection for the year 2000 in 1990 is less than a quarter of that of 1977 (IAEA 1981; WNIH 1992).

A further effect of resistance is the relocation of innovations (Rucht, Chapter 13). Resistance to new technology may have spatial and cluster effects. Successful resistance in one location is no guarantee of success in other locations. This results in concentration of the technology in one region or in one country, while in neighbouring areas the technology is hardly present (Hagerstrand 1967). The effect will be the spatial distribution of new technology. Nuclear power is an example: some countries, such as France or Japan, have realized a large nuclear power programme while other countries have abandoned it. In the debate on new biotechnology dislocation is a much anticipated effect. If companies are prevented from building research facilities in some areas by local resistance or because of strict safety regulations, they may move into other countries, and finally to the Third World, where conditions seem to be 'favourable'.

Often the definition of the problem is at stake. Resistance may oppose the particular way a problem is 'framed' in terms of issues, themes and concerns. Such a frame may mean a lack of choices, or present an unacceptable choice. For

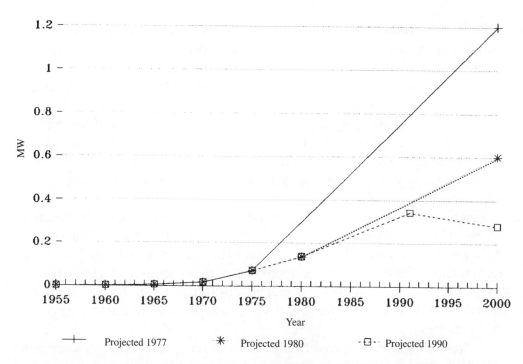

Figure 1.2. Changing expectations of the nuclear industry 1977 to 1990. The graph shows the production and projections of nuclear power production for the year 2000 by 1977, 1980 and 1990. Source: Nuclear Industry Handbook 1992.

political decisions issues are often reduced to a binary option corresponding to the polarities of an adversarial political system. This dichotomy is achieved in various ways. The role of resistance may be to indicate how inadequate the binary option is and to reopen the debate beyond the simplifying either/or. A variety of perspectives get voiced where formerly experts dominated the debate. As an outcome the parameters of a problem may multiply. The environmental movement introduced a set of new criteria (environmental impacts) on existing problems such as resource management and waste disposal. The environmental agenda has penetrated the rhetoric of most political parties to the extent that the movement has lost its distinct political profile; the environmental movement becomes a victim of its own success. The success of a resistance movement may be a condition of its dissolution.

Indirect and foreseen effects

Indirect effects are instrumental in achieving the objectives of resistance activities; as such they are intended. Of more tactical than strategic interest are media effects. Media coverage is important for setting the public agenda, mobilizing public opinion and forcing governments to act (see Kepplinger, Chapter 17).

Activities of resistance intend to be newsworthy and spectacular. Greenpeace and anti-abortionist activities, such as chaining protesters to a tree or to furniture or camping on the top of a chimney that pollutes the air, epitomize this quest for publicity. Depending on how the media are taking up the issue media coverage can be supportive or can backfire. Access to the media is important for defining an issue for public debate. Kepplinger shows how the anti-technological movements were able to mobilize major parts of the German press for their causes. Media coverage is also important for recruiting new members, stabilizing a public profile and legitimacy, gaining coalition partners, and challenging the reactions of opponents (Schmitt-Beck 1990).

In democratic societies a **referendum** can be an effect of resistance, particularly where such a procedure is not normally foreseen. The referendum on the Austrian nuclear power plant at Zwentendorf in 1978 is such a case. The Swiss referendum on biotechnology and reproductive technologies, part of the normal political procedure, is analysed by Buchmann (Chapter 10). Legal regulations set an important context for the development of new technology such as biotechnology.

Resistance often leads to new regulations to prevent misuse of knowledge and related practices. **Getting the legal system moving**, i.e. the production of new laws or new interpretations, can be attributed to resistance, while the kind of regulation that emerges from the process follows different constraints. Jasanoff (Chapter 15) will show how legal traditions condition the regulatory response to resistance to biotechnology in Germany, Britain and the USA: process versus product regulations, or a combination of both.

Unforeseen effects

Unforeseen effects are effects of resistance on the technological process that have not been intended nor anticipated by resistance actors. These effects share an element of surprise for actors, as they are unexpected. The analogy of evolution (mutation, selection and retention) on technological change (Mokyr 1992) stipulates resistance as an external factor that selects products and processes over time. Retrospectively these appear as lines of progression. At the time the selection process is surprising and subject to attempts to gain control. Any present range of technical devices and ideas is the outcome of past resistance which works as a filter; screening out some items, letting through others, pushing the path of 'progress' into a particular direction. Technological development is at times gradual, at times it jumps from one 'punctuated equilibrium' to the other, depending on large-scale environmental events. Within the temporary state of equilibrium small changes accumulate; fine tuning based on existing knowledge contributes massively to overall gains in productivity; revolutions from one equilibrium to another alter the basics of design with totally new ideas.

The resistance of consumers to new products or of investors to new processes make innovations disappear from the market. These products or processes are retrospectively the waste of evolutionary selection. Innovative activity produces variants of products and processes, from which consumer preferences and investment rationales select viable ones.

Another unforeseen outcome of resistance to new technology consists of institutional innovations and new political opportunities (Tarrow 1988). It seems paradoxical that social innovation is attributed to a movement that resists innovations. Resistance movements may bring new forms of political and social negotiation that did not exist before. Procedural outcomes are new forms of debate in new 'arenas' and with access to information that was previously not available. Evers and Nowotny (1987, pp. 279ff) have shown how formerly unknown technological risks are contained by new social institutions such as insurance, technology assessment in various forms, public consultations, hearings and other fora for public participation. However, more of a good thing is not always better. Participatory institutions create fora for discussions between the individual, groups, corporations and the state. Simultaneously they may produce an overload of democratic decision-making capacity. Too many democratic proceedings tire the citizens, slow the decision process, and if a new bureaucracy is established, institutional paralysis may be the consequence.

Once locked in a particular frame of definition, a problem may need to be opened again to find different solutions. Institutional proceedings that close a debate need the complement of those that open them, otherwise the social system becomes paralysed in divisions and inactivity. The 'reinvention of politics' is predicted for issues of new technologies (Beck 1993).

Private business contributes to institutional innovations; new forms of consultancies emerge, dealing with risk communication, public perceptions, public relations, and organizing fora for public debates on new technology. Consultancies provide a kind of 'therapy'; consultants are being well paid for telling highly unpleasant news to decision makers.

Another unforeseen effect is what social psychologists call 'silent conversions', 'delayed effect' or 'sleeper effect' (Mugny & Perez 1991, p. 63). A consistent and persistent minority will put its point across in an unexpected way. People are changing their attitudes silently; while publicly admitting to be in favour of some contested technology, some of these people are changing their attitude in private; the determinants of the old attitude weaken and a sudden change becomes possible, which will materialize during periods of public debate. For a consistent minority the potential support is often much larger than is publicly visible.

In the nuclear debate resistance managed to raise the safety standards to higher levels than were ever envisaged by the nuclear industry. The industry, by complying with safety standards to render nuclear power more acceptable, incurs increasing costs. Safety expectations in the public have increased the costs of building, running and waste disposal for nuclear power plants to an extent that makes its economic viability doubtful in competition with other energy sources (see Rucht, Chapter 13). In 1980 the unit costs of nuclear power generated electricity had risen to 571% of the 1968 level (Radkau 1983, p. 585). The recalculation of the economic rationale of nuclear power is at least in part due to the public scrutiny of anti-nuclear movements and its effects on public expectations and government regulations.

Social differentiation is an unforeseen effect of resistance. In taking part in a controversy on new technology public opinion is likely to polarize along the lines of that controversy if the issue is kept in the public eye for long enough. The segmentation of opinions and attitudes in social and geographical space may cut right across traditional lines of social differentiation, such as political party affiliation, religion, social class or income, as the environmental movement has shown. The alignment that anti-technology movements have introduced into Western societies in the last 25 years defies an explanation in terms of economic categories such as income or property ownership.

The paradox of resistance is that it may **increase the rate and the depth of change**. Crozier (1963) argued that bureaucratic institutions change in a cycle of cataclysms. Periods of small or no changes alternate with periods of fundamental upheaval. This may lead to the effect that the total rate of change and its range may be higher than otherwise. A similar characteristic of scientific progress was suggested by Kuhn (1962, p. 64) who observed that 'normal science' leads to a refinement of instruments and concepts and to institutional rigidity which will be shaken more fundamentally the more rigid it is. In the catastrophe model of organizational change by Bigelow (1982), resistance is the splitting factor that creates the cusp of **unpredictability** in the system. Depending on the degree of resistance in a system, the rate of change will be gradual or unpredictable. The linear correlation between forces for change and resulting changes becomes chaotic with high levels of resistance in the system. Resistance leads to abrupt, unpredictable and hence uncontrollable changes in a system. From the point of view of interest in fundamental change, resistance in the system is ultimately conducive to innovation.

Recursive effects of resistance actions

An outcome of any action is to inform future actions of a similar kind. All the effects previously discussed are contributions to the socio-technical process of which resistance is a subprocess; recursive effects strengthen or weaken the viability of future resistance. It makes sense to conceive the interaction of social actors, innovators and resistance as a learning process of structural change. Protest is its own school, and resistance feeds on past resistance for strategy, tactics, cognition, motivation, and its cultural basis (see Randall, Chapter 3). Conceptually one may argue that we are not dealing with **social analogies** – different social processes that solve similar problems – of resistance across technologies, but with **social homologies** – social processes that are similar due to common origin and tradition.

Resistance activity provides a social identity and fosters social cohesion. Scattered individuals or small groups merge in a larger network with an image and awareness of themselves in contrast to others, giving social significance to their members. The post-war history of European nations provides many examples: national identities are redefined based on the experience of resistance against fascism and Soviet–Russian hegemony. The experience and practice of resistance provides the bridge to span the gap between political parties and their ideological divisions, and locks them into an historic consensus (Wippermann 1983). As soon as the memory of that common resistance fades away, the structure that is built on it weakens, and the post-war political system is in crisis.

Official recognition is another recursive effect of resistance. An initially marginal core actor may gain official recognition to speak for the people in resistance. This **legitimation** has an internal and an external dynamic. Internally several actors and factions may compete for leadership; the external recognition of leadership may solve at least temporarily an internal conflict. The legitimation of leadership may provide a distinct profile, direction, coordination and personalized image of its causes. Personalization of ideas is important to obtain effective media coverage. A representative can join the formal political process in parliament or in commissions to negotiate face-to-face with other parties. On the other hand the basis of leadership is fragile as the increasing distance between figurehead, formal organization and grassroots may undermine its legitimacy (the iron law of oligarchy). The fate of the German Green movement, torn between basic democracy ('*Fundis*') and political effectiveness ('*Realos*') tells a story in that respect. A similar story is told about Greenpeace, where the increased attempt to legitimize claims by invoking scientific research and authority opens up the gap between the leadership of the movement and its basis, and undermines the basis of the movement (Yearley 1992).

Motivational effects need to be taken into account. Motivation describes the processes of beginning, ending, and sustaining resistance activity in the face of difficulties. Media coverage, favourable or unfavourable, provides a public profile and identity, which is likely to strengthen the social cohesion and determination of collective action. Success in some geographic location against a particular technological development may increase the determination and confidence of the actors involved, and by the spread of news stimulate actors in other areas. Repression deters people from action, but may increase the determination of a small group of actors to go underground and to fight with all means available, which may lead to extremism and to violence.

Measuring effects of resistance

Several measures of effect that are used in empirical research will be discussed in this volume. Rucht (Chapter 13) and Rudig (1991) estimate the **rate of completion** of national nuclear power programmes by comparing the status quo, current plans and initial plans of energy capacity. The difference between plan and reality after the occurrence of resistance is an index of its effect. Similarly projected figures of future (past futures) production of nuclear energy measure the **changing expectations** of the industry in reaction to resistance. The number of **orders of nuclear power plants cancelled** is another index of effect. However, the link between these effect measures and resistance requires detailed case analysis, because contingencies other than popular resistance influence the development of the nuclear industry.

Content analysis of media coverage measures the **attention** that resistance activities mobilize, favourable and unfavourable, for their causes. Media coverage is a crucial inroad to study social reality, because of the central role of media in reducing the complexity of modern reality for the citizens. What, to what extent, and how resistance is covered, are basic data for measuring the effects of resistance on public discussion. Intensity measures allow us to assess the strength and the structure of the 'resistance signal' and to relate it to the changes in the technological process. Time series data are essential for this purpose.

Another source of measures is diffusion research. Resistance in a system will lead to **clusters of innovations** in certain regions (Hagerstrand 1967). The degree of regional disparity, the number of clusters, and their size may be taken as indicators of the effect of resistance. Delays in the timing of the take-off of a diffusion process, and the rate of diffusion are indicators of resistance to innovation in a social system (Botelho, Chapter 11).

An overview of the book

The contributions to this book are grouped into five parts to compare resistance and its effects according to several criteria: comparisons by (a) time, between historical events before 1945 and after 1945, (b) across different technologies in the post-war period; and (c) cross-national. This three-dimensional grid of comparisons shows the complexity within which one attempts to generalize empirical findings; abstraction may prove to be more promising than induction.

The first part of the book introduces **conceptual issues of resistance**. Alain Touraine (Paris, France) opens up a wide horizon with reflections upon the origins of the present 'crisis of progress' and the modern dilemma of universal rationality and particular identities: we still believe in science but no longer in 'Progress' writ large. The progress of science brought the rise of 'big science', 'technocracy' and, dialectically, of anti-technocratic attitudes. He pleads for an extension of democracy to bridge the increasing gap between functionally differentiated social activities. Adrian Randall (Birmingham, UK) presents a detailed reassessment of three forms of Luddite resistance and their impacts on the early nineteenth century English textile industry: strike, violence and appeal to authority. His historiographical comments situate traditional views of 'Luddism' in Whiggish accounts of technological progress and labour history. Dancker Daamen and Ivo van der Lans (Leiden, Netherlands) discuss the measurement problem of 'context effects'. In measuring present-day resistance with surveys questioning public attitudes to technology, we incur measurement effects. The way we sequence survey questions in questionnaires and interviews affects significantly the results of the survey. The 'assimilation effect' is demonstrated and strategies to increase the validity of such survey data are discussed. Finally, Bauer (London, UK) will trace the periodic revival of the concept of 'technophobia' to assess people's reaction to new technology. Cyberphobia epitomizes the clinical view of resistance which attributes a pathological deficit to the computer non-user. We may more usefully conceive resistance as a signal of the mismatch of expectations between users and designers. With data from a study I show how this resistance signal affects a software development project over eight years in a Swiss Bank. I shall develop this idea of a 'signal' in the last chapter on the functionality of resistance.

Part II brings together **five national and regional studies** to show the **historical diversity** of resistance in form and effect. Kristine Bruland (Oslo, Norway) shows how, in the context of nineteenth century Scandinavia, resistance to new technology is not confined to labour. Governments, regional administrators, civil servants and industrialists have vested interests in old technologies. Long-term conflicts over new technologies in Scandinavia concern new fishing nets and the introduction of the steam saw in the timber industry. The private automobile in Norway and nuclear energy in Sweden are examples of public resistance forcing a selection on the technological process in the twentieth century. Staudenmaier (MIT, United States) shows how the name 'Ford' evokes images which are equally ambiguous as Henry Ford's personal reaction to his own achievements. Ford himself, in his later years indulging in an idyllic escapism while simultaneously bringing his technological visions to a climax, exemplifies the ambiguity with which humans confront technology. Roy MacLeod (Sydney, Australia) unfolds a narrative of the Australian debate on nuclear issues: of 'Nuclear Knights', of the official secrecy, of international relations, and of the resistance from various sources that to date has prevented nuclear power from producing a significant amount of energy on that continent. Roderick Martin (Glasgow, UK) recalls the first round of industrial conflict on the introduction of new computer based production technology in Fleet Street, London, where, between 1975 and 1980, highly unionized staff successfully resisted attempts to introduce computerized photocomposition. This was possible because neither proprietors nor unions were united in their strategies. New ownership and new industrial relations were capable of implementing the new technology ten years later. He discusses the functional role of resistance in 'voicing' issues that are neglected. Finally, Marlis Buchmann (Zürich, Switzerland) analyses voting behaviour in a Swiss referendum, where in May 1992 the electorate decided upon the legal framework for biotechnology and reproductive technology. A public debate gave strong signals to interested institutions and expressed the relevant concerns. Survey data show that supporters of the new regulations, young, urban and well educated, express a value pattern of 'limited progress' which is different from the 1950s' unconditional quest for modernization.

Part III compares **resistance to a new technology across different countries** and shows how both the occurrence of resistance and its effects are contingent upon contexts. Antonio Botelho (MIT, United States) analyses the diffusion of semiconductor technology until the mid-1960s in France and Japan. Resistance in the diffusion process is the product of a political arena where the state, industry and professional cultures meet. He unravels how France and Japan differentiate their involvement in the semi-conductor industry despite similar starting positions in 1945. Miles and Thomas (Manchester and London, UK) show how resistance manifests itself in the context of information technology such as videotex and audiotex with differential success rates in France, Germany and the United Kingdom. They order recent developments of interactive information technologies and distinguish forms of consumer resistance to explain the relative failure of videotex in Britain. They discuss the effectiveness of UK regulatory bodies in voicing concerns on behalf of the public. Moving to nuclear power, Dieter Rucht (Berlin, Germany) reviews the history of anti-nuclear resistance and presents an elaborate attempt to assess both resistance and its effects across sixteen countries. The strength and effectiveness of anti-nuclear resistance varies around the world. The emerging picture is complex. The effects of resistance on nuclear policy are contingent upon economic contexts, the way political decisions are made, and the attitudes of elites in different countries.

On new biotechnology, Robert Bud (London, UK) reviews the actions of various government agencies in Japan, Europe and the USA between 1970 and 1986. Trying to accommodate both public concerns and industrial interests, different concepts of biotechnology are promoted at different times. An undercurrent of public distrust survives all regulations and may resurface with new issues such as the human genome project. Jasanoff (Cornell, United States) compares the legal cultures of the USA, the UK and Germany and shows how they accommodate the risks of new biotechnology in different ways. She identifies three paradigms of control – product, process and programme – as contingent forms of public reassurance with regard to physical, social and political risks. Forms and effects of public resistance are both contingent upon these cultural contexts. The definition of what is 'new' is at the heart of the matter.

Part IV compares resistance **across different technologies in one country** and explores the problems of drawing analogies between technologies. Joachim Radkau (Bielefeld, Germany) compares the debate on nuclear power and biotechnology in Germany with a stage model and discusses the 'risks' of the analogy by risk. Nuclear power and biotechnology have mobilized the German Green Party; a parliamentary commission on new biotechnology preferred analogies with information technology to analogies between nuclear power and biotechnology. Both the risks and the chances associated with these developments require a more calm and more critical assessment. Hans-Mathias Kepplinger (Mainz, Germany) compares the press coverage of nuclear power and biotechnology in Germany and its institutional context. The long-term trend towards critical commentary on new technology seems to bear out a paradox: attention and negative attitudes increase while actual damage decreases. This is explained with a two-culture model where the scientific and technological elite confronts a 'reflective-elite' of literary and social scientific provenence. The mentality of the reflective elite changes the culture of journalism to a focus on 'negative' events. The agenda setting function of some newspapers for public opinion is assessed quantitatively. Dorothy Nelkin (New York, United States) finally compares information technology and biotechnology in the USA by their impact on people's privacy. She equally points to the contradiction between real impact and actual concern. Information technology, more intrusive into privacy than biotechnology, does not mobilize as much public concern. This contradiction reveals the hierarchy of values and exposes a gap between rhetoric and reality in American society. What makes people move are health risks, organized interests and religious agendas, while lip service is paid to the protection of privacy, freedom and democracy.

In Part V, Bauer will attempt to integrate the various contributions with the help of the 'pain analogy' of resistance by abstracting from various contexts: resistance is a signal that things go wrong. Resistance works as the 'acute pain' of the technological process. Social system theory provides the framework to elaborate the self-monitoring functions of resistance: it allocates attention, evaluates and alters the technological progress. Resistance is primarily a functional process that is constituted in communication about it; dysfunctions are likely, but are secondary. Several implications for research on resistance within socio-technical progress are suggested: the analysis of resistance shifts from causes to effects; and to the analysis of the symbolic encoding of events over long time periods and in the contexts of activity systems.

Notes

1 Many of the following ideas were developed during discussions at the conference workshop on 7 April 1993. I would like to express thanks to the many contributors, equally to Jane Gregory, George Gaskell, Sandra Jovchelovitch and Alan Morton for helpful comments on the chapter. However, I take full responsibility for the way ideas are presented here.

2 For a view of the managerial bias in the resistance literature in the Lewin tradition which I described as the 'forced feeding paradigm' see Bauer (1991, 1993); for a sharp critique of notions of resistance in modernization theory see Reverendi (1975).

3 Other comparisons were suggested: compulsory vaccination was withdrawn on public demand in Britain in 1907 after 37 years of resistance. Mazur (1975) has compared public resistance to fluoridation of water and to nuclear power in the USA. Mazlish (1965) explored analogies between railway building in the 19th century and the space programme in the 1960s with regard to social impact.

4 The debate on the empirical evidence for and the explanation of these cycles continues (Freeman 1984; Hall 1988). I use the idea of long waves descriptively to demarcate periods of historical events (van Roon 1981).

5 I assume that the OECD reports on particular developments indicate the timing of the giving of political attention by key member states to significant new developments; some states may have identified the issue earlier at a national level and drawn the attention of others through the OECD activity. For example, the German DECHEMA report on biotechnology dates back to 1974 and may have served as an international agenda setter (see Bud, Chapter 14).

6 I disregard here earlier reports on microprocessor developments in 1961, and on the automation debate in the 1950s and 1960s. These issues were temporary concerns which did not initiate a continuous focus of attention.

7 'New materials' are often being identified as a new base technology, but they do not give rise to public debate, nor to Utopian images of future societies such as the 'nuclear society', the 'information society', or the 'biosociety'.

8 Strictly speaking the demonstration of this serial media attention would require data from one single country; however, longitudinal data on all three issues are not available from any one country. One can expect that countries peak in different years, but the overall series in Figure 1.1 remains the same in the light of the available evidence. We are collecting comparative longitudinal data on press coverage of all three technologies in Britain between 1945 and the present, but the data are not ready yet (for preliminaries see Bauer 1994).

9 The coverage of nuclear power issues went beyond the 1979 peak to cover the Chernobyl disaster in 1985. For evidence of an international peak of nuclear power news in 1979, see Saxer et al. (1986, p. 169) for Switzerland; for the USA see Gamson and Modigliani (1989, p. 17); also Mazur (1984, p. 106), who indexed articles on 'nuclear power plants'; Weart (1988, p. 387) finds the news peak on civil nuclear power four years earlier in 1975 by following the

keywords 'atomic', 'nuclear' and 'radioactivity'. Weart equally shows an earlier peak in coverage on civil nuclear power in 1955 and 1956. Contrary to the late 1970s, the discussion in the 1950s was mainly in a positive tone.

10 Back in 1963 an American journalist working in London, Harvey Matusov, founded the 'International Society for the Fight Against Data Processing Machines' (ISFADPM) with the aim 'to destruct man's overdependence on the computer' (1968, p. 7). Matusov (1968) published a record of computer atrocities. At the end of the book he lists a series of 'guerilla warfare' tips to confuse and challenge data processing machines; on request the ISFADPM would send more detailed instructions. How far this was a spoof or a serious endeavour is difficult to reconstruct without further evidence about the society's activity.

11 The topical distribution of public opinion surveys may be another indicator of public concern. To establish an inventory of such surveys across different countries is a topic for future research. See Jaufmann & Kistler (1986) for an international secondary analysis of public opinion on information technology. They managed to diffuse the idea of a particular syndrome of 'Technikfeindschaft' (technophobia) in Germany by international comparison.

12 In fact the British judge, John Presser QC, has rejected a claim for compensation for RSI, declaring that the term RSI was 'meaningless' and 'had no place in the medical dictionary' (The Independent, 23 November 1993).

13 On such visions of nuclear power as a mobile source of energy for various small-scale purposes, see Radkau (1983, pp. 79f) in his discussion of the 'nuclear myths' of the 1950s.

14 This idea is taken from the discussion of what constitutes 'resistance' in the context of Fascism and National Socialism in Germany (Kershaw 1985, p. 781).

15 In the discussion on resistance against Nazism in Germany, a functional definition had the effect of shifting the focus of research from 'elite groups' and a few spectacular events to more widespread daily forms of non-conformism, which demonstrated the diverse ways in which limits were set to a horrendous political project. The symbolic significance and real effects of such actions remain controversial (Kershaw 1985, p. 780).

16 The resistance of 18th century French peasants as well as of present day Malaysian farmers to tax demands of central government is silent, and neither coordinated nor self-conscious as Scott (1985) has described in his Weapons of the Weak.

17 Newspapers have recently reported violent attacks on research institutions in the USA. Letter bombs seriously injured Carles Epstain, at Berkeley University, and exploded in the Computer Centre at Yale University. The FBI stresses that this is likely to be the act of an individual (e.g. The Guardian, 25 June 1993; NZZ, 26 June 1993). In Britain some scientists receive police protection due to threats from animal protection terrorists.

18 Psychotherapists call this 'interactional resistance' (Petzold 1981).

References

AYERS, R. U. (1990). Technological transformations and long waves, Part II. *Technological Forecasting and Social Change* **36**, 111–37.

BAUER, M. (1991). Resistance to change – a monitor of new technology? *Systems Practice* **4** (3), 181–96.

BAUER, M. (1993). Resistance to change. A functional analysis of responses to technical change in a Swiss bank. PhD thesis, London School of Economics and Political Science.

BAUER, M. (1994). Science and technology in the British press, 1946–1986. In *When science becomes culture*, Vol. II, ed. B. Schiele, M. Amyot and C. Benoit. Boucherville: University of Ottawa Press.

BAUER, M., DURANT, J. & EVANS, G. (1994).   European public perceptions of science, *International Journal of Public Opinion Research* **6** (2), 163–86.

BECK, U. (1993).   *Die Erfindung des Politischen.* Frankfurt: Suhrkamp.

BIGELOW, J. H. (1982).   A catastrophe model of organisational change. *Behavioural Science* **27**, 26–42.

BRODDE, K.(1992).   *Wer hat Angst for DNS? Die Karriere des Themas Gentechnik in der deutschen Tagespresse von 1973–1989.* Frankfurt: Peter Lang.

BUD, R. (1993).   *History of Biotechnology.* Cambridge: Cambridge University Press.

CARLOPPIO, J. (1988).   A history of social psychological reactions to new technology. *Journal of Occupational Psychology* **61**, 67–77.

COVELLO, V. T. (1991).   Risk comparison and risk communication: issues and problems in comparing health and environmental risks. In *Communicating risks to the public,* ed. R. E. Kasperson and P. J. M. Stallen, pp. 79–124. Dortrecht: Kluver Academic Publishers.

CROZIER, M. (1963).   *Le phenoméne bureaucratique.* Paris: Seuil. (Reprint 1971.)

DEGREENE, K. B. (1988).   Long wave cycles of sociotechnical change and innovation: a macropsychological perspective. *Journal of Occupational Psychology* **61**, 7–23.

DERTOUZOS, M. L. & MOSES, J. (ed.) (1979).   *The computer age: twenty-year view.* Cambridge, Mass: MIT Press.

DURANT, J., HANSEN, A. & BAUER, M. (1995).   Public understanding of new human genetics. In *The benefits and hazards of new human genetics,* ed. T. Marteau and M. Richards. Cambridge: Cambridge University Press.

EUROPEAN COMMISSION (1993).   *Biotechnology and genetic engineering,* Direction Generale 12. Brussels: INRA and European Commission.

EVERS, A. & NOWOTNY, H. (1987).   *Ueber den Umgang mit Unsicherheit; die Entdeckung der Gestaltbarkeit der Gesellschaft.* Frankfurt: Suhrkamp.

FREEMAN, C. (ed.) (1984).   *Long waves in the world economy.* London: Pinter.

FREEMAN, C. (1985).   Long waves of economic development. In *The information technology revolution,* ed. T. Forester, pp. 602–16. Oxford: Blackwell.

GAMSON, W. A. & MODIGLIANI, A. (1989).   Media discourse and public opinion on nuclear power: a constructivist approach. *American Journal of Sociology* **95**, 1–37.

HAGERSTRAND, T. (1967).   *Innovation diffusion as a spatial process.* Chicago: University of Chicago Press.

HALL, P. (1988).   The intellectual history of long waves. In *The rhythms of society,* ed. M. Young and T. Schuller, pp. 37–52. London: Routledge.

HALTINER, K. W. (1986).   Die Widerstandsmotivation in der Schweizer Bevölkerung. *Allgemeine Schweizerische Militärzeitschrift* **7/8**, 403–9.

HOBSBAWM, E. J. (1976).   Economic fluctuations and some social movements since 1800. *Labouring Men,* ed. E. J. Hobsbawm, pp. 126–57. London: Weidenfeld and Nicholson.

HULTBERG, J. (1991).   *A tale of two cultures. The image of science of C. P. Snow.* Report No. 165, Department of Theory of Science, University of Gothenburg.

INGLEHART, R. (1990).   *Culture shift in advanced industrial society.* Princeton, NJ: Princeton University Press.

INTERNATIONAL ATOMIC ENERGY AGENCY (1981).   *Power reactors in member states.* Vienna: IAEA.

JAUFMANN, D. & KISTLER, E. (1986). Technikfreundlich? – Technikfeindlich? Empirische Ergebnisse im nationalen und international Vergleich. *Politik und Zeitgeschichte* **48**, 35–53.

JOERGES, B. (1989). Soziologie und Maschinerie. *Technik als sozialer Prozess*, ed. P. Weingart, pp. 44–89. Frankfurt: Suhrkamp.

JOHNS, E. A. (1973). *The sociology of organisational change.* Oxford: Pergamon Press.

JUNGK, R. (1979). *The nuclear state.* London: John Calder. (First published in German, 1978.)

KEPPLINGER, H. M. (1988). Die Kernenergie in der Presse, *Kölner Zeitschrift für Soziologie und Sozialpsychologie* **40**, 659–83.

KEPPLINGER, H. M., EHMIG, S. C. & AHLHEIM, C. (1991). *Gentechnik im Widerstreit.* Frankfurt: Campus.

KERSHAW, I. (1985). 'Widerstand ohne Volk'. Dissens und Widerstand im Dritten Reich. In *Der Widerstand gegen den Nationalsozialismus*, ed. J. Schmädeke and P. Steinbach, pp. 779–98. Munich: Piper.

KLEIN, J. A. (1984). Why supervisors resist employee involvement. *Harvard Business Review* **62**, 87–95.

KLING, R. & IACONO, S. (1988). The mobilization of support for computerization: the role of computerization movements. *Social Problems* **35** (3), 226–43.

KNORRE, H. (1992). 'The star called Wormwood': the cause and effect of the Chernobyl catastrophe. *Public Understanding of Science* **3** (1), 241–50.

KUHN, T. S. (1962). *The structure of scientific revolutions.* Chicago: University of Chicago Press.

LATOUR, B. (1988). The prince for machines as well as for machinations. In *Technology and social process*, ed. B. Elliot, pp. 20–43. Edinburgh: Edinburgh University Press.

LAWRENCE, P. R. (1954). How to overcome resistance to change. *Harvard Business Review* **32** (3), 49–57. (Classics reprint 1969.)

LITTEK, W. & HEISIG, U. (1986). Rationalisierung von Arbeit als Aushandlungsprozess. Beteiligung bei Rationalisierungprozessen im Angestelltenbereich. *Soziale Welt*, **37** (2/3), 237–62.

LYON, D. (1988). *The information society.* Cambridge: Polity Press.

MARLIER, E. (1992). Eurobarometer 35.1. In *Biotechnology in public. A review of recent research*, ed. J. Durant, pp. 52–108. London: Science Museum.

MATHES, R. & PFETSCH, R. (1991). The role of the alternative press in the agenda building: spill-over effect and media opinion leadership. *European Journal of Communication* **6**, 33–62.

MATUSOW, H. (1968). *The beast of business. A record of computer atrocities.* London: Wolfe.

MAZLISH, B. (1965). Historical analogy: the railroad and the space program and their impact on society. In *The railroad and the space program. An exploration in historical analogy*, ed. B. Mazlish, pp. 1–52. Cambridge, Mass: MIT Press.

MAZUR, A. (1975). Opposition to technological innovations. *Minerva* **13**, 58–81.

MAZUR, A. (1984). Media influences on public attitudes toward nuclear power. *Public reactions to nuclear power: are there critical masses?* ed. W. R. Freundenberg and E. A. Rosa, pp. 97–114. Boulder, Colo: Westview Press.

MOKYR, J. (1990). *The lever of riches. Technological creativity and economic progress.* Oxford: Oxford Univeristy Press.

MOKYR, J. (1992). Technological inertia in economic history. *Journal of Economic History* **52** (2), 325–38.

MUGNY, G. & PEREZ, J. A. (1991). *The social psychology of minority influence.* Cambridge: Cambridge University Press.

PAKULSKI, J. (1993). Mass social movements and social class. *International Sociology* **8** (2), 131–58.

PETZOLD, H. (ed.) (1981). *Widerstand: ein strittiges Konzept in der Psychotherapie.* Paderborn.

RADKAU, J. (1983). *Aufstieg und Krise der deutschen Atomwirtschaft 1945–1975. Verdrängte Alternativen in der Kerntechnik und der Ursprung der nuklearen Kontroverse.* Reinbeck: Rowolt.

RASCHKE, J. (1988). *Soziale Bewegungen. Ein historisch-systematischer Grundriss.* Frankfurt: Campus.

REVERENDI, J. C. (1975). Les resistances et les obstacles au developpement: l'envers d'une theorie. *Recherches Sociologiques* **6** (1), 62–97.

RHINOW, A. R. (1985). *Widerstandsrecht im Rechtsstaat?* Staat und Politik 30. Bern: Haupt.

ROON, G. VAN. (1981). Historians and long waves. *Futures*, October, 383–8.

RÜDIG, W. (1990). *Anti-nuclear movements: a world survey of opposition to nuclear energy.* London: Longmans.

RUHRMANN, G. (1992). *Das Bild der biotechnischen Sicherheit und der Genomanalyse in der deutschen Tagespresse 1988–1990*, TAB–Diskussionspapier No. 2, Bonn.

SAXER, U., GANTENBEIN, H., GOLLMER, M., HATTENSCHWILER, W. & SCHANNE, M. (1986). *Massenmedien und Kernenergie.* Bern: Huber.

SCHMÄDEKE, J. & STEINBACH, P. (ed.) (1985). *Der Widerstand gegen den Nationalsozialismus.* Munich: Piper.

SCHMITT-BECK, R. (1990). Ueber die Bedeutung der Massenmedien für soziale Bewegungen, *Kölner Zeitschrift für Soziologie und Sozialpsychologie* **42** (4), 642–62.

SCHOT, J. (1991). Constructief technology assessment as hedendaags Luddism. Dissertation, University of Twente.

SCOTT, J. C. (1985). *Weapons of the weak. Everyday forms of peasant resistance.* New Haven, Conn: Yale University Press.

SCREPANTI, E. (1984). Long cycles in strike activity: an empirical investigation. *British Journal of Industrial Relations* **22**, 99–124.

SENSALES, G. (1990). *L'informatica nella stampa italiana.* Milan: Franco Angeli.

SHARP, G. (1973). *The politics of nonviolent action*, 3 vols. Boston: Porter Sargent.

SHOTTON, M. (1989). *Computer addiction? A study of computer dependency.* London: Taylor and Francis.

SIMON, H. A. (1981). *The science of the artificial*, 2nd edn. Cambridge, Mass: MIT Press. (First edition 1969.)

SNOW, C. P. (1959). *The two cultures and the scientific revolution.* Cambridge: Cambridge University Press.

STAUDENMAIER, J. (1985). *Technology's storytellers: reweaving the human fabric.* Cambridge, Mass: MIT Press.

TARROW, S. (1988). National politics and collective action. *Annual Review of Sociology* **14**, 421–40.

TOURAINE, A. (1992) *Critique de la modernité.* Paris: Fayard.

WEART, S. R. (1988). *Nuclear fear. A history of images.* Cambridge, Mass: Harvard University Press.

WEIZENBAUM, J. (1976). *Computer power and human reason. From judgement to calculation.* San Francisco: W. H. Freeman and Co.

WIPPERMANN, W. (1983). *Der Europäische Faschismus.* Frankfurt: Suhrkamp.

World Nuclear Industry Handbook (1992). Sutton, UK: WNIH.

WRIGHT, S. (1986). Recombinant DNA technology and its social transformation 1972–1982. *Osiris* **2**, 303–60.

WYNNE, B. (1993). Public uptake of science: a case for institutional reflexivity. *Public Understanding of Science* **2** (4), 321–38.

YEARLEY, S. (1992). Green ambivalence about science: legal–rational authority and the scientific legitimation of a social movement. *British Journal of Sociology* **43** (3), 511–32.

PART I
Conceptual issues

2

The crisis of 'Progress'

ALAIN TOURAINE

It would be misleading to speak of an anti-scientific mood in public opinion today. Most people support advanced technology or scientific medicine, but it is true that criticism of economic modernization or hospital life is growing. Science is not widely criticized, but the idea of a scientific society is often rejected by science-educated people. We still believe in science, but no longer in progress. I would like to make some comments on this general statement.

Progress means that scientific and technological achievements trigger welfare, freedom and happiness. Condorcet, at the end of the eighteenth century, during the most violent period of the French Revolution and a few weeks before he died, wrote a most enthusiastic hymn to progress, announcing a future of abundance, emancipation and peace. The idea of eternal peace was central to eighteenth-century British, French and German political thought.

Social analysis, from the beginning of the nineteenth century to recent years, has been dominated by the opposition between modernity and tradition, categories which were practically synonymous with good and evil. Modernity was defined as rationalization, both in economic and administrative fields, and as secularization or disenchantment, to use Max Weber's terms, as far as culture was concerned. It was a global view of collective as well as individual life. Modernity was identified with the creative use of reason in all fields, and conceived as domination and mastery of nature, to use Descartes' and Bacon's ideas; also as an instrument of liberation from customs, privileges and prejudices.

What we are experiencing now is the waning of this 'progressist' or modernist view of the world and of human life. Its decay is not new. It began during the last third of the nineteenth century. For sure, during most of this century, some anti-modernist thinkers, from Carlyle and the Pre-Raphaelites on and from traditionalists to social critics, had criticized the spread of the black country and the destruction of the green country. But, during the last part of the nineteenth century, the enlightenment idea of the parallel progress of modern civilization and self-realization of the individual was brutally rejected and replaced by an analysis of the conflicts between social trends and personal needs. First Nietzsche and then Freud emphasized the conflict between moralization and will of power (*Wille zur*

Macht) or between pulsions and law or social controls. All kinds of '*Lebens-philosophien*', from Simmel to Bergson, emphasized this duality and opposition between reason and life.

Modernity: discontents and resistance

In our time, the rupture of the global concept of progress is complete. On one side, rationalization has been transformed into instrumentalism and objective reason into subjective reason, to use Horkheimer's terms, which were followed by other members of the Frankfurt School. Our world appears to be integrated as a world market, but the counterpart of this globalization is the more and more aggressive defence of personal and collective identity. Instead of living in a cosmopolitan world, as some people pretend we do, we live in a dualized world in which not only North and South are more and more distant from each other, but where rich and poor districts of the cities are more and more separated universes and in which most individuals are split between their participation in a globalized world and their consciousness of individual and collective identity.

Rationalization and science have destroyed the traditional correspondence between social organization and personal life. We no longer internalize social norms and roles. We as individuals resist a mass society, in which we participate actively at the same time. The traditional patterns of social and cultural organization by which personal identity was identified with social status and role, are disappearing, so that the integration of individual behaviour into a social system is declining. Religion was the main institution by which rational thought and non-rational creeds were associated with each other. Secularization on the contrary means, on one hand, the development of scientific research. No Galileo can be stopped by a church in a secularized country, while biology has been or is hampered by political or religious prohibitions in various authoritarian and non-secularized regimes. But, on the other hand, it means that individuals and groups are no longer protected by the mechanisms of social and cultural integration. Individuals are politically free, but they are still exposed to pain, grief and death, and their individual experience appears to be separated from their situation, in the same way as the patient in a hospital is separated from his illness and from the instruments of diagnosis and therapy. We no longer can understand, or explain, our individuality in social terms, while most of us nevertheless feel the necessity to discover an explanation of our individual character and experience. So the very success of a scientific approach to human life stimulates indirectly the de-velopment of the rational concept of a personal destiny, for example of a strict astrological determinism. Among hospital patients, this search for individualized and non-scientific treatment is not necessarily contradictory to biology based medical care. Few are the patients who reject scientific medicine and trust only 'soft' treatments. Much more numerous are the people who feel it necessary to combine the universalistic methods of science and technology with a direct reference to their own individual life history. It has been observed that, in most cases, patients who suffer from very serious diseases, while they use non-scientific resources such as placebos or prayers, nevertheless behave rationally in the hospital and strongly believe in scientific medicine, even if they are hopeless about themselves. Non-scientific behaviour in these cases is not contradictory to scientific medicine; it answers a different kind of need. Patients try to rely on a personal authority which helps them to bear the burden of their sufferings and anxiety. The objective and subjective universes are more distant from each other than in the past.

What is observable at the individual level is even more clearly visible at the level of collective life. No human group, national, ethnic or religious in particular, identifies itself completely with its participation in world markets. We all feel threatened in one way or another by a loss of collective identity. Our subjective situation and our subjective experience part company. We belong to a world-wide mass culture but our behaviour corresponds less and less to the internalization of norms coming from this culture through processes of socialization. David Riesman, long ago, analysed the processes through which inner-directed behaviour was replaced by other-directedness. We imitate, we are attracted by mass consumption patterns, by the messages of advertisement or propaganda agencies, more than we act in conformity with norms and what used to be called values. What we do and what we feel are more and more disconnected. The more we live in a world full of objects, signs and stimuli, the more our Self needs to be rooted into a specific culture or tradition. National, ethnic or religious identifications offer strong supports to a Self which is separated from social behaviour.

An uprooted urban population, which has lost traditional social and cultural patterns in which individuals were educated, feels the need, sometimes, to come back to tradition – although this is in general impossible – and, much more often, to participate in wrongly named fundamentalist movements which re-create, in a non-traditional way and generally in a politically radical way, a collective identity. This image of the nation is not modernist, but on the contrary is opposed to imported modernization, and imposes cultural and social norms which reject the patterns of modern life.

This reaction is most violent when modernity is directly identified with foreign or social domination. But everywhere, from Third World countries to the so-called Western world, rationalization and modernization are more and more often identified with power. This is a direct consequence of the triumph of instrumental over substantive rationality. When rationalization meant the creation of a rational world and of rational individuals, it could appear to be a defence against irrational powers and privileges. On the contrary, when it refers only to instruments which allow attainment of ends which are no longer defined as rational, as is the case in a mass consumption, hedonistic society, rationalization appears to be an instrument in the hands of groups and individuals who control and govern the social and economic system. Rationalization, during the second half of the nineteenth century, became synonymous with industrial management, that is with methods which were considered by radical critics as instruments of manipulation of industrial workers by managers who were more interested in profits than in rational methods of production. More recently, this image of social life as dominated by a power elite has been partly replaced by a more systemic view of society, which is even more destructive of the old correspondence between norms and motivations, system and actors. Society is described then as an auto-poietic, self-transforming system which adjusts itself to a changing environment, increases its level of internal differentiation, its own complexity, and is not influenced by actors' values and intentions. Actors actually are only agents. Following this interpretation, it is possible to describe the development of science as an internal process, so that public opinion feels more and more alien to a scientific development which can no longer be identified with social needs or with progress. The widening gap between autonomous systems and social demands explains the process of delegitimization of science and of economic institutions as well.

'Big science' and 'technocracy'

To this idea, of a growing separation between instruments and meanings, must be added a more traditional analysis which observes that the very progress of science replaces the traditional image of pure and disinterested scientific activity by two complementary images. The first one is 'big science', that is the existence of power structures in science. The second is technocracy, that is the domination of society by people who control large rationalized organizations. The control of complex technological resources tends to increase the degree of control exerted by systems – or by their centres – over ordinary members of the system or over its environment. In all these cases, anti-technocratic attitudes grow, which challenge the management of science and technology more directly than science itself.

Very recently in many European countries, a large number of patients who had received blood transfusions were contaminated by the AIDS virus. From the spring of 1985 onwards, this danger was well known by specialists, but administrative authorities, including many physicians, were slow in stopping the use of dangerous blood products. In the French case, groups of haemophiliacs sued the main officers of the public blood transfusion system and some of these were condemned to jail amidst manifestations of public emotion and anger. Public opinion accused these doctors of sacrificing the lives of many patients to bureaucratic, economic or national interests. In a parallel way, in several countries, governments and health institutes are accused of insufficient support for research and welfare on behalf of AIDS victims. In the case of blood transfusion as, half a century earlier, in the case of the nuclear bomb, science is not directly condemned but scientists cannot just pass the responsibility to politicians because they participate in the decision-making process; they mobilize resources, they react as individual beings and their laboratories can be analysed by sociologists as large organizations with complex decision-making processes. Science is no longer a small community of distinguished minds: it is part of the economic and of the political system and decisions about science are not entirely scientific.

Today, debates about biology and medicine are more dramatic than the physicists' misgivings about the effects of their discoveries, because this last debate was held *a posteriori*, while today anti-AIDS policies are criticized in the hope that dramatic problems can be solved and that the discovery of new therapies can be accelerated. But this last example shows clearly that these campaigns against scientific institutions are not based on anti-scientific orientations. On the contrary, they are rather similar to anti-bureaucratic criticisms which attacked post-revolutionary parties and governments as 'traitors' to a revolution which was considered in itself as positive and necessary.

Even when science and rationalization appear to be instruments of a social or national domination, there is no widespread resistance to new technologies and medical care. Moreover the resistance, when it exists, does not come from the population itself but, most of the time, from political leaders who mobilize people against what is considered as a foreign invasion. Sometimes the resistance comes from people who are suffering negative consequences, for example of better agricultural methods which induce a higher concentration of land and the elimination of small farmers, but what is rejected is socially, nationally or religiously defined, not cognitively or technically.

Outcomes of resistance

Scientists answer what they consider as attacks against science, by complaining bitterly about false information which is diffused by media and by political movements. They accuse pseudo-scientists of supporting statements which are not scientifically correct and are part of a political campaign more than normal steps in the process of scientific discovery. Many scientists have recently condemned the weakness of the scientific basis of many statements which were uttered during the Rio Conference. These members of the Heidelberg group are probably right in some cases and it is dangerous to consider as demonstrated causal relationships which have not been well enough analysed. Nevertheless, we must recognize that the ecological movement has drawn attention to many problems which were not sufficiently observed and analysed by scientists, and it seems perfectly normal that social demand be one of the ways by which scientific research is orientated. The progress of science is not entirely endogenous, even if it is so to a large extent; it results from political, social, economic and military demands as well.

It is understandable that many scientists stick to the idea of progress as it has been elaborated by the Enlightenment. But the very positive defence of science cannot justify the very superficial idea that determinants and consequences of scientific activities are part of the scientific process itself. In the same way, churches identify themselves with a sacred principle, but historians of religion cannot be satisfied with their arbitrary explanations. They recognize that patterns of religious organization which give a concrete social form to religious creeds are related to historical circumstances and not just to a sacred message.

Even if most scientists rightly want to devote all their time and preoccupations to their scientific work, it is more and more difficult for them not to participate in public opinion debates. That is true not only for the social sciences but for the natural sciences themselves. These debates can have positive consequences for science itself. Public demand can help innovative individuals or groups to overcome the resistance of vested interests. When the access of the public to scientific results is difficult, the danger of the influence of non-scientific interests on science is more limited; it is much larger for social sciences in which prejudices or ideologies have often seriously threatened or paralysed scientific research. Especially in these fields, public debate is a way of isolating what is strictly scientific from what is socially or politically determined in the development of scientific activity. Moreover, the non-scientific statements of voluntary associations are probably less dangerous for science itself than the limits imposed on the diffusion of scientific results by governments and companies.

When science had few practical uses, it could be easily protected by academic institutions, although it is far from demonstrated that universities have always had positive attitudes toward scientific innovation; but today science plays such an important role in society that it is no longer possible to maintain a complete autonomy of science as a subsystem of society. Politics, culture and science overlap in public opinion debates and it is impossible to condemn scientists who take part in political life with the prestige that their scientific activity gives them, even if there is obviously no scientific answer to political problems.

Rationalization and individual freedom

The main result of the growing importance of social determinants and consequences of scientific activity is that we no longer believe in the entirely positive and emancipatory effect of science. We no longer believe in progress, as I defined it, as parallelism between scientific and technological achievements on one hand, and abundance, freedom and happiness on the other. Moreover, we no longer believe in the final triumph of reason and in the creation of a transparent world of technological and legal artefacts.

From the Enlightenment to the philosophies of Progress, the idea prevailed that modern societies were guided by reason and that rational individuals would build a reasonable society thanks to the resources created by science and technology. This correspondence between individuals, society and nature was progressist because it meant the elimination of traditions and privileges, as well as scarcity and authoritarian power. For this social philosophy, education had a central role, not only to diffuse scientific knowledge and rational social thought, but to separate public from private life and to prepare young people – at least young males – to participate in public life and to acquire different gratification patterns which were indispensable for subordinating consumption to saving and investment.

This Western 'classical' view of modernity implied a polarized view of society: it opposed reason to feelings, traditions and creeds as much as adulthood to childhood and, in general, male rational culture to feminine irrational psychology or bourgeois rationality to working class brutality or laziness, without forgetting the opposition between rational Westerners and irrational natives. The direct association between economic growth and peace or freedom was linked with the opposition between reason and passion and, at least in the case of Latin countries, between reason and religion.

This ambitious and integrated view of modernity collapsed during the last third of the nineteenth century. I already mentioned that the image of the rational or educated individual was criticized by Nietzsche and Freud who opposed unconscious pulsions to repressive law or 'moralization', so that progress appeared to have a high psychological cost, to be repressive more than liberating.

On the political level, the British, American and French idea of the nation as *res publica* based on people's sovereignty, was largely replaced by the Völkisch concept of nation which gives more importance to a community of culture or to ethnic specificity (*Schicksalsgemeinschaft*) than to Rousseau's *Contrat Social*, to Locke's *Trust* or to Hobbes' *Covenant*. Cultural and political particularisms rejected a universalism which was identified with the domination of ruling classes or nations.

From this period on, and during most of the twentieth century, in many parts of the world, the distance constantly increased between reason and personality or culture, between the universe of manufactured objects and the search for personal or collective identity. The ideology of progress has been not only abandoned but rejected as destructive of national or cultural identity. Even in the Western world, liberal or radical ideas are often linked with the acute consciousness of a basic conflict between the logic of the systems and the defence of the '*Lebenswelt*', the socio-cultural environment and values of the individual or of collective subjects.

Science is not directly threatened by radical neo-communitarian movements; these, on the contrary, find their supporters and leaders among professionals who are educated in the scientific, medical and engineering departments of the universities. The Islamists, both in Iran and Egypt, are anti-traditionalist and favourable to modern technology as an instrument necessary to build an authoritarian and theocratic power which uses science and technology as pure instruments of power politics.

In Western countries and liberal circles, science is not rejected either, but in both cases, the integrated image of a scientific world which was introduced by the Enlightenment and by the positivists is strongly rejected and replaced by a dual view of modernity: rationalization on one side, individual freedom on the other. That corresponds to a complete change from the eighteenth and nineteenth century views of modernity.

Functional differentiation and public opinion

Modernity has often been defined as the substitution of a unified and transparent view of the universe for a *Weltanschauung* which was dominated by the distance between God and humankind or grace and sin. It seems to me, on the contrary, that the pre-modern view of the world emphasizes the unity of the creation which was considered as both rational and sacred, while modernity means the separation of nature and humankind, of inner life and natural existence, of consciousness and structure. This interpretation of modernity is supported by the fact that strong anti-modernist tendencies are constantly struggling against it in the name of the unity of the universe. Structuralists at the theoretical level, ecologists or at least deep ecologists at a more political level, want to come back either to the materialism or to the utilitarianism of the eighteenth and the nineteenth centuries to defend a unified view of nature and to reject the subjectivism of modern culture.

On the contrary, we give more and more importance to ethical debates. They contradict the traditional trust in scientific progress. They consider that some moral principles are independent of scientific achievement in the same way as natural law was considered to be independent from and superior to the *Raison d'Etat*, to the interest of the State. For example, in many countries, for ethical reasons, the human body or parts of it cannot be bought or sold. For this reason, blood is donated free for transfusion and organs can be transplanted only in public hospitals, to avoid priority being given to the richest patients. These rules are based not on utilitarian principles or on rational calculation but on purely moral principles, completely independent of scientific arguments. There is more distance from a positivist or utilitarian view to these ethical principles than from these positive or utilitarian views to the naturalist doctrine of the Catholic Church, because they all refer themselves to the same general concept of natural laws which must be respected, while ethics is based on value judgements and on the idea of fundamental rights.

Two opposite trends are visible in our political culture. On one side, we adopt a more and more integrated view of economic growth and we recognize that public administration, education and even protection of the environment are essential aspects of a self-sustained growth and even more of a sustainable development. But at the same time, and for deeper reasons, we want to limit the consequences of rationalization and economic rationality and we defend individual freedom and cultural diversity of minority rights. We refuse more and more to give a total privilege either to universalism or to particularism: we want to combine science and culture, open markets and protection of cultures as well as of species.

We no longer call modern a society which imposes the same rational rules upon everybody; on the contrary, we consider as modern societies which are able to combine science or rationality with cultural diversity. And this combination appears to be one of the main *raisons d'être* for democracy. If we dream of a rational world, we are tempted to accept a king philosopher. Such was the Bolshevik concept of democracy: the avant-garde's often violent action was legitimized by its goal – to free progressive forces from the irrational power of private profit. In the opposite way, if we accept the complete specificity of each culture, we can too easily favour policies of cultural homogenization and of ethnic cleansing within a national territory, policies which are clearly anti-democratic.

If we reject, on the contrary, any central principle of unification, if we accept a basic separation between technical rationality on one side, and moral principles or cultural identity on the other, we are favourable to an open political system, which is able to elaborate compromises and looks for a partial integration between independent and opposite principles. Democracy in its very nature combines plurality of interests and values with the unity of law and power: *ex pluribus unum.*

A concrete consequence of this dualism is that the effects of scientific research cannot be controlled only by scientists or be decided by scientific arguments. Management of science is not only scientific and the effects of science and technology cannot be accepted or rejected only for scientific reasons.

It is difficult to draw a clear frontier between the endogenous development of science and social political or moral debates. Sometimes, scientists feel hindered in making new discoveries by political regulation or moral resistance. Sometimes, on the contrary, moral authorities and public opinion itself are shocked by techniques or decisions which appear to them to contradict moral values. There is no rational solution to such a conflict and there must be an open and permanent debate between scientists, moral authorities and politicians, leading to a careful and limited intervention of law and administrative regulations in scientific matters. In our time, when science often transforms so deeply our image of ourselves and of our social life, the respectable idea of pure science can easily be transformed into a technocratic attitude.

Extending the democratic process

In a parallel way, at the political level, as I already mentioned, the national State is in deep crisis, at least in Europe. It was an extraordinary combination of particularism and universalism, of '*Rechtsstaat*' and local life in all Western countries, in Spain, Italy or Germany, almost as much as in Britain and France. Today, this synthesis is crumbling because of the rapid process of globalization and mass culture. Privileged social categories can recreate active social participation through voluntary associations and control their own self through education; but weak groups either reject universalism and accept authoritarian forms of defence of a national or a cultural identity, or become a mass of uprooted consumers of material or symbolic goods, without being able to create original cultural patterns. Europeans in general are afraid to become part of a world-wide cultural market more than a plurality of original and creative cultures.

In industrial societies, we emphasized the importance of poverty, exploitation and class conflict. They resulted sometimes in revolutions, but more often, at least in the most industrialized countries, in reform movements which slowly created industrial democracy. But in today's societies, at the world level as well as at the national or the individual level, alienation is more visible than exploitation.

While rationality becomes a pure instrument which is subordinated to power or profit, subjectivity can easily be transformed into an aggressive defence of identity. The triumph of the market disorganizes most societies and increases the number of marginalized or excluded groups of individuals, while the defence of cultural specificity often leads to authoritarian or totalitarian regimes.

Both universes, the universalist, scientific and technological world, as well as the particularistic national or religious regimes, lead to massive desocialization.

We need to restore some kind of communication between these two opposite universes. That means creating new forms of control and regulation of economic or even scientific activity. Modernity has been identified with the growing autonomy of subsystems: politics has been separated from religion since Machiavelli, economy from politics since Locke and private life from public life since Rousseau or the arts from social life since Baudelaire. This functional differentiation has had very positive consequences, especially for scientific knowledge and arts and, according to many people, for religion itself, since the American Constitution has rejected the idea of an official church. And very few people would accept, at least in the Western world, a process of desecularization, of re-enchantment of the world, to use Weberian terms. But this differentiation supposes the extension of political processes through which conflicting values are combined. If these processes do not exist, our world will be dominated by a latent civil war between markets and tribes, mass society and closed cultures, imperial liberalism and aggressive nationalism.

Science is neither above society nor completely autonomous. Its financial resources, the careers of the scientists, sometimes the goals of the laboratories themselves, depend on non-scientific factors, so that, if scientists must defend scientific objectivity against external pressures, they must at the same time combine the endogenous development of science with external demands and moral principles. A century ago, the democratic idea penetrated into the industrial world with great difficulty, first in Britain. Today, the democratic process must be extended to all fields of social life, to avoid a complete segmentation and fragmentation of society and to limit as much as possible irrationalist or anti-scientific attitudes which result from a complete differentiation of the scientific and technological activities from the rest of personal and collective life.

If we try to maintain the eighteenth and nineteenth centuries' trust in a global progress, if we keep believing in a rationalized society, we can only accelerate the rupture between powerful systems and powerless actors which destroys the creative capacity of individuals and societies.

3

Reinterpreting 'Luddism': resistance to new technology in the British Industrial Revolution

ADRIAN RANDALL

The historiography of resistance to new technology, generically referred to as Luddism, in the British Industrial Revolution has been a curious one. Machine breaking plays little part in most economic history textbooks, labour's reaction to technological displacement, when noted at all, being seen as little more than an irritating and futile minor impediment to progress. Political and social historians have found the riotous crowd rather more interesting but they have tended to see Luddism as symptomatic of some other problem, economic depression or high food prices, rather than as direct hostility to technological change. Thus Briggs refers to the Luddites as the 'helpless victims of distress' (Briggs 1959, p. 182). Labour historians, though sympathetic to the problems faced, have often found that Luddism ill-accords with that Whiggish development of a 'proper' labour movement characterized by orderly trade unions, deemed the mark of progress. Machine breaking and riot are often seen as being very different and separate from orderly collective bargaining (see, for example, Cole & Postgate 1949, pp. 184–5; Thomis 1970, pp. 133–4; or Hunt 1981, p. 195). Even Hobsbawm, whose pioneering essay on the machine breakers showed how pre-industrial labour utilized machine breaking as a weapon, also saw industrial violence as an anachronism by the early nineteenth century (Hobsbawm 1968, pp. 5–22). Thus the Luddites have been too easily absorbed into popular notions as backward-looking, blinkered obstructionists, men who failed to see the ineffable benefits of the Industrial Revolution and who were therefore justly and legitimately defeated. The term Luddism has consequently become both an epithet to disparage anyone deemed hostile to the march of 'progress' and a self-evident explanation for why they must fail.

This chapter will suggest that we might do well to recognize that both these assumptions are built upon a particular reading of our history and one which our contemporary experience would suggest might be flawed. In an era of horrendous unemployment statistics, not just in the United Kingdom but across Europe, we should note the universality of the blow which the threat and the reality of technological redundancy can mean. Psychologists tell us that the loss of one's work, in particular the loss of a trade or career, is a devastating psychological blow to the individual's self-esteem, exceeded only by the shock of bereavement. We might wonder why this blow should be deemed less severe for a craftsman in 1793 than for one in 1993. We might also wonder why we suppose that the experience of the massive social and economic change which we refer to as the Industrial Revolution should have been welcomed with universal open arms any more than the similar painful process which Eastern Europe is now undergoing. Change is a necessary feature of life. 'Change' may ring well from the lips of aspirant politicians. But much fundamental change is painful and that which destroys old ways of life more painful still. We should approach our analysis of those who resisted change with more humility and with more sympathy.

Historiographical contexts

The historiographical context of this analysis also needs examination since the history of the Industrial Revolution has itself undergone much re-writing in the last 30 years. Long gone are the accounts of the heroic age of the rise of the factory and of rapid take-off into self-sustaining growth. Their place has been filled by the effusions of the New Economic History, anxious to demonstrate that, in aggregate terms, the Industrial Revolution, like hurricanes in Hertford, Hereford and Hampshire, hardly happened. While advocates of the proto-industrialization thesis among others have been concerned to show that the process of economic change was a protracted one, stretching back far into the eighteenth century, economic historians concentrating upon the 'classic period' of the Industrial Revolution have increasingly come to conclude that the rate of change then was slow indeed, 'emphasising gradualism and continuity and playing down the possibility of major discontinuities in either economic or social life' (Hudson 1989, p. 1). One historian reviewing the recent historiography of this period was in fact moved to conclude that 'the British Industrial Revolution is now depicted ... as a limited, restricted, piecemeal phenomenon, in which various things did *not* happen or where, if they did, they had far less effect than was previously supposed' (Cannadine 1984, p. 162).

One aspect of this reductionism has been the diminution almost to the point of inconsequence of the roles of machinery and the factory. Once seen as the engine of growth and the forger of new social relationships respectively, their importance is now seen as at best marginal. Now we are told technological change played only a minor role in generating economic acceleration. Growth was as much about a multiplication of handworkers as of machines, as much a consequence of increased domestic production as of large-scale factories. Only in textiles and the pottery industry did the factory mark significant change and even in the cotton industry, the 'leading sector', half the work-force in 1830 still worked outside the factory (see, for example, Crafts 1983, pp. 84, 87, 88; Mathias 1983, pp. 2–4, 14–16; or Crouzet 1982, p. 199). Social historians likewise have reduced the formerly perceived role of the machine as an agency of social change. Thus Price notes, 'The process of proletarianisation was not primarily a matter of subordination to a labour process that was technologically driven. Rather it was a process that occurred more in the sphere of market relations and involved increased exposure to the vagaries of market forces' (Price 1986, pp. 21–7; also Reddy 1984; Rule 1986, pp. 384–93).

All of which is very well. The broadening of our perspective upon the period, still resolutely referred to as the Industrial Revolution, provided by the researches into proto-industry, into change in the artisan trades, into the significance of demographic expansion and consumer preference and the corresponding move away from a simplistic equation of change with technology and the factory, marks a valuable development of our historical understanding. However, we should resist entirely throwing out the baby with the bath water. A principal factor in the rapid expansion of production in the late eighteenth century was technological innovation. None were especially dramatic but their impact was frequently considerable. No one industry was completely transformed by technological innovation. None saw the complete imposition of a factory system in place of workshop or cottage. But machinery was a major agency in the rapidly increasing pace of change and it, together with the factory, provided the impetus for those who wished to cast aside existing barriers to more wholesale industrial reorganization.

The impact of the machine and the factory can be seen in three ways. First, they directly displaced some key labour groups. Secondly, their impact was felt at second hand by the families of those so displaced and by the trades of others into which the displaced flocked. And finally, most remotely but most extensively, the machine and the factory provided a role model for other industries and trades as yet untouched by change. The organizational role of the factory had been appreciated long before the Industrial Revolution but pre-industrial labour, as in the textile industries, proved unremittingly hostile to attempts to group them together in such structures. The advent of powered machinery made a factory system necessary. But, once established from need, the factory rapidly became established as a production mode to which other industrialists without direct need for powered machinery began to aspire. The factory system, with its regularity, order, control and discipline became not just a model for modern industrial efficiency. It also became a metaphor for society and social relations as a whole. The fear of the factory, with its connotations of the workhouse, hung over many old handicraft trades for decades before they were absorbed. The machine and the factory therefore constituted and were seen to constitute major threats to old forms of production. Those so threatened did not always give up without a fight.

Places of resistance

Looked at quantitatively, it is clear that most of the new technologies in the British Industrial Revolution did not encounter significant resistance.[1] But if resistance was not the norm, it was in many ways remarkably widespread. The most famous episodes are of course the Luddite disturbances which convulsed the East Midlands framework knitting industry, the West Riding woollen industry and the Lancashire cotton industry in 1811–12 as wide frames, gig mills, shearing frames and power looms were attacked and destroyed (for Luddism, see Peel 1968; Thompson 1968, Ch. 14; Thomis 1970). But Lancashire had witnessed machine-breaking riots in 1767, in 1769, spectacularly in 1779 and again in 1792 prior to Luddism (Wadsworth & Mann 1931, pp. 380, 478, 480, 496; Berg 1985, p. 262). And Yorkshire Luddism, which followed earlier conflicts in the county over new machinery, was a successor to the Wiltshire Outrages of 1802 which themselves followed a long and protracted battle against new machinery in the West of England woollen industry which dated from 1767 (Randall 1982, 1991). Nor did machine breaking end with Luddism. Lancashire Luddism was followed fourteen years later by more extensive anti-power loom riots (Hammond & Hammond 1979, pp. 99–100) while the Scottish cotton spinners were prepared to utilize violence against the self actor mules in the 1830s.[2] By the early 1840s the wool combers' contest with the combing engine, which in different forms had excited resistance since the 1790s, was finally resolved in favour of the machine.[3] And while Luddism in 1811–12 affected four English counties, the most geographically extensive manifestation against new technology took place in 1830 among those who were normally the least riotous of the labour force, the agricultural labourers. The Swing Riots, occasioned at least in part by a grim hatred of new threshing machinery, swept from Kent across southern England to Dorset and Wiltshire and northwards into East Anglia and the Midlands (Hammond & Hammond 1966, Ch. 11 and 12; Hobsbawm & Rude 1969). This widespread disorder also allowed paper makers in Buckinghamshire to exhibit their hostility to new machinery which threatened their trades and sawyers in Southampton to destroy new and extensive steam sawmills, just as London sawyers had destroyed the first steam sawmill to be erected in London some 62 years earlier in 1768 (Hobsbawm & Rude 1969, pp. 121, 144–5; Rude 1970, pp. 249–50; Shorter 1971, p. 106). And as late as the 1860s brick makers in and around Manchester were attacking new brick-making machines which threatened their jobs in styles and forms which echoed the Luddites (Price 1975).

Such episodes, particularly Luddism and Swing, were events of major political importance and consequently have been well documented. However, resistance to new technology was not necessarily limited to the spectacular. Powerful labour groups might for long periods delay the adoption of new machinery simply by covert threats and sanctions. Wiltshire cloth dressers or shearmen, for example, were assiduous in bringing various pressures to bear to dissuade employers from introducing new gig mills and shearing frames for many years before 1802 (Randall 1982, pp. 286–8). These forms of quiet coercion rarely leave a record. Equally, we can know little of delays in the take-up of machinery caused by the vested interests of capital. Mokyr rightly points out that capitalists may have even greater reasons to resist new technologies than labour since their capital investment in old methods might thereby be rendered redundant (Mokyr 1990, p. 256). The catalogue of resistance to new technology in the British Industrial Revolution is therefore extensive. What can we learn from it?

What was resisted?

We should initially note that many of the innovations which excited protest in the early Industrial Revolution were not strictly *new*. Gig mills, over which the Wiltshire and Yorkshire cloth dressers clashed so violently with their employers in 1802 and 1812, had been in use for over a century in parts of both counties on coarse white cloths. It was only when gig mills were applied to fine cloths in the 1790s that trouble really developed (Randall 1991, p. 124). The wide frames, which in part occasioned the initial Luddite outbreaks in Nottinghamshire and Leicestershire in 1811, were far from new, having been used for decades to produce cheap knitted cloth. It was their increasing adaptation to produce 'cut ups', hosiery which was not 'full wrought', which occasioned the violent response (Hammond & Hammond 1979, pp. 184–5). The power loom was certainly not a new invention when it occasioned riots in Lancashire in 1812, still less was it when in 1826 they were repeated.[4] Combing engines had been around since the 1790s[5] and the threshing machine in growing use in the east of England long before the Swing Riots of 1830 (Chambers & Mingay 1966, pp. 72, 119; Hobsbawm & Rude 1969, p. 359; Macdonald 1975, 1978). The introduction of the flying shuttle loom, widely taken up in Lancashire from the 1760s and in

Yorkshire from the 1770s, was the occasion of extensive protests in Frome in Somerset as late as 1822 (Mann 1971, p. 161; Hammond & Hammond 1979, pp. 39–40; Randall 1991, pp. 98–9). Protest therefore stemmed from the *application* of these new technologies, not their invention. The timing of that application varied from place to place.

Why did some new technologies engender resistance and not others? And why did the level and vehemence of that resistance vary from place to place? It is necessary to examine two aspects if we are to make sense of these problems: the structure of work and impact of the machine upon work itself and upon the livelihoods of workers and petty and large capitalists; and the culture of work and the culture of the society within which that work takes place.

A crucial factor which influenced whether or not a new technology encountered resistance was the structure of work and the impact of the machine upon workers' autonomy. Some machines could be incorporated into the existing work structure without challenging the employment or the autonomy of the worker. This was true of the many small innovations which transformed the Birmingham trades in the Industrial Revolution. The stamp, the press, the draw bench and the lathe were for the most part hand powered and easily adopted by the small-scale masters and their journeymen without serious disruption (Berg 1991, pp. 197–8). The new technologies of the textile industry proved more potent in their labour-saving potential but here too the structure of work was crucial in determining the response of labour. This can be seen by comparing the structure of the woollen industry in Yorkshire with that in the West of England (see Figure 3.1). In the West of England, the industry was dominated by large-scale merchant capitalists, the gentlemen clothiers, who organized a highly capitalized outworking system (or, in the parlance of the proto-industrialization debate, a *Verlagsystem*[6]) which was based upon a high degree of subdivision of the labour force. Thus the wool went from sorter to scribbler to spinner to weaver, each of whom performed their specialized task in their own home or in a workshop. This specialization brought advantages of superior quality and standardization, enabling the region to dominate the fine cloth trade. In Yorkshire the industry was organized on the Domestic System (a *Kaufsystem*) around master clothiers, petty capitalists who made up cloth from sorting to weaving in their own homes and small workshops with the assistance of their family and living-in journeymen, each of whom might assist at any stage of production. Because of this small-scale basis of production and because of the lack of specialization, the product of the Yorkshire industry was of very inferior standard to that found in the West of England. However, the specialist skills of the West Country workers made them much more vulnerable to new technology. There the new scribbling engine, carding engine and the spinning jenny displaced entire trades. In

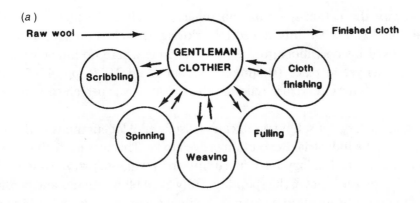

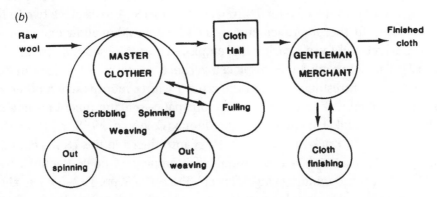

Figure 3.1. The organizational structure of (*a*) the West of England woollen industry, (*b*) the Yorkshire woollen industry. From Hudson 1989, pp. 180, 181.

Yorkshire, while causing temporary disruption, the lack of specialism meant that labour could relocate their energies within the master clothiers' workshops since the new technologies threatened only part of their work. However, where the structure of work organization was the same in both regions as was the case with cloth finishing, the resultant response to new labour-saving technology was identical, as the Wiltshire Outrages and Yorkshire Luddism showed.

Where new technologies did not threaten established labour groups, resistance was rare. Thus the carding engine and the spinning jack seem to have encountered no hostility in the woollen industry since they were not seen to take the work of any existing labourer (Mann 1971, pp. 288, 289–90). The same is true of the new pug mills which were introduced into the brick-making industry from the 1840s. 'These kinds of machines do not seem to have occasioned any resistance because they threatened no significant change in the work structure of the trade and were often ... merely aids to basic skills' (Price 1975, p.

120). Likewise the cotton-spinning machinery, which had triggered riots in Lancashire in 1779 when introduced by Arkwright, excited no such response when he moved his operations to the East Midlands where no major spinning industry was as yet in place (Hammond & Hammond 1979, pp. 43–5). The response to machinery therefore reflected the threat it was perceived to pose to established trades.

We should not forget just how significant a displacer of labour many of the new technologies of the Industrial Revolution proved to be. This was particularly true of the preparatory technology introduced into the textile industry. Thus in the woollen industry, the effect of the spinning jenny, scribbling engines and carding engines, flying shuttle looms and gig mills combined was to displace three in four of those previously employed. Women workers in particular suffered major technological redundancy (Randall 1991, pp. 50–61). The self actor cotton-spinning mules introduced after 1826 were said in 1834 to displace two in three spinners while further improvements by 1842 were said to allow one man to do the work which eight years earlier had employed four (Musson & Kirby 1975, pp. 302–12). Threshing machines displaced perhaps as many as fourteen in fifteen labourers in the threshing barns in the crucial winter months when other work was hard to obtain.[7] Fourdrinier continuous paper-making machines made the old hand paper maker completely superfluous (Coleman 1958, p. 191; Mokyr 1990, pp. 106–7). Brick-making machinery introduced in the 1860s cut labour costs by *c.* 40% and replaced 'the moulder as employer and craftsman ... by the moulder as machine-minding wage-earner' (Price 1975, pp. 122–3). The gig mill and shearing frame displaced completely the old skills of the cloth dresser (Randall 1991, pp. 120–2). And the combing engine, when eventually perfected by Lister and Donisthorpe in 1841, rapidly displaced the wool combers (Burnley 1889, p. 183). All these machines were targets for the machine breakers.

A critical factor, therefore, in determining the incidence of resistance to new technology in the British Industrial Revolution was whether the new machine could be accommodated within existing work structures and patterns. However, resistance was far from an automatic response even where new machinery constituted a major threat to employment. Some places witnessed resistance. Others facing similar displacement saw none. How can we account for this?

A significant clue in ascertaining where resistance occurred has to be sought in the character of the community which was under threat of technological redundancy. Some communities proved far more ready to resist than others since they had both a previous history of active resistance to detrimental change and the indigenous cohesion and leadership to sustain new resistance. For example, the long tradition of riotous protest found in the West of England woollen towns and districts over consumer and producer conflicts informed their protests against machinery (Randall 1991, pp. 86–90). Other areas had no such tradition and here machinery took over without a fight. This can be seen if we examine the response to the advent of the spinning jenny in the West of England. At Shepton Mallet and at Frome in 1776 and 1780 respectively, the jenny encountered violent resistance from the whole woollen-working community, threatened by the loss of a major source of family income. Indeed, in the main woollen towns few clothiers dared to contemplate introducing the machine until the 1790s. The rapid spread of the jenny thereafter was greatly aided by the expansion of trade and the opportunities created for the wives of textile workers to obtain compensatory employment in the loom or working the new machines themselves. The impact of the jenny, however, was devastating to the numerous agrarian communities around the textile districts where the wives of agricultural labourers found their staple by-employment disappear at a stroke. Here there was no protest since, unlike the much more pugnacious textile workers with their long history of collective resistance to threats to their trades, the agricultural labourers had no such experience of protest on which to draw. Only one non-textile community in the West of England possessed the tradition of and capacity to protest at the loss of work threatened by the jenny. That was Keynsham, where miners' wives had long supplemented family income by spinning. The miners were well versed in the arts of the riot but their distance from the woollen towns made their protests easier to ignore by entrepreneurs than those of the woollen workers (Randall 1989, pp. 189–90). Therefore if we are to make sense of the geographic variations in response to new technologies in the Industrial Revolution, we must pay careful regard to the previous histories of community protest which existed from place to place.

Forms of resistance

Resistance to the introduction of new technology in the British Industrial Revolution took three forms: the use of orthodox industrial strategies such as the use of strikes; industrial violence; and appeals to both the local and national authorities for protection. These forms were not mutually exclusive. The cloth dressers, for example, made use of all three in their campaign against the gig mill in 1802. They can, however, conveniently be separated for analytical purposes. How effective was each weapon?

Strikes

Resistance which took the form of orthodox labour sanctions such as strikes, blacking of employers' work or shops and attempted negotiations, clearly required a high degree of labour organization. Cloth dressers in both the West of England

and in Yorkshire fitted this category, having a remarkable level of organization by 1800. In both regions their trade union, the Brief Institution, provided common membership cards, unemployment and sickness benefit and proved a powerful force safeguarding wages and conditions (Randall 1991, pp. 131–9). The paper makers likewise had a long and impressive trade union history before 1800, their local organizations linked, as was usual for artisan trades, by the tramping system (Shorter 1971, pp. 95–7; Rule 1981, pp. 182–3). Wool combers formed some of the oldest combinations on record and such was their power that their autonomy was challenged only rarely by their employers (Burnley 1889, pp. 164–5; Dobson 1980, pp. 20–1, 25; Rule, 1981, pp. 164–5). Brick makers had a more intermittent history of combination in the eighteenth century but, by the mid-nineteenth century, they too were a powerful group (Price 1975, pp. 113–15). And cotton mule spinners, a labour group created by the new technology of the Industrial Revolution, were the archetype of the new factory-based trade unions of the nineteenth century (Lazonick 1979, p. 238; Price 1986, p. 79; Rule, 1986, 270–1). All utilized their undoubted industrial muscle to resist the new technologies which challenged their trades.

The foundations of these combinations lay in the very skill which machinery was threatening. It was the capacity of these trades to limit recruitment, the difficulty of acquiring the complex hand skills – dexterity, strength, 'feel' – and their determination to remain an elite which were the foundations of their industrial strength. It was therefore not surprising that these strengths were brought to bear upon innovators. But how successful was orthodox trade union action when faced with new labour-displacing technologies?

The power of vested labour groups could certainly deter many innovators from risking conflict. Particularly when that new technology was untried, the uncertain profit from innovation could be weighed against the certainty of industrial disruption and found wanting. Thus the early attempts to set up gig mills in Leeds soon fell victim to shows of strength by the cloth dressers or croppers. With scarcely a man out of the fifteen hundred cloth dressers in the town not a member of the Institution, few merchants, apart from Benjamin Gott, thought the risk worth taking. Gott's attempt incurred summary intimidation and the sanction of industrial action until he took the machine down. The strike of Gott's croppers in 1802, although ostensibly over the issue of apprenticeship, was also a show of strength directed at rumours that Gott was thinking of trying once more to bring in gig mills. When clothiers in Wiltshire tried to introduce the same machine into their shops in 1802, they were immediately the victims of strikes and found that they could not get their cloths finished in the area, having to transport them out of the district to be dressed and then back before they could be sold. This made their cloths vulnerable to attack as the Warminster clothiers found to their cost in 1802 (Randall 1991, pp. 130, 146–8, 155). The famous six month long strike of the wool combers in Bradford, Yorkshire, in 1825 was principally over the issue of wages but, as in subsequent disputes, the threat of the introduction of machinery was present. The problems of developing the combing engine to compete with manual labour in the finest wools clearly frustrated some manufacturers who continued to look to new technology as a means to break the power of the over-mighty combers. Others showed much less enthusiasm to challenge the combers and were slow to introduce combing engines before the machine was eventually perfected in 1841 (Burnley 1889, pp. 125–6, 166–74). Manchester brick makers too used strike action in an attempt to resist the introduction of brick-making machinery in the 1860s (Price 1975, pp. 125–9).

However, orthodox industrial sanctions could succeed only where the strength of labour was very high and the trade highly concentrated. Innovators outside the main centres of production might take up the machinery with less fear of reprisals and, once they had proved the technology and discovered the profitability of reducing skilled hands, it was only a matter of time before the pressure of their competition drove employers in the labour strongholds to grasp the nettle. This was what happened in Yorkshire in 1811–12. The Huddersfield master clothiers' utilization of gig mills had by the late 1790s been threatening the trade of merchants in the major cloth dressing centres but the risk of conflict with their croppers meant that few chose to follow. However, in the years from 1809 the rise in the eastern Yorkshire woollen districts of a few large-scale and well-fortified factories which housed shearing frames precipitated the increasing use of frames across the district in 1811 (Thomis 1970, pp. 63–4). The cotton spinners found likewise that when technology began to get a hold, as with the self actors after 1826, the weapon of strike action proved much less effective (Musson & Kirby 1975, pp. 32, 143). To go on strike against a machine which was both proven and would permanently displace you was not likely to prove successful. With the stakes high, few labour groups of any tradition tamely succumbed. Violence offered a clearer method of resistance.

Violence: anti-machinery riot and machine breaking

It is possible to identify two distinct characteristic forms of violence against new technology, the anti-machinery riot and the more narrowly focused machine-breaking attack. Luddism in 1812 revealed both forms of response with the mass assaults upon the weaving factories in Lancashire which involved many hundreds and the small machine-breaking 'gangs' who broke frames in both Yorkshire and Nottinghamshire. The community based anti-machine riots, which typified many of the protests against textile machinery in Lancashire, Yorkshire and the West of England, have much in common with the food riot which was so characteristic of eighteenth-century protest (for eighteenth-century food riots, see Thompson 1971). The crowd summoned by horns, the pressing of all to leave their work to take part, the strong assertion of legitimacy, all echoed the actions of the food rioters. Thus, the ceremony with which the rioters protesting at the establishment of the first scribbling engine to be introduced into the West of England at Bradford on Avon in 1793 'arraigned' and tried the machine before burning it on the town bridge betokens the clear sense of popular legitimation of crowd intervention so often seen in the popular retribution handed out to millers and bakers found to be withholding stocks or adulterating flour (Charlesworth & Randall 1987, pp. 205–6; Randall 1991, p. 80). The collective empowerment of these community protests did not necessarily lead to violent conflict. Just as farmers and middlemen, confronted by the reality of mass protest, might often back down and sell grain at the 'just' price, so too did innovators often seek to forestall violence by retreat. Thus numerous farmers sought to save their ricks and property by voluntarily dismantling the hated threshing machines whenever bands of Swing rioters approached in 1830 (Hobsbawm & Rude 1969, pp. 233–5). The systematic assaults of machine-breaking gangs often reflected a more deliberate and systematic policy of destruction, echoing the trade union base of these forms. This form characterized the actions of the shearmen in Wiltshire in 1802 and the Yorkshire croppers in 1812, as well as those of the brick makers in the 1860s. These groups saw machine breaking as a calculated and effective means to safeguard their trades. The attacks on the wide frames in the East Midlands in 1811 likewise revealed careful planning and effective leadership. It was because these Luddite gangs proved so successful and so difficult to intercept or to suppress that the government dedicated far more troops to chase Luddites at home than it gave to Wellington to fight the French in the Peninsular War (Thomis 1970, pp. 144–5).

The violence utilized in the various campaigns to resist new technologies, whether employed by the rioting crowd or by the machine-breaking gang, was, however, far from being mindless or uncontrolled. Violence usually occurred only after some other indication of popular hostility had been manifested towards the innovating capitalist. An escalating scale of violence was often utilized as labour sought to intimidate and coerce innovators into compliance. The Wiltshire Outrages in 1802 show this clearly (Randall 1982, p. 297). Anonymous threatening letters usually opened the 'negotiations', supplemented by some minor piece of destruction to the innovator's property, often a hay rick or, in one case, a dog kennel. Failure to comply would see increasing activity including actual attacks on the innovator's cloths, house and factory or workshops. Guns might be discharged outside his home. Only in the last resort were outright attacks mounted to destroy factories. Violence was also sometimes instigated by workers from outside the immediate area. Clothiers at Shepton Mallet, mindful of likely trouble, had persuaded their workers to accept mutually monitored trials of the newly-introduced jenny in 1776 but this agreement was not acceptable to the workers in the neighbouring towns. Thus it was the workers of Frome who marched into the town later that night and demolished the machines. Such workers felt that they needed to act to prevent the establishment of a technology which would in time oust them from their work. They also no doubt felt that their chances of being identified were somewhat less than for the locals.[8]

The protesting anti-machinery crowd rarely 'went on the rampage'. Indeed, the restraint of the crowd, often under very considerable provocation, is remarkable. For example, when a large crowd of textile workers at Bradford on Avon, protesting at the establishment there of the first scribbling engine in the county in 1793, were fired upon by the owner, killing three people outright, they offered no retaliatory violence to him personally although all his windows were smashed and the hated machine seized and destroyed (Randall 1991, p. 80). The same was true in Lancashire in 1826 when guns were fired into the crowd besieging power loom factories (Hammond & Hammond 1979, pp. 99–100). Thus, while property was often destroyed and many entrepreneurs claimed to be in fear of their lives, violence against the person of the innovators was rare. The murder of Horsfall in 1812 and the attempted murders of Jones in 1808 and of Cartwright in 1812 came only after other forms of protest had proved failures. Such personal attacks clearly did not receive the community support accorded to attacks upon mills and machinery. While the authorities still found evidence hard to obtain, it is significant that, after these events, mass popular support for the Luddites fell away (Thompson 1968, pp. 624–5; Thomis 1970, p. 80; Randall 1991, p. 180).

This raises the other characteristic of anti-machinery protest which deserves attention, namely the widespread public support which the machine breakers often enjoyed. The participants in many anti-machinery riots frequently included a cross-section of working class occupations, not merely those trades immediately at risk. Thus among those taken up for Luddite attacks in Yorkshire in 1812 were not only croppers, whose jobs the machines were threatening, but also weavers, shoemakers, tailors and clothiers (Thompson 1968, p. 643). Those involved in the power loom riots in Lancashire in 1826 included workers from all the main textile trades as well as others in no way connected to cotton. Thus in Blackburn, among those arrested for the attacks upon mills were spinners, labourers, a farmer, a confectioner and a butcher, together with power loom weavers and manual weavers. And at Chorley, an eyewitness noted, 'there can be no doubt a great multitude of the townspeople were their friends. The women supplied the rioters with stones, concealing the missiles under their aprons.'[9] The numerous anti-machinery disturbances in the West of England were likewise characterized by a wide cross-section of textile trades taking part. And the Swing rioters included in their ranks not only agricultural labourers but also members of many rural crafts (Hobsbawm & Rude 1969, pp. 243–6; Randall 1991, p. 102). The resistance to machinery was therefore a genuinely community response, not merely a trade response, reflecting the general hostility towards those who sought to overthrow customary economic relations. Further, in Luddism, as in the disturbances in the West of England and Lancashire before, the central authorities were both shocked and angered by the way in which ratepayers, shopkeepers, small employers and even some magistrates sided with the machine breakers or gave help in the form of alms. Thus a Headingley magistrate informed the Home Office in 1812, 'it is surprising how much the opinions and wishes even of the more respectable part of the Inhabitants are in unison with the deluded and ill-disposed of the population with respect to the present object of their resentment, Gig Mills and Shearing Frames' (Hammond & Hammond 1979, p. 251). While this was often ascribed to fear, there is good reason to believe that hostility towards the innovators extended well up into the respectable classes. In 1779 following the anti-jenny riots, magistrates at Wigan suspended the use of all carding, roving and spinning machines worked by water until Parliament had been consulted as to their propriety (Hammond & Hammond 1979, pp. 43–4; Stevenson 1979, p. 118). In Wiltshire in 1802 some magistrates 'refused to act' with any vigour against the riotous cloth dressers (Randall 1991, p. 167). And John Anstie, an innovating clothier, admitted in 1803, 'there are a considerable number of persons of respectability who...still continue to consider the introduction of machinery into the woollen industry as unfriendly to the general interest and peculiarly injurious to the poor' (Anstie 1803, pp. 7–8). Small shopkeepers feared that new large factory masters would, by truck, destroy their trades. The same was true of the many small manufacturers and master workers who correctly recognized that their trades were under siege from those with large capitals. And ratepayers had long memories of the way in which they had picked up the bill for unemployment in the past and for the ubiquitous way in which in all manufacturing areas the manufacturers seemed to pay less than their 'fair share' of poor rates (Mann 1971, p. 117; Randall 1991, p. 106).

Appeal to authority

The other recourse of those seeking to resist the advent of machinery which threatened their trades was to appeal to the authorities, both locally and nation-ally, for protection. In Lancashire in the years from 1776 to 1779 (Wadsworth & Mann 1931, pp. 497–500; Hammond & Hammond 1979, pp. 43–4), in the West of England from 1776 until 1809, in Yorkshire from 1793 to 1809, labour groups sought to impress the magistracy and Parliament that new machinery should be prohibited or closely controlled (Crump 1931, pp. 59–63; Randall 1991, pp. 74, 93, 96, 124–7). In the years from 1802 to 1806 the woollen industry in particular was subject to careful scrutiny and an extensive campaign sustained by both the forces for innovation and those for the status quo for the hearts and minds of the legislators.[10] In this, and from the campaigns of other groups in other industries, we can discern something of the alternative political economy of the machine breakers to that of *laissez-faire* advanced by the advocates of change.[11]

While the innovators argued that machinery represented progress which could not and should not be impeded, that the state had no legitimate role interfering in the productive process and that the workers represented only their narrow self-interest and not the interests of the country as a whole, labour argued a very different case. Unregulated machinery they saw as a major threat to the economic well-being of the community since it might displace hands, throwing them upon the scrap heap in exchange merely for the enhanced profit of the few. Labour had no conception of the machine as the agent to extend markets. In their view, markets were already supplied by the old hand production methods. All that machinery might do was to fill orders more quickly. Thus an under-consumptionist crisis must occur in which fewer labourers earned a living from the manufacture, thereby curtailing the market and serving only to increase the burden of poverty upon the ratepayers. This essentially mercantilist view of trade had been a commonplace view less than a century before and even in the mid-eighteenth century was still being expounded by economists of significance (Berg 1985, pp. 54–6). But the new gospel of free trade, *pace* Smith, and the success of the cotton industry in expanding both a domestic and a world market with machine-made cheaper products, was already undermining the old view. Nonetheless, anti-machinery petitioners continued to portray the innovating machine owners as selfish would-be monopolists, concerned only with their own short-term profits at the expense of the long-term interests of the community as a whole.[12] In this way, the odium attached to the forestaller and regrater in the food market was attached to the machine innovator.

Additionally, the anti-machine petitioners continued to argue that the strength of the British economy lay in the indigenous skills of its workers. In that machinery would render these skills worthless, it was, they asserted, folly to allow them to be destroyed. Besides, they argued that the labourer's skills were his property and as such merited every bit as much protection and regard as the innovator's property in the machine. This notion that skill was a hand craft is interesting. The machine was deemed to produce both worse quality products (a claim that in many cases, such as that of the combing engines before 1840, was true (Burnley 1889, pp. 165–6) but which ignored the fact that it was in the cheapest products that the greatest market growth was occurring) and required less skill to control it. Only with the spinners in the nineteenth century would this argument be modified, as definitions of 'skill' became increasingly associated with machine minding and as complex machine minding skills came to define the characteristics of a new labour aristocracy. But it is worth noting that the spinners went out of their way to 'tune' their mules to their own individual specification, thereby ensuring that their own hand skills reacquired significance and a degree of irreplaceability.[13]

The solution advanced most usually by those seeking interventionist action by the state was not the outright banning of new machinery but its regulation and control. The cloth dressers in 1802 and 1806 vociferously argued that the gig mill should indeed be banned because it was so prohibited by the old Act 5 & 6 *Edward VI, c22* but even here they accepted that, as it had for years been used on coarse cloth, it should only be banned from the fine trade (Randall 1991, p. 243). Other groups, however, sought either some sort of phased introduction of machinery which could both allow a gradual re-employment of workers or a regulation of the numbers or financial advantage of the machine. Thus Lancashire spinners demanded a series of incremental taxes upon carding and spinning machines in 1780, woollen workers and rate payers in Saddleworth in 1795 and in Leeds in 1803 sought to phase the introduction of preparatory and finishing machinery by taxes, while handloom weavers continued to campaign for a tax upon power looms until the 1830s.[14]

In all these views we can discern that 'moral economy of the crowd' which Thompson so brilliantly delineated in his rescue of the eighteenth-century food riot (Thompson 1971). The same view that the customary rights of the consumer and producer should be preserved against the rapacity of entrepreneurial capital, the same view that the state and its local agents had no choice but to intervene to protect the community from oppressive economic power, come through strongly.

The moral economy of the crowd, however, presumed the paternal authority and motivation of the state. Unfortunately for the machine breakers, the old moral economy values of paternal regulation were being eroded from the mid-eighteenth century as a growing national market in grain, enclosure, turnpikes and the success of the unfettered cotton industry offered hard evidence to support the theories of Dr Smith. Thus Mokyr argues that 'The case of the workers [against machinery] was politically hopeless.' (Mokyr 1990, p. 258). Perhaps so, but the reasons he advances to explain this are not entirely convincing. Thus his belief that 'The argument that stopping the new machinery would only lead to its flight abroad was persuasive' does not square with the state's continued efforts to prevent the export both of machinery and of artisans until well into the nineteenth century (Berg 1980, pp. 205–7, 209). Mokyr also argues that, since the landed did not see machinery as a threat to their own property, they saw no reason to resist it. Yet, as many were aware, machinery could well result in rapidly increasing poor rates, a very present concern for the landed up to 1834 and one which induced many country gentlemen to contemplate some alternative to the unfettered adoption of machinery (Randall 1991, p. 235). Nor will the argument that 'not all workers in the traditional sector were initially made worse off' explain the attitudes of the authorities to machinery since in no case was the well-being of threatened labour groups seen as a major determinant of policy.

The workers' case against machinery *was* doomed but the major reasons were political and perhaps also cultural. Increasingly the English upper classes saw the problem of machinery as part of a wider and growing problem of law and order. To some degree this preoccupation with social control was the consequence of the rise of a new sort of industrial society in the late eighteenth century which was not responsive to the old methods of paternal authority. But the major reason lay with the events in France in 1789 and the years which followed. That old self-confidence of the English landed classes in the security of their power, a vital factor which had underpinned the paternal model of government, was increasingly eroded by the perceived menace of revolution from below. Thereafter violent protest against machinery or about food prices smacked too much of the *sans culottes* for comfort. Likewise, the rise of an indigenous and vital radical movement after 1791, coupled with the growing concern that organized labour harboured subversive intentions (Thompson 1968, pp. 545–7; Wells 1986, pp. 51–3), served only to tilt the balance abruptly and decisively against paternal intervention. In this climate it was property, not labour, which seemed in need of protection. And, significantly, in defining that property, the state chose not to recognize the property rights of skill, even though it continued to attempt to prevent its export. Additionally, it may well be that the landed classes' ignorance of technology, a consequence of a gentlemanly culture lacking in any significant scientific education, ill-equipped them to see the problem of machinery in any way other than as an issue of law and order.[15] Thus the state sided with the machine, with capital and with the free market and against labour and regulation. The debate about the 'Machinery Question' which developed in the 1820s and 1830s never looked likely to shift this fundamental new balance (Berg 1980, pp. 313–14).

Conditions of success and failure

How successful was machine breaking? Clearly it was not successful in that machinery ultimately came to triumph and to establish a new form of manufacturing economy. In some cases this triumph was rapid. In other areas, however, the ascendancy of the machine was only slowly established and in large measure this delay was due to resistance. Even in the cotton industry, determined labour protest slowed the early adoption of the large jenny and the power loom (Bythell 1978, p. 45; Hammond & Hammond 1979, pp. 43–4). In Yorkshire too, new machines were delayed by protest and the croppers' violent resistance in 1812 slowed the adoption of the shearing frame until the post-war years (Thomis 1970, p. 164; Randall 1991, pp. 79, 175–6). The Wiltshire shearmen's Outrages in 1802 bought them even longer as innovators were reluctant to reintroduce finishing machinery in large numbers until well after the war ended (Randall 1982, p. 302). Agricultural labourers suffered heavily for their temerity in smashing threshing machines in 1830 but significantly it was not for many years that these machines began to be reintroduced (Hobsbawm & Rude 1969, pp. 298–9). The Manchester brick makers bought themselves many years of continued employment by their sustained and coordinated machine-breaking violence in the 1850s and 1860s (Price 1975, pp. 127–8). More work is needed before we can make the same judgements about the Buckinghamshire paper makers who destroyed the paper mills between Loudwater and Wycombe in 1830. Did they succeed in delaying the advent of continuous paper making there?

We might also note that 'success' in resisting new technology might in the end prove expensive. The continual battle Wiltshire entrepreneurs encountered at every turn may well have significantly delayed the development of the woollen industry there and ensured that it could not match its rivals either in Yorkshire or in Gloucestershire. By 1826 the industry in the county was in terminal if protracted decline (Mann 1971, p. 168, 170). Yet the hostility to machinery remained. When Gurney drove his prototype road-going steam engine into the town of Melksham in 1829, he was met by a large and angry crowd of textile workers who stoned his horseless carriage, shouting, 'We are starving, let's have no more machinery' and 'Down with machinery'.[16]

Finally, we might ask, why were the machine breakers unsuccessful? In part, of course, their defeat was a consequence of the perceived benefits which machinery made possible. The opposition of weavers in Lancashire, Yorkshire and even in the West of England to spinning machinery eventually diminished when they saw the advantages the new more even machine-spun yarn had for weaving. But it was not often that those displaced saw any of the benefits. The consumer may have enjoyed more choice and more attractive prices (Mokyr

1990, p. 256). For most of those displaced, their concerns were more immediate and concerned finding alternative work. It was rare they could find anything as well paid or conferring equal status to the trade they had lost (Thompson 1968, pp. 342–3). The defeat of the machine breakers may also legitimately be ascribed to the 'heroic' innovators and entrepreneurs who, in spite of all problems, boldly persisted and ultimately won through. This, of course, was very much the story which the new manufacturing factory owning class wished to leave to posterity (Thompson 1968, p. 613). But we would be wrong to deny that, just as it took courage to attack Rawfolds Mill, it also took courage to build, equip and fortify it in the face of widespread community hostility.

The major determining factor in the defeat of the machine breakers, however, was the policy of government. The triumph of *laissez-faire* philosophy in the arena of industrial regulation was a remarkably abrupt one. Though the paternal model was already fatally flawed by the late eighteenth century and though governments were from the 1730s reluctant to embark upon any new regulatory legislation, under some circumstances Parliament could be persuaded to intervene on labour's side and, even as late as 1803, the Cotton Arbitration Acts offered hope to the workers that the advent of the machine age could be controlled.[17] The great campaign over the role of legislative regulation in the woollen industry between 1803 and 1809 proved to be the turning point. The old national staple, the woollen industry was hedged around with over 70 regulatory statutes which labour groups sought to utilize to impede the development of a machine- and factory-based industry. Innovators in 1802 mounted a determined campaign to remove them from the Statute book and, after two major enquiries and much procrastination, in 1809 Parliament obliged with wholesale repeal (Randall 1991, pp. 9, 221–7). Thereafter labour stood little chance of securing any stay of execution from the state when it came to machinery. The 1806 Report of the Select Committee on the Woollen Trade made this clear, asserting 'the right of every man to employ the capital he inherits, or has acquired, according to his own discretion, without molestation or obstruction'.[18] The state took the side of 'progress' and in a remarkably short period of time moved decisively from the role of arbiter to that of enforcer for capital (Thompson 1968, pp. 595–6). Its role in enabling rapid mechanisation in the first half of the nineteenth century was far from insignificant, both in the removal of further legislative obstacles and in the punishment of those who sought to resist.[19] As the machine breakers were to discover, the dice were heavily loaded against them.

Notes

1 Thus Mokyr 1990, p. 258: 'Most innovations were adopted without significant trouble, and not all trouble necessarily reflected resentment toward the new machines.'
2 Berg 1985, p. 263. Lancashire spinners too attacked new machinery in 1830 and 1834. See Musson & Kirby 1975, pp. 128, 294.
3 Burnley 1889, pp. 165, 175; *Journal of the House of Commons* 1793, Vol. 49, pp. 322–3; Randall 1991, p. 105.
4 Early power looms were invented, among others, by Rev. Edmund Cartwright in 1784 and by William Radcliffe in 1803. Hammond & Hammond 1979, pp. 55, 222.
5 Cartwright's 'Big Ben' was first introduced in Doncaster in 1790 to apparent enthusiasm though it was not welcomed in Bradford. Burnley 1889, pp. 124–6.
6 On the debate on proto-industrialization see Mendels 1972; Berg, Hudson & Sonenscher 1983, Ch. 1. For an evaluation of the usefulness of the proto-industrial model to an understanding of the English woollen industry, see Randall 1991, pp. 22–6.
7 Hobsbawm and Rude argue that, while the cost advantages of the threshing machine were limited, its great advantage was that 'it saved precious time'. Hobsbawm & Rude 1969, p. 362.
8 Mann 1971, p. 123: Randall 1991, pp. 72–3. The clothiers' choice of venue for the experimental trials of the jenny, the workhouse, could hardly have soothed the fears of displacement and poverty which the jenny's advent excited.
9 *Blackburn Mail*, 10 and 24 May 1826: *Preston Chronicle*, 29 April 1826. I am grateful to Dr David Walsh for these references.
10 *B. P. P.* 1802/3, Vol. 5 (H. C. 30 and 95); *B. P. P.* 1806, Vol. 6 (H. C. 268).
11 For this debate, see Randall 1986 and 1991, Ch. 7, from which the following two paragraphs are derived.
12 Thus the opponents of the large jennies in Lancashire in 1780 complained that the 'machines appeared as a mere monopoly "for the immense Profits and Advantage of the Patentees and Proprietors"'. Wadsworth & Mann 1931, p. 500.
13 Price 1986, pp. 73–5, 78–9; Catling 1970, pp. 74–9. See also Lazonick (1979). For discussion of the definition and acquisition of skill in the nineteenth century, see More 1980, pp. 15–26.
14 Wadsworth & Mann 1931, pp. 500–1; Nield 1795, p. 8; *Leeds Mercury*, 29 January 1803; Thompson 1968, pp. 335–6; Bythell 1969, p. 175.
15 I owe this interesting suggestion to Prof John Harris. See Harris 1993, pp. 93, 99–100.
16 *Devizes and Wiltshire Gazette*, 30 July 1829.
17 For example, in 1756 petitions and pressure from the Gloucestershire weavers resulted in the passing of an act which empowered magistrates to fix, and incidentally raise, piece rates, while petitions and violence secured wide-ranging protection of their trade for the Spitalfields silk weavers in 1773.
18 *B. P. P.*, 1806, Vol.3 (H. C. 268), Report, 12.
19 Rule 1992, pp. 313–14. It is interesting to compare the British experience in this with that of France where governments remained far more cautious about siding so obviously with the innovators and against the crafts. Mokyr 1990, p. 260.

References

ANSTIE, J. (1803). *Observations on the Importance and Necessity of Introducing Improved Machinery into the Woollen Manufactory.* London.

BERG, M. (1980). *The Machinery Question and the Making of Political Economy 1815–1848.* Cambridge: Cambridge University Press.

BERG, M. (1985). *The Age of Manufactures: Industry, Innovation and Work in Britain 1720–1820.* Fontana.

BERG, M. (ed.) (1991). *Markets and Manufacture in Early Industrial Europe.* London: Routledge.

BERG, M., HUDSON, P. & SONENSCHER, M. (ed.) (1983). *Manufacture in Town and Country before the Factory.* Cambridge: Cambridge University Press.

BRIGGS, A. (1959). *The Age of Improvement, 1783–1867.* London: Longmans.

British Parliamentary Papers, B. P. P., 1802/3, Vol. 5 (H. C. 30 and 95). Report of and Minutes taken before the Select Committee on the Woollen Clothiers' Petition.

British Parliamentary Papers, B. P. P., 1806, Vol. 6 (H. C. 268). Report of and Minutes taken before the Select Committee to Consider the State of the Woollen Manufacture.

BURNLEY, J. (1889). *The History of Wool and Woolcombing.* London.

BYTHELL, D. (1969). *The Handloom Weavers. A Study in the English Cotton Industry During the Industrial Revolution.* Cambridge: Cambridge University Press.

BYTHELL, D. (1978). *The Sweated Trades. Outwork in Nineteenth-Century Britain.* London: Batsford.

CANNADINE, D. (1984). The present and the past in the English Industrial Revolution. *Past and Present* **103**, 131–72.

CATLING, H. (1970). *The Spinning Mule.* Newton Abbot, UK: David and Charles.

CHAMBERS, J. D., & MINGAY, G. E. (1966). *The Agricultural Revolution, 1750–1880.* London: Batsford.

CHARLESWORTH, A. & RANDALL, A. J. (1987). Morals, markets and the English crowd in 1766. *Past and Present* **114**, 200–13.

COLE, G. D. H. & POSTGATE, O. (1949). *The Common People, 1746–1946.* London: Methuen.

COLEMAN, D. C. (1958). *The British Paper Industry 1495–1860.* Oxford: Clarendon Press.

CRAFTS, N. (1983). British economic growth, 1700–1831: a review of the evidence. *Economic History Review* **36**, 177–99.

CROUZET, F. (1982). *The Victorian Economy.* London: Methuen.

CRUMP, W. B. (1931). *The Leeds Woollen Industry, 1780–1820.* Leeds.

DOBSON, C. R. (1980). *Masters and Journeymen: A Prehistory of Industrial Relations, 1717–1800.* London: Croom Helm.

HAMMOND, J. L. & HAMMOND, B. (1966). *The Village Labourer.* London: Longmans. (1911, 1966 edn.)

HAMMOND, J. L. & HAMMOND, B. (1979). *The Skilled Labourer.* London: Longmans. (1919, 1979 edn., ed. J. Rule).

HARRIS, J. R. (1993). French industrial policy under the ancien régime and the pursuit of the British example, *Histoire, Economie et Société*, 1e Trimestre.

HOBSBAWM, E. J. (1968). *Labouring Men: Studies in the History of Labour.* London: Weidenfeld and Nicholson. (1964, 1968 edn.)

HOBSBAWM, E. J. & RUDE, G. (1969). *Captain Swing.* London: Lawrence and Wishart.

HUDSON, P. (ed.) (1989). *Regions and Industries: a Perspective on the Industrial Revolution in Britain.* Cambridge: Cambridge University Press.

HUNT, E. H. (1981).   *British Labour History 1815–1914*. London: Weidenfeld and Nicholson.

Journal of the House of Commons (1793), Vol 49.

LAZONICK, W. (1979).   Industrial relations and technical change: the case of the self-acting mule. *Cambridge Journal of Economics* **3**, 231–62.

MACDONALD, S. (1975).   The progress of the early threshing machine. *Agricultural History Review* **23**, 63–77.

MACDONALD, S. (1978).   Further progress with the early threshing machine: a rejoinder. *Agricultural History Review* **26**, 29–32.

MANN, J. de L. (1971).   *The Cloth Industry in the West of England from 1640 to 1880*. Oxford: Oxford University Press.

MATHIAS, P. (1983).   *The First Industrial Nation*. London: Routledge. (1969, 1983 edn.)

MENDELS, F. F. (1972).   Proto-industrialisation: the first phase of the industrialisation process. *Journal of Economic History* **32**, 241–61.

MOKYR, J. (1990).   *The Lever of Riches: Technological Creativity and Economic Progress*. Oxford: Oxford University Press.

MORE, C. (1980).   *Skill and the English Working Class, 1870–1914*. London: Croom Helm.

MUSSON, A. E. & KIRBY, R. (1975).   *The Voice of the People, John Doherty, 1798–1854*. Manchester: Manchester University Press.

NIELD, D. (1795).   *Addresses to the Different Classes of Men in the Parish of Saddleworth Showing the Necessity of Supporting the Plan for Augmenting the Price of Labour in the Woollen Manufactory*. Leeds.

PEEL, F. (1968).   *The Risings of the Luddites, Chartists and Plug Drawers*. Frank Cass. (1880, 1968 edn.)

PRICE, R. (1975).   The other face of respectability: violence in the Manchester brick-making trade, 1859–1870. *Past and Present* **66**, 110–32.

PRICE, R. (1986).   *Labour in British Society*. London: Croom Helm.

RANDALL, A. J. (1982).   The shearmen and the Wiltshire Outrages of 1802: trade unionism and industrial violence. *Social History* **7**, 283–304.

RANDALL, A. J. (1986).   The philosophy of Luddism. *Technology and Culture* **27** (2), 1–17.

RANDALL, A. J. (1989).   Work, culture and resistance to machinery in the West of England woollen industry. In *Regions and Industries: a Perspective on the Industrial Revolution in Britain*, ed. P. Hudson. Cambridge: Cambridge University Press.

RANDALL, A. J. (1991).   *Before the Luddites. Custom, Community and Machinery in the English Woollen Industry, 1776–1809*. Cambridge: Cambridge University Press.

REDDY, W. (1984).   *The Rise of Market Culture: the Textile Trade and French Society, 1750–1900*. Cambridge: Cambridge University Press.

RUDE, G. (1970).   *Paris and London in the Eighteenth Century*. London: Fontana.

RULE, J. (1981).   *The Experience of Labour in Eighteenth-Century Industry*. London: Croom Helm.

RULE, J. (1986).   *The Labouring Classes in Early Industrial England, 1750–1850*. London: Longmans.

RULE, J. (1992).   *The Vital Century. England's Developing Economy, 1715–1815*. London: Longmans.

SHORTER, A. H. (1971). *Paper Making in the British Isles: an Historical and Geographical Study*. Newton Abbot: David and Charles.

STEVENSON, J. (1979). *Popular Disturbances in England, 1700–1870*. London: Longmans.

THOMIS, M. I. (1970). *The Luddites: Machine Breaking in Regency England*. Newton Abbot: David and Charles.

THOMPSON, E. P. (1968). *The Making of the English Working Class*. Harmondsworth, UK: Penguin. (1963, 1968 edn.)

THOMPSON, E. P. (1971). The moral economy of the English crowd in the eighteenth century. *Past and Present* **50**, 76–136.

WADSWORTH, A. P. & MANN, J. de L. (1931). *The Cotton Trade and Industrial Lancashire, 1660–1780*. Manchester: Manchester University Press.

WELLS, R. A. E. (1986). *Insurrection: the British Experience, 1795–1803*. Gloucester: Alan Sutton.

4

The changeability of public opinions about new technology: assimilation effects in attitude surveys

DANCKER D L DAAMEN and IVO A VAN DER LANS

Introduction

As pointed out in other chapters in this volume, resistance to new technologies can take several forms. A prerequisite for most of these forms is a negative attitude towards the technology. Therefore, to predict public resistance to new technologies, it is crucial to know people's attitudes (e.g. Williams & Mills 1986; Eurobarometer 1989; Daamen, van der Lans & Midden 1990; Miller 1991). Surveys are important instruments for assessing public attitudes to new technologies. Unfortunately, responses to specific items may be changed dramatically by the characteristics of the questionnaire. For instance, it has been demonstrated in numerous studies that answers to a survey item may either assimilate towards or contrast away from (the central tendency in) responses to preceding items (for reviews see Schuman & Presser 1981; Hippler & Schwarz 1987; Tourangeau & Rasinski 1988). The topic of this chapter is assimilation effects in surveys on technology perceptions.

An early example of an assimilation effect in a survey comes from Salancik & Conway (1975). They demonstrated that attitudes towards being religious were affected by responses to earlier questions about religious behaviour: Subjects in two conditions had to indicate whether statements of pro- and anti-religious behaviours applied to them (e.g. 'I attend a church or synagogue'; 'I use the expression "Jesus Christ"'). In one condition, Salancik and Conway inserted 'on occasion' in the pro-religious behaviour statements and 'frequently' in the anti-religious statements. In the other condition this was reversed. By this they successfully manipulated the subjects on the independent variable, i.e. the probability of endorsement of the behaviour statements, in condition 1 in a pro-religious and in condition 2 in an anti-religious direction. As predicted, respondents in condition 1 responded more favourably to the subsequent religious attitude question than those in conditon 2. Referring to the self-perception theory (Bem 1967), Salancik and Conway's interpretation of these results is that subjects partly infer their attitudes from easily accessible and relevant information about their own behaviour, i.e. their responses to the religious behaviour statements. If their responses are predominantly pro-religious, respondents infer that they must have a positive attitude towards religion. On the other hand, if their responses to the behaviour statements are in an anti-religious direction, respondents decide that they must have a more anti-religious attitude.

Tourangeau and Rasinski (1986) provide a second example. They found that respondents in condition 1, who were asked questions about women's rights subsequently had a more favourable attitude towards legalized abortion, compared with those in condition 2, who answered prior questions about traditional values. They provide the following explanation: women's rights items more readily activate related 'pro-choice' beliefs than 'pro-live' beliefs, whereas for traditional value items it is the other way around. Subsequently, these activated related beliefs, pro-choice in condition 1 and pro-life in condition 2, are more easily retrieved when respondents are asked for their attitude to abortion and this influences their responses (see Tourangeau & Rasinski 1988).

These studies have a lot in common. In both studies the preceding context influences the responses to the attitude question in a direction that is consistent with (the responses to) the preceding questions. In the interpretation of results in both studies, the traditional view on attitudes is left. An attitude towards a topic is not conceived as a ready-for-use predisposition to respond to that topic, but as a satisfying answer to the question, 'What is your attitude about topic x?' When respondents are confronted with such a question they will search in their memory for relevant information about the topic. This search process is assumed to be 'quick and easy': most respondents do not go through the trouble to retrieve from their memory **all** relevant information, but mainly retrieve easily accessible information. Preceding questions about aspects of topic x – or responses to these questions – raise the accessibility of specific information units and by this the chance respondents base their attitude towards x upon these units. All else being equal, attitudes will be more positive, when preceding questions raise the accessibility of predominantly positive information about the topic.

Both studies are the first in a series:[1] The findings of Tourangeau and Rasinski (1986) are replicated for eight out of ten political issues (Tourangeau et al. 1989a,b). Results similar to those of Salancik and Conway (1975) are reported by Bishop and his colleagues (Bishop, Oldendick & Tuchfarber 1982, 1984; Bishop 1987).

Apart from their similarity, there is an important difference between the two studies. In the account of Salancik and Conway, respondents' attitudes are influenced by preceding **responses** (the preceding questions are essentially the same in both experimental conditions). Respondents retrieve their responses to the earlier questions. There is a **direct** relation between the 'parts', the responses to the specific behaviour questions, and the 'whole', the response to the general attitude question. The only assumption to be made is that respondents remember the central tendency in their preceding responses and infer their attitude from this central tendency (cf. Smith 1982, 1992).

In the studies of Tourangeau *et al.* the assimilation results from differences in the **questions** that precede the attitude question. In their research the general attitude question is preceded by a series of items that are assumed to be associated with predominantly favourable beliefs about the attitude object in one condition, and a series of supposedly unfavourable beliefs in another condition. Referring to recent studies in the social judgment literature, Tourangeau and Rasinski (1988) point out that beliefs that are primed by prior questions may in turn activate related beliefs and the latter may be retrieved from memory and used to answer the general attitude question. Thus, the relation between the 'parts', the prior beliefs items, and the 'whole', the attitude item, is less direct and the assimilation effect need not depend on responses to the part items.

The present study concerns public perceptions of modern technologies. The focal item in our survey concerns the attitude to 'technology in general'. We investigated to what extent this general attitude assimilates towards the **responses** to prior attitude questions about nine specific areas of technology. We hypothesized that respondents at least in part infer their attitude to 'technology in general' from the central tendency in their responses to the specific attitude questions. Thus, our study must be placed in the tradition started by Salancik and Conway (1975).

Some researchers hypothesized that respondents who are more involved with an issue should be less susceptible to context effects (e.g. Rugg & Cantril 1944; Converse 1974). Regarding assimilation effects the argumentation for the hypothesis is triple: (a) Respondents who are more involved with an issue – compared with those who are less involved – have gathered and processed more information about that issue and have at their disposal enough accessible inputs to generate satisfying responses to questions about that issue. They simply do not need the information contained in preceding items for their response to the general question because that information (and/or enough other relevant information) is already available and easily accessible to them (cf. Feldman & Lynch 1988). (b) Issue involvement will heighten the motivation to retrieve from memory all information that is relevant to the general question, not only information that is primed by prior items. The more thorough this retrieval process, the less influence one may expect from prior items (or from responses to these prior items). (c) More involved respondents may more often have stored in memory a ready-for-use evaluative judgment about the issue (a chronically accessible attitude) and they simply retrieve it from memory when the general question is asked (cf. Fazio *et al.* 1986). Although these arguments sound reasonable, there is hardly any evidence for a simple relation between issue involvement and susceptibility to assimilation effects in surveys (Krosnick & Schuman 1988; Bishop 1990).[2] In the present study we will nevertheless examine this relation.

Method of the study

In 1990 we published results of a large survey ($N = 2037$) and its pre-test ($N = 197$) conducted to investigate attitudes of the Dutch public toward main areas of technology. For details of the questionnaire and the method of these surveys, we refer to Daamen *et al.* (1990). The present study concerns assimilation effects in the pre-test.

We asked respondents for their attitude to new 'Technology in general' at the start of the interview (abbreviated T.genstart). This attitude question was repeated verbatim at the end of the interview (T.genend). In between, attitudes were asked towards nine areas of technology (T.area). The attitudes to 'technology in general' were measured as a global evaluative judgment about the extent to which advantages exceeded disadvantages (seven-point scale anchored with 'advantages are...very much smaller (-3) / very much larger ($+3$)...than disadvantages'). The attitudes to the nine areas of technology were measured on a five-point evaluative scale ranging from 'very bad' (-2) to 'very good' ($+2$).

The areas were military technologies, nuclear energy, automation, computer technologies, communication technologies, genetic engineering, *in vitro* fertilization, transplantation, and environmental technologies. Each area was introduced by a short description together with three examples of the area. For instance, the examples of military technologies were cruise missile, spy satellite and laser weapon. If respondents said they had at least some notion of two or all three examples, they were asked for their attitude to that area. If they did not pass this filter no more questions were asked about that area.

Obviously, the 'whole' question in the present study concerns the attitude to 'technology in general'; the 'part' questions concern the attitudes to the nine areas of technology. In this within-subjects design we can study the effect of context (the attitudes to T.area) on the 'post-test' (the attitude to T.genend) controlling for the 'pre-test' (the attitude to T.genstart). Obviously, this analysis is informative only if significant results are obtained. Non-significant results may be due to the short time lag between 'pre-test' and 'post-test': respondents remember their response to the general attitude question at the start of the interview and try to be consistent when this question is repeated verbatim at the end of the interview.

Involvement with new technology was measured with three items, selected out of six on the basis of a reliability analysis. These items concerned the frequency of being fascinated by technological innovations, the frequency of reading about technological innovations in magazines and papers, and the frequency of watching TV programmes on science and technology (these three items were averaged into an involvement index with a Cronbach's alpha of 0.70).

Respondents were sampled from the population of the city of Leiden ($N = 197$). Their ages ranged from 17 to 76 years. Respondents were interviewed by professional interviewers face to face in their homes.

Results and discussion

The mean of attitudes to the technology areas was $+0.5$, indicating that the central tendency of the attitudes to the technology areas was somewhat positive. Consistently, the attitude towards 'technology in general' was significantly more positive at the end of the interview than at the start (means of T.genstart and T.genend are $+1.41$ and $+1.63$ respectively, paired t-test is 2.89, $p \leqslant 0.01$). More importantly, respondents who were on average more positive towards the specific technology areas (compared with those who were less positive), appeared to have a significantly more positive attitude to 'technology in general' at the end of the interview (controlled for their attitude at the start). This became apparent from an analysis of covariance with the mean of attitudes to T.area as a factor ('low' vs 'high', median split), the attitude to T.genend as the dependent variable, and the attitude to T.genstart as a covariate, giving a significant result ($F = 35.08$, $p \leqslant 0.001$). The two groups ('low' vs 'high' T.area) differ by 0.3 in mean attitude to 'technology in general' at the beginning of the interview (T.genstart). At the end, this difference is more than doubled (the difference in mean T.genend between the two groups is 0.7). Apparently, if respondents changed their general attitude, they changed it in the direction of the central tendency in (here the mean of) their attitudes to specific technology areas. Correlational analysis yielded a comparable result: the part correlation between attitude to T.genend and the mean of attitudes to T.area with the attitude to T.genstart being partialled out from the attitude to T.genend, equals 0.49 ($p \leqslant 0.001$).

Thus, we find support for our assimilation hypothesis. Forty-five per cent of the respondents did not change their attitude to 'technology in general' during the interview (so their T.genstart is equal to T.genend). Possibly, they had a stable

Table 4.1. *Pearson correlations between attitudes. Attitudes to the specific areas of technology and the attitude to 'technology in general' at the beginning of the interview or the same judgment at the end*

	Attitudes to 'technology in general'	
Attitudes to areas	At the start	At the end
Computer technologies	0.21^b	0.36^b
Automation	0.21^b	0.40^b
Communication technologies	0.12	0.23^b
In vitro fertilization	0.12	0.21^b
Genetic engineering	0.11	0.30^b
Transplantation	0.04	0.07
Nuclear power	0.03	0.36^b
Military technologies	−0.04	0.16^a
Environmental technologies	−0.03	−0.01

$^a p \leqslant 0.05$, $^b p \leqslant 0.01$.

attitude to 'technology in general', or they just remembered their attitude rating at the start of the interview and wanted to be consistent when the same question was asked at the end. Of those who did change their attitude to 'technology in general', 72% changed their T.genend in the direction of their mean T.area. This is a clear indication of the assimilation effect.

Another indication of the assimilation effect is provided by Table 4.1. The attitudes to only two areas correlated significantly with the attitude to 'technology in general' at the beginning of the interview (T.genstart). The number of significant correlations between specific attitudes and general attitude at the end of the interview (T.genend) was considerably higher. Moreover, the correlations between the general attitude and the specific attitudes were on average lower at the beginning of the interview than at the end: the mean of attitudes to the areas (T.area) correlated 0.16 with the general attitude at the start (T.genstart) and 0.51 with the general attitude at the end (T.genend).

Confronted at the beginning of the interview with a rather vague question like 'What is your attitude to new technology in general?' probably few respondents are willing and capable to hold in mind all relevant technological innovations, to judge advantages and disadvantages of all these various innovations and to combine these pros and cons into one overall evaluation. This cognitive task is rather too demanding for an average respondent. We argued (Daamen *et al.*

1990) that the average respondent will 'satisfice' by using judgments about just a few salient specific technologies as the basis for their attitude to 'technology in general'. We provided evidence that the majority of the Dutch public sponta- neously associate 'technology in general' with automation and in particular its computer applications.[3] At the beginning of the interview most respondents probably base their attitude to 'technology in general' on these spontaneous associations. This explains the significant correlations between T.genstart and the attitudes to computer technologies and automation. At the end of the interview, after answering a number of questions about the nine areas of technology, the judgments about these areas are more accessible and respondents probably use these judgments to adjust their attitude to 'technology in general'. This explains why the correlations between T.genend and (the central tendency in) the attitudes to the technology areas are higher at that moment.

A higher involvement with technology covaried with more spontaneous associations with 'technology in general' at the beginning of the interview (a significant but weak correlation existed between the involvement index and the number of examples of technological innovations: $r = 0.20$, $p \leqslant 0.05$). Con- sistent with this result, a significant correlation was found between T.genstart and T.area for respondents 'high' on the involvement index ($r = 0.24, p \leqslant 0.05$), while this correlation was not significant for respondents 'low' on this index ($r = 0.05$, n.s.).[4] Thus, before they are primed by questions about specific technology areas, more highly involved respondents probably include in their attitudinal judgment about 'technology in general' more technologies than respondents who are less involved. However, no differences were found between 'low' and 'high' involved respondents in the extent their T.genend assimilated towards T.area controlling for T.genstart. Analysis of covariance as well as correlational analysis yielded non-significant results. Therefore, we conclude that in this study higher involvement did not render respondents less susceptible to assimilation effects.

One could maintain that the assimilation effect in the present study is merely a result of the ambiguity of our target item. Because the meaning of 'technology in general' is unclear to respondents, they use the prior questions about the technology areas as cues to infer what the target item is about. One could argue that if our target item had been unequivocal no assimilation effects would have emerged. However, recently we replicated the assimilation effect in a national survey ($N = 1065$) with a target item that was far less ambiguous (Daamen & Kips 1993). This target item concerned the attitude to the use of nuclear power to generate electricity. The design of this survey was similar to the present study: in a pre-test respondents gave their attitude to nuclear power, three weeks later they participated in the actual survey in which first beliefs were asked about advantages and disadvantages of nuclear power and directly after that the general attitude (this question was identical with the one in the pre-test). The results were similar to those in the present study: respondents who were on average more positive to the belief items (compared with those who were less positive) appeared to have a significantly more positive general attitude to nuclear power at the end of the interview with the variance accounted for by their general attitude at the pre-test partialled out. Because of this replication, we are inclined to generalize our results to surveys concerning other specific technologies. We think that it goes for most respondents that they use information provided by the questionnaire to adjust their attitudes to new technologies.

Item context effects: points of view, implications, remedies

Now that we have shown that assimilation effects do emerge in the judgments about technology, the question arises of what the implications are for research on attitudes towards new technologies. Because the implications of assimilation effects are not essentially different from those of context effects in general, we will discuss them from that broader perspective.

At least two points of view can be adopted with respect to context effects in surveys. On the one hand context effects can be interpreted as 'bias' or 'error' in the judgments in a survey,[5] as artefacts of the questionnaire that distorts the assessment of respondents' 'true' attitudes. On the other hand context effects can be interpreted as unavoidable and integral constituents of judgments both in a survey and in real life. We will discuss below some implications of these points of view.

In the first point of view – context effects as biases – people may hold the conviction that survey judgments are completely inspired by the unpredictable, momentary thoughts induced by the questionnaire context. Those people believe that there is hardly any relation between the respondent's opinion and the overt survey response and they may conclude that one might as well give up survey research entirely. This extreme version of the first point of view is in our opinion too pessimistic. All the existing research into context effects in surveys indicates that there is a large extent of constancy in judgments over contexts: the part of the variance in judgments which can be ascribed to the opinions held by the respondents concerning the content of the item is often greater than the part that can be ascribed to the context. Schuman takes a less rigorous standpoint: he does not believe that survey results are meaningless. He does, however, consider judgments in surveys to be so amenable to the reference framework that the reporting of answer percentages for separate opinion items is generally misleading. On the basis of empirical data (see, for example, the review of Schuman & Presser 1981) he has considerably more confidence in survey results concerning the relationships. 'Even where marginals are easily changed, it is rare for the direction of a relationship involving the same variable to change, and even the magnitude of the relationship is unlikely to shift appreciably' (Schuman 1988, p. 580). So, if one is convinced that context effects influence answer percentages of separate items to an unacceptable extent, then a 'solution' is not to note these marginals in survey reports and to restrict reporting to relationships between judgments. There are, for that matter, exceptions to the 'rule' that relationships between variables are better able to withstand context effects than marginals (e.g. Budd 1987; Strack, Martin & Schwarz 1988; or see Table 4.1).

If one interprets context effects as errors then another way out is to try to avoid or reduce them. The most rigorous attempt to avoid context effects is by offering *no* context. In psychophysical experiments, Poulton (1973, 1979) suggests avoiding certain context effects by offering each subject (or group of subjects) only one stimulus, and by giving no examples in the instructions. Such attempts to create a psychological vacuum are doomed to failure: subjects compare that one stimulus with their experiences with similar stimuli outside the laboratory. Birnbaum (1982) provides arguments and empirical data which demonstrate that 'contextless' judgments can produce misleading results (see also Stephenson 1953). Moreover, only one judgment per subject rules out the possibility of studying correlations between judgments. In addition to the undesirability of 'contextless' judgments, it is also difficult to realize in surveys: one question per respondent is impractical and expensive.

A more practical approach to reduce the influence of context effects on survey results is keeping the context (i.e. the questionnaire) constant. Often in studies on technology attitudes one is interested in trends (e.g. Midden & Verplanken 1990). When the questionnaire is kept constant (questions, question order, instructions, and so on), then one may assume that the **trends** in longitudinal data are not 'distorted' by the questionnaire context. It may also be assumed in that case that in cross-sectional research **differences** between groups are not caused by context effects. For instance, one may assume that differences between countries found in the Eurobarometer on science and technology (1989) are not due to context effects.

If one wants to reduce assimilation effects in part–whole question sequences (as described above) then the research of Tourangeau *et al.* (1989a) is of importance. They established that the effect of contextual items on the ratings of a target item is considerably weakened when these contextual items are spread throughout the questionnaire, as opposed to being presented in a block directly before the target item. Thus, if you have some idea which items affect each other you can spread these items throughout the questionnaire, i.e. you mix these items with questions about other issues with the aim of 'breaking' the context.[6] If you have no notion which items affect each other you can rotate or randomize the order of the items in the series. That is to say, you can make different versions of a questionnaire in which the order of the items varies (cf. Perreault 1975). Using this procedure context effects cannot be banished, but are equally spread over all items.

As we have seen, the second point of view states that the inclusion of context information is an integral part of the generation of a response to a survey item. Those who defend this point of view question the claim that judgments are in 'real life' stable and would be distorted in a survey context. Thus Singer (1988) proposed: 'All human communication, not only survey response, is subject to frame-of-reference effects... In everyday life, as in research settings, people do not consider all information that potentially bears on the judgment but draw on the information that comes to mind most easily at that particular time. At a different time, in a different context the judgment will also be a different one' (Singer 1988, pp. 576–7). In other words, people need a frame of reference when making judgments, and both in ordinary life and in surveys, this frame is influenced by the context of the moment. From this point of view, any attempt to suppress context effects is naive, because there is no **one** true judgment which will remain stable throughout all contexts. If one tries to 'break' the questionnaire context then the judgment is also 'broken'. There is no reason why the context which is made available by the questionnaire is worse than the comparisons which the respondent makes spontaneously. This point of view implies, as far as survey methodology is concerned, that the interpretation of the ratings for a particular item depends on the questionnaire context.

Survey researchers who interpret context effects as integral constituents of survey responses have at least three options: (1) they can vary the context systematically to learn its effect on judgments, (2) they can attempt to make the questionnaire context representative of the judgment context in real life, or (3) they can design a questionnaire context which respondents should involve in their judgments if they are to reach well-considered opinions.

The first option, systematic variation of the context, is called systextual design by Birnbaum (1982). He claims that provided one has developed a theory concerning context effects one is able to generalize across contexts. Based on Parducci's range–frequency theory (1965, 1983), we performed two survey experiments in which we studied to what extent category ratings were affected by range and skewness of series of preceding items (Daamen 1991; Daamen & de Bie 1992). We did know, in the case of the target items, the margins between which the ratings moved as a result of manipulation of the questionnaire context. If these studies had served as pre-tests for a large nationwide survey, then we would have been able to indicate the limits between which the ratings probably lie. Such pre-testing on the lines of systextual design is time-consuming and expensive, but when the precise interpretation of the results is what counts, then it is worthwhile.

The second option is called **representative** or ecological design (Brunswick 1956; Petrinovich 1979): one attempts to have the context in the investigation match the context in 'real life' to which the research results will be generalized. Representative design is only performable in survey research in a weakened form: items about aspects of an object are accompanied by questions on 'natural' comparative objects. For instance, it may appear that people bring judgments about nuclear power to the fore especially in conversations about other forms of energy. Then, a relatively representative survey context for questions about nuclear power would be a series of questions about advantages and disadvantages of nuclear energy, fossil fuels, alternative energy sources, energy saving, and so on (e.g. Midden, Daamen & Verplanken 1984). Because the questionnaire context is similar to the judgment context in 'real' life, it is hoped that the judgments made on the basis of items in the questionnaire will be representative of similar judgments in 'real' life. Obviously, the problem with representative design is that survey researchers do not know the median 'real live' judgment context in most cases. Nevertheless, the trend in opinion research to administer omnibus questionnaires (questions about all sorts of different topics) is abhorrent to those who consider representative design the best approach to context effects in surveys.

In addition to this representative design (aimed at description), there is also a more prescriptive approach in which the researcher designs a survey context which respondents **should** involve in their judgments if they are to reach well-considered opinions. In these 'information questionnaires' the respondent is provided with information about a topic which, in the researcher's view, respondents should be aware of if they are to come to a responsible point of view; only then are questions about that topic asked. If the information provided is not compiled with the greatest care, this can lead to crude manipulation (see Lagendijk 1982, for a disastrous example with respect to energy technology). When due care and attention are invested, such manipulation is avoidable (see Neijens 1987 or Neijens, De Ridder & Saris 1992, for an excellent example of an information questionnaire with respect to energy technology).[7] According to Tourangeau & Rasinski's review (1988), the lower the familiarity of respondents with the topic of a survey the more they will be influenced by context effects. **New** technology is a survey topic that is inherently unfamiliar: an average respondent will lack sufficient (valid) information. Under such conditions an information questionnaire can be a valuable help to respondents in defining their view.

In summary, there are a number of approaches to the questionnaire context and its effect on the responses to certain items. One can try to avoid context effects (e.g. omit the context by asking only one question per respondent), to evade them (e.g. keep the questionnaire constant and study only trends or group differences), to reduce them (e.g. randomize item order), or to accept and to study them (e.g. vary context systematically in embedded survey experiments to learn its effect on the responses, or make the questionnaire context representative of the judgment context in 'real' life). Whatever approach one chooses, a first and important step is to realize that respondents often use information provided by the prior items to interpret a later item, to think about it, and to select an adequate response.

Notes

1 Assimilation effects between *two* items in surveys have been reported before (e.g. Duncan & Schuman 1980; Schuman & Presser 1981; Strack, Martin & Schwarz 1988). The present studies concern assimilation in longer series of items.
2 See Tourangeau *et al.* 1989a, b for a more complex relation.
3 When asked to generate examples of new 'technology in general' 75% mentioned one or more examples of computer technologies; 30–40% mentioned an example drawn from the areas of automation and communication. Examples of the other technology areas were seldom mentioned.
4 Division in 'low' and 'high' involvement according to median split. We performed this analysis and all other involvement analyses with more fine-grained divisions in levels of involvement with comparable results.
5 From this point of view, researchers who study context effects are often seen as harbingers of bad news. When his congress contribution was scheduled for 9 a.m. on Saturday morning, J. G. Bachman (1987) assumed jokingly that this unfavourable time must be his punishment for accidently discovering another questionnaire context effect.
6 Possibly it is merely passage of time between the part questions and the whole question that reduces the assimilation effect (cf. Srull & Wyer 1980).
7 The information provided in the questionnaire of Neijens *et al.* was compiled by various energy experts and translated for lay people by the researchers. An independent committee of representatives from various societal sections judged whether the information was well balanced and complete. Moreover, a procedure was designed and tested to help respondents to process the information provided in the questionnaire.

References

BACHMAN, J. G. (1987). Friends may disapprove of drug use, but not as much as parents do: Another lesson in questionnaire context effects. Paper presented at the annual meeting of AAPOR, Hershey, Pennsylvania, May 1987.

BEM, D. J. (1967). Self-perception: an alternative interpretation of cognitive dissonance phenomena. *Psychological Review* 74, 183–200.

BIRNBAUM, M. H. (1982). Controversies in psychological measurement. In *Social attitudes and psychological measurement*, ed. B. Wegener, pp. 401–85. Hillsdale: Lawrence Erlbaum.

BISHOP, G. F. (1987). Context effects on self-perceptions of interest in government and public affairs. In *Social information processing and survey methodology*, ed. H. J. Hippler, N. Schwarz and S. Sudman, pp. 179–99. New York: Springer.

BISHOP, G. F. (1990). Issue involvement and response effects in public opinion surveys. *Public Opinion Quarterly* **54**, 209–18.

BISHOP, G. F., OLDENDICK, R. W. & TUCHFARBER, A. J. (1982). Political information processing: question order and context effects. *Political Behavior* **4**, 177–200.

BISHOP, G. F., OLDENDICK, R. W. & TUCHFARBER, A. J. (1984). What must my interest in politics be if I just told you 'I don't know'? *Public Opinion Quarterly* **48**, 510–19.

BRUNSWICK, E. (1956). *Perception and the representative design of experiments*. Berkeley: University of California Press.

BUDD, R. J. (1987). Response bias and the theory of reasoned action. *Social Cognition* **5**, 95–107.

CONVERSE, P. E. (1974). Comment: the status of nonattitudes. *American Political Science Review* **68**, 650–60.

DAAMEN, D. D. L. (1991). Range and skewness effects in survey interviews. Dissertation, Leiden University.

DAAMEN, D. D. L. & DE BIE, S. E. (1992). Serial context effects in survey interviews: subjective probability ratings affected by range and skewness of series of preceding items. In *Context effects in social and psychological research*, ed. N. Schwarz and S. Sudman, pp. 97–113. New York: Springer.

DAAMEN, D. D. L. & KIPS, J. (1993). *De energiemonitor: Publieksoordelen over kernenergie, kolen en andere energiebronnen* [Monitoring public opinions on nuclear power, coal and other energy sources]. Werkgroep Energie- & Milieuonderzoek, University of Leiden.

DAAMEN, D. D. L., VAN DER LANS, I. A. & MIDDEN, C. J. H. (1990). Cognitive structures in the perception of modern technologies. *Science, Technology and Human Values* **15**, 202–25.

DUNCAN, O. D. & SCHUMAN, H. (1980). Effects of question wording and context: an experiment with religious indicators. *Journal of the American Statistical Association* **75**, 269–75.

Eurobarometer (1989). Brussels, Commission of the European Communities.

FAZIO, R. H., SANBONMATSU, D. M., POWEL, M. C. & KARDES, F. R. (1986). On the automatic activation of attitudes. *Journal of Personality and Social Psychology* **50**, 229–38.

FELDMAN, J. M. & LYNCH, J. G. (1988). Self-generated validity and other effects of measurement on belief, attitude, intention, and behaviour. *Journal of Applied Psychology* **73**, 421–35.

HIPPLER, J. J. & SCHWARZ, N. (1987). Response effects in surveys. In *Social information processing and survey methodology*, ed. J. J. Hippler, N. Schwarz and S. Sudman, pp. 102–22. New York: Springer.

KROSNICK, J. A. & SCHUMAN, H. (1988). Attitude intensity, importance and certainty and susceptibility to response effects. *Journal of Personality and Social Psychology* **54**, 940–52.

LAGENDIJK (1982). *Schaduwparlement en energiebeleid* [Shadow parliament and energy policy]. Apeldoorn: Lagendijk Opinieonderzoek.

MIDDEN, C. J. H., DAAMEN, D. D. L. & VERPLANKEN, B. (1984). Personal attitudes towards large scale technologies. In *Linking economics and psychology*. Proceedings of the 1984 IAREP Annual Colloquium, Tilburg: Tilburg University.

MIDDEN, C. J. H. & VERPLANKEN, B. (1990).   The stability of nuclear attitudes after Chernobyl. *Journal of Environmental Psychology* **10**, 111–19.

MILLER, J. D. (1991).   *The public understanding of science and technology in the United States, 1990: a report to the National Science Foundation.* DeKalb, Ill.: Public Opinion Laboratory, Northern Illinois University.

NEIJENS, P. (1987).   *The choice questionnaire: design and evaluation of an instrument for collecting informed opinions of a population.* Dissertation, Amsterdam, Free University Press.

NEIJENS, P., DE RIDDER, J. A. & SARIS, W. E. (1992).   An instrument for collecting informed opinions. *Quality and Quantity* **26**, 245–58.

PARDUCCI, A. (1965).   Category judgment: a range-frequency model. *Psychological Review* **72**, 407–18.

PARDUCCI, A. (1983).   Category ratings and the relational character of judgment. In *Modern issues in perception*, ed. H. G. Geissler, H. F. J. M. Bulfart, E. L. J. Leeuwenberg and V. Sarris, pp. 262–82. Berlin: VEB Deutsche Verlag der Wissenschaften.

PERREAULT, W. D. (1975).   Controlling order–effect bias. *Public Opinion Quarterly* **39**, 544–51.

PETRINOVICH, L. (1979).   Probabilistic functionalism. A conception of research method. *American Psychologist* **34**, 373–90.

POULTON, E. C. (1973).   Unwanted range effects from using within subject experimental designs. *Psychological Bulletin* **80**, 113–21.

POULTON, E. C. (1979).   Models for biases in judging sensory magnitude. *Psychological Bulletin* **86**, 777–803.

RUGG, D. & CANTRIL, H. (1944).   The wording of questions. In *Gauging public opinion*, ed. H. Cantril, pp. 23–50. Princeton, NJ: Princeton University Press.

SALANCIK, G. & CONWAY, M. (1975).   Attitude inferences from salient and relevant cognitive content about behaviour. *Journal of Personality and Social Psychology* **32**, 829–40.

SCHUMAN, H. (1988).   Rejoinder. *Public Opinion Quarterly* **52**, 579–81.

SCHUMAN, H. & PRESSER, S. (1981).   *Questions and answers in attitude surveys: experiments on question form, wording and context.* New York: Academic Press.

SINGER, E. (1988).   Letter to the editor. *Public Opinion Quarterly* **52**, 576–9.

SMITH, T. (1982).   *Conditional order effects* (GSS Technical Report No. 33). Chicago: NORC.

SMITH, T. W. (1992).   Thoughts about the nature of context effects. In *Context effects in social and psychological research*, ed. N. Schwarz and S. Sudman, pp. 163–84. New York: Springer.

SRULL, T. K. & WYER, R. S., Jr (1980).   Category accessibility and social perception: some implications for the study of person memory and interpersonal judgments. *Journal of Personality and Social Psychology* **38**, 841–56.

STEPHENSON, W. (1953).   *The study of behaviour: Q-technique and its methodology.* Chicago: University of Chicago Press.

STRACK, F., MARTIN, L. L. & SCHWARZ, N. (1988).   Priming and communication: social determinants of information use in judgments of live satisfaction. *European Journal of Social Psychology* **18**, 429–42.

TOURANGEAU, R. & RASINSKI, K. A. (1986).   Context effects in attitude surveys. Unpublished manuscript.

TOURANGEAU, R. & RASINSKI, K. A. (1988).   Cognitive processes underlying context effects in attitude measurement. *Psychological Bulletin* **103**, 299–314.

TOURANGEAU, R. & RASINSKI, K. A. BRADBURN, N. & D'ANDRADE, R. (1989a).   Belief accessibility and context effects in attitude measurement. *Journal of Experimental Social Psychology* **25**, 404–21.

TOURANGEAU, R. & RASINSKI, K. A. BRADBURN, N. & D'ANDRADE, R. (1989b). Carryover effects in attitude surveys. *Public Opinion Quarterly* **53**, 495–524.

WILLIAMS, R. & MILLS, S. (ed.) (1986).   *Public acceptance of new technologies*. London: Croom Helm.

5

'Technophobia': a misleading conception of resistance to new technology[1]

MARTIN BAUER

In the history of technology the concept of 'technophobia' seems to undergo a periodical revival to deal with people's reactions to innovations. This tendency to detect symptoms of pathology in people's experience of new technologies reappears in public debates. According to the historian Goffi (1988) we may distinguish a universal from a particular form of technophobia. Universal 'phobia' is expressed in ancient myths such as Prometheus, The Golem, Dr Faustus or the Greek notion of Hybris. The particular form is the anti-scientific attitudes in the recent industrial age. Often technophobia is part of the larger concept of 'neophobia' which refers to people's general aversion against all things new.[2]

In the nineteenth century, 'Siderodromophobia' subsumed adverse reactions to railway work and railway journeys: fever in the aftermath of journeys; the '*delirium furiosum*', a mental agitation caused by the mere sight of a locomotive steaming by; and a hysterical aversion to work among locomotive and wagon personnel (Fischer-Homberger 1975, pp. 40f).[3] Mitchell (1984) reports how at times the nuclear debate in the USA was conducted under the heading of 'nuclear phobia': images of nuclear power, spread by the media, touch on unconscious motivations, and give rise to an emotional over-reaction which brought the nuclear power industry to a virtual stop.

To use the notion of 'phobia' to describe people's experience of and behaviour towards new technology is pragmatically not neutral; the psychopathological classification presents the problem through the 'clinical' eye. In the following I analyse the most recent and vociferous revival of technophobia as 'cyberphobia', 'computer phobia', or 'computer aversion'. A relatively confined body of literature – around 300 recorded publications since 1980 – employs 'anxiety' and 'phobia' as core concepts for understanding resistance to computers at school, at work and at home. This growing body of literature probably has limited influence, but it nevertheless shows the misleading implications of this approach.

My concern with 'cyberphobia' is critical. Diagnosis and treatment of 'cyber-phobia' constitute the clinical gaze on the problem of resistance in the context of computers and information technology. The idea is rooted in a clinical and educational setting and has popular appeal and news value. I challenge the 'clinical model of resistance', the diagnosis of a psychological deficiency, and the provision of psychotherapy for those who resist computers at work, at school or at home. First, the diagnosis of cyberphobia is simplistic and not up to professional standards; and secondly, the clinical eye leads to a therapeutic intervention centred on the individual that is inadequate for dealing with people's resistance to new technology and does not ensure technological effectiveness.

The clinical eye on resistance

Communication systems operate with particular distinctions that define their identity, and which allow actors to make sense of the world in their own terms. Clinical terms define reality under the main distinctions of illness and health, normal and abnormal. Illnesses are classified and constitute the 'medical code' of communication (Luhmann 1990, pp. 183ff). In a health context problems of the 'world' are likely to be translated into the code of health/illness, its diagnosis and its treatment.

Mental, emotional or physical discomfort makes people seek medical or psychological advice. The provision of such advice is controlled by experts and their institutions in a variable historical form. Attempts to define normal/ abnormal, and mental health/mental illness are controversial issues (Tiles 1993). Without entering this debate directly, it should be possible to observe how the idea of illness/health is extended into a new area. Sociologically orientated historians have shown how medicalization, the expansion of medical expertise to areas of life formerly not within that expertise, functions overall to control, to discipline and to constrain deviance within society at large (Conrad 1992). Medicalization may be regarded as a modern equivalent of religion with its partial function of controlling people's thought and behaviour.

The rationale for this kind of question is social constructivist. It highlights how institutions gain the power to define and to give reality to phenomena by classifying observations and by administering expertise and skills. Institutions are relatively autonomous in so doing. The health professions, including medicine, psychiatry and clinical psychology, define for themselves what constitutes a clinical case, and who has the expertise to treat it. Extending or reducing such definitions means social change. Abstraction, respecification and institutional-ization are processes of 'social construction of knowledge' (Berger & Luckmann

1967). Abstraction refers to defining concepts beyond a particular context, respecification applies such concepts in a context different from the original one, and institutions endow this activity with legitimacy. A particular institution will see problems and offer solutions according to its practices. Expertise, beneficial as it is in many cases, may overreach itself by applying its resources to diagnose and to treat in inadequate contexts.[4] Material and cognitive resources define the range of possible problems. To make an analogy: if only hammer and nails are at hand, all problems of fixation become problems of hammering and nailing, to the effect that more elegant or more adequate solutions, such as glueing or screwing, become unthinkable. This is a general problem of cognition which Duncker (1935) called the 'functional dependence of practical thinking'. I shall explore the following thesis.

> The concept of 'cyberphobia' epitomizes the clinical eye on resistance to new technology. Besides providing media hype, the concept prioritizes personal therapy over technological design. It individualizes the problem when the technology requires more adequate (re)design according to social and psychological criteria.

I track the diffusion of the 'cyberphobia' concept since 1980 by counting publications; then I describe the institutional boundaries of 'cyberphobia', its definition, and empirical results on its prevalence and its etiology. The 'cyberphobia' concept is inadequate for at least three reasons: (a) the diagnosis of cyberphobia is technically a weak diagnosis; (b) the idea of resistance as 'phobic' is based on a statistical definition of pathology – an assertion which I would like to question; and (c) the approach succumbs to the attribution error of unilaterally blaming the user. My criticism of the clinical eye leads to an alternative view of resistance in systemic terms.

The diffusion of the concept 'cyberphobia'

Diagnosis, screening and therapy of cyberphobia make up a full package which a group of people developed, sought funding for, defined their academic identity with, and gave a label to, to sell on the health market (see, for example, Weil, Rosen & Sears 1987; Greenly 1988). To map the diffusion of the concept I counted the publications on 'cyberphobia' and related terms in electronic data bases.[5] The word 'cyberphobia' is often used interchangeably with 'computer anxiety', 'computerphobia', 'user resistance', 'computer aversion' or 'negative attitudes to computers'.

Figure 5.1 shows the frequency of publications on 'cyberphobia' and related terms since 1980. The thick line shows the total number; the other lines show the

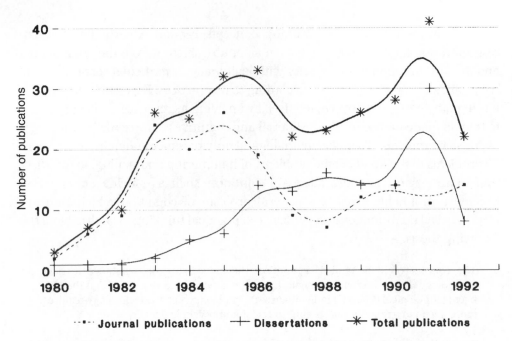

Figure 5.1. Publications on cyberphobia and related terms 1980–92. Four sources of data bases were used: ABI/INFO, SSCI, PsychAbstract, and DissAbstracts. Keywords used are 'cyberphobia', 'computer phobia', 'computer anxiety', 'computer aversion', 'computer attitudes'. N = 298.

number of journal articles and PhD dissertations. The curve is smoothed to show the trend. A total of 298 publications were counted between January 1980 and December 1992, 173 articles and 125 dissertations, on average 23 publications per year.[6] Before 1980 the term 'cyberphobia' is not used, although studies on negative attitudes to computers date back to the 1960s (e.g. the pioneering IBM study by Lee 1970). Jay (1981), a psychologist at North Adams State College, Massachusetts, seems to have introduced the term to hype a story in a popularly written article for educationalists entitled 'computer phobia: what to do about it?'. By 1985 he had elaborated the concept with a psychometric scale in a presentation to a US ergonomics conference. Weinberg, a clinical psychologist and information scientist, claims to have treated problems that came to be called 'cyberphobia' since 1978. Together with Mark Fuerst, a science journalist, he wrote a book entitled *Computer phobia* (Weinberg & Fuerst 1984).[7]

Three observations can be made. First, the concern for 'cyberphobia' comes in two waves. The first wave peaks between 1984 and 1986, at the time when public opinion studies on computers and their impacts cluster internationally (Bauer 1993). The year '1984' had gained notoriety from Orwell's dystopia of an omni-present social control. These were also the years when the computer in the form of the personal computer (PC) reached out into homes and offices on a larger scale. The second peak follows six or seven years later, in 1991. Secondly, the first peak is based on popular writings; articles and editorials in management and educational magazines peak in 1985. Cyberphobia personalizes the user's problems and may therefore suit the logic of popular writing. The second peak around 1991 is based more on reporting empirical research in scholarly work. We find a shift to scholarly journals and PhD theses after 1986. Thirdly, coverage of cyberphobia in popular magazines precedes the scholarly concern. Dissertations significantly increase after 1984 to peak in 1991. After 1984 one recognizes also the impact of funding bodies: the three-year US government 'Computer Phobia Reduction Program' started in 1985 (Weil *et al.* 1987). Cyberphobia is a case of popular agenda setting that leads on to a scholarly exercise. This is interesting in the context of popularization. The popularization of science normatively assumes the direction of influence from scientific communication to public communication; however, the health and social sciences may be more reactive to public demands than other sciences.

Institutional settings

Definitions and expertise are practised in institutions. Analyzing who investigates and provides therapies for cyberphobia shows how the expertise is socially located. Four questions are relevant: who conducts the research; who funds the research; where are the results published; and who is the audience?

The most vociferous protagonists of 'cyberphobia' are situated in the Department of Psychology, at California State University, San Dominguez, where Michelle Weil is a clinical consultant and Larry Rosen is professor of psychology; and in the Catholic University of America, Washington, DC, where Carol R. Glass is at the Department of Psychology. They offer the most elaborated intervention programme for cyberphobia. About 15% of dissertations on cyberphobia are educational doctorates. Technical universities are less frequently involved. The 'Computerphobia Reduction Program 1985–' (Weil *et al.* 1987) is funded by the US Department of Education Fund for the Improvement of Post-Secondary Education (FIPSE). This must be seen in the context of concerns about declining standards of US schools in mathematics and science in the early 1980s. In 1985 the US Phobia Society supported a study that screened a representative sample of the American population for 'cyberphobia' (Meier 1985). Another study reports how the president of the US Phobia Society validated 'phobic' response patterns in survey interviews (Patton Gardner *et al.* 1989). Publications with empirical data on 'cyberphobia' are published primarily in clinical and educational journals, and only secondarily in the journals of the engineering related human factor or ergonomics community.[8] Research is presented at the annual meetings of the American Association for the Advancement of Behaviour Therapy, for example in Houston in November 1985. Much research draws conceptually on concerns with mathematics and test anxiety (Glass & Knight 1988), which are represented by the Society for Test Anxiety Research. Regular conferences provide the network for an international comparative study on cyberphobia in 1989 (Weil & Rosen 1991).

Cyberphobia is defined, diagnosed, and 'healed' within a complex of clinical psychological, educational, and behaviour therapy involvement, mainly centred in Southern California. The main group is clinical with a cognitive–behavioural orientation, working in educational settings with affinities to the research on test anxiety. The Human Factors community is only marginally involved. Research is mostly, if not exclusively, conducted in the USA.[9] The attempts of Rosen and Weil to recruit international partners for comparative studies seem to have triggered interest in other parts of the world, but hardly before 1989. I will focus on the work of the group around Rosen and Weil at California State University, Dominguez Hill – the most sophisticated and most complete programme on 'cyberphobia'. It includes a review of literature, the development of a diagnostic instrument, population screening, international comparisons, a programme of therapy, and its evaluation, and an ongoing 'conference' via electronic mail.

Diagnosis and correlates of 'cyberphobia'

Cyberphobia is a particular case of technophobia (Jay 1981; Rosen & Weil 1992). Fear of computers at home, at school or at work is a particular form of aversion to technical devices such as railways, aircraft, cars, electric light, nuclear power stations, video recorders, computer games, microwave ovens and other household gadgets.

The prefix 'cyber' shares its source with 'cybernetics', the Greek word for the helmsman who controls the ship. Phobias are forms of anxiety and aversions; phobic people dissociate between their cognition of a harmless situation and their extreme emotional reaction to it, a dissociation of which they are aware. The composite word specifies the object of fear: zoophobia is fear of animals; agoraphobia is fear of public places. Definitions of anxiety, phobia and cyberphobia are given in Table 5.1. Jay, and Weil et al. highlight the attitude component of cyberphobia; Gardner stresses the bodily symptoms of anxiety. Authors often distinguish, on a hypothetical continuum of attitudes, between cyberphobia, an intense and extreme fear, and computer anxiety, a more moderate fear.

Most empirical research works with operational definitions of computer phobia using self-report rating scales.[10] A rating measure consists of a number of statements; respondents are asked to agree or disagree, to estimate or to rate some experience as shown with examples in Table 5.2. Overall test scores locate individuals within a defined population. A normal distribution of test scores is

Table 5.1. *Definitions of anxiety, phobia and cyberphobia*

Anxiety	'An unpleasant feeling of generalized fear and apprehension accompanied by increased physiological arousal ... anxiety can be assessed by self-report, by measuring physiological arousal, and by observing overt behaviour' (Davison & Neale 1978, p. 634)
Phobia	'Extreme fear and avoidance of an object or situation which the person is able to recognize as harmless' (Davison & Neale 1978, p. 133).
	'A persistent fear of a circumscribed stimulus ... exposure to the specific stimulus almost invariably provokes an immediate anxiety response ... the person recognizes that his or her fear is excessive or unreasonable ...' (DMS-III-R 1987, p. 144)
Cyberphobia	'... appears generally in the form of a negative attitude toward technology. The negative attitude takes the form of (a) resistance to talking about computers or even thinking about computers, (b) fear or anxiety toward computers, and (c) hostile or aggressive thoughts about computers' (Jay 1981, p. 47)
	'... one or more of the following: (a) anxiety about present or future interaction with computers or computer-related technology; (b) negative global attitudes about computers, their operation, or their societal impact; or (c) specific negative cognitions or self-critical internal dialogues during present computer interaction or when contemplating future computer interaction' (Weil *et al.* 1990, p. 362)
	'... acknowledging a combination of reported symptoms including avoiding using the computer, panicking when facing a computer, feeling of unreality, fear of losing control, sweating palms, pounding heart, tightness or pain in the chest, trembling or shaking, shortness of breath and dizziness or lightheadedness when faced with a computer' (Patton Gardner *et al.* 1989, p. 93)

constructed for statistical reasons by item selection. In pre-tests uncorrelated items are eliminated; highly correlated items form consistent test scores measuring the same hypothetical phenomenon. A standardized instrument may have several dimensions. The test instrument is supposed to identify pathological

Table 5.2. *Examples of questions used to diagnose 'cyberphobia'*

General Attitudes toward Computers Scale (GATC: 20 *items*)

How much do you agree or disagree with these statements?

   (1) strongly agree —— strongly disagree (5)

Computers can save people a lot of work;
Computers can increase control over your own life;
Some ethnic groups are better with computers than others;
There is an overemphasis with computer education in this society;
Computers will never be smarter than people;

Computer Anxiety Rating Scale (CARS: 20 *items*)

How anxious would these situations make you?

   (1) not at all —— very much (5)

Thinking about taking a course in a computer language;
Watching a movie about an intelligent computer;
Visiting a computer centre;
Thinking about buying a new personal computer;
Reading a computer manual;

Computer Thought Survey (CTS: 20 *items*)

When you use a computer or think about using a computer, how often do you think:

   (1) not at all —— very much (5)

I am going to make a mistake;
Everyone else knows what they are doing;
I like playing on the computer;
This will shorten my work;
I hate this machine;

Source: Rosen and Weil, 1992.

cases of 'cyberphobia' at the negative extreme of an unknown distribution of attitudes to computers in a population; cases at the opposite extreme are often called 'computer addicts'. To match test scores and 'real' cases of cyberphobia and to define a cut-off point on the test distribution is the problem of external validation. Once a cut-off point is defined we are able to diagnose pathological cyberphobia with the test score.[11] However, a test score is an insufficient criterion for real 'cyberphobia'; it can at best efficiently reproduce a pre-existing expert judgement, which, as it seems, is more suggested by popular expectations, than by insisting on phenomenological evidence.

According to Rosen and Weil (1990, p. 181) the differential diagnosis of cyberphobia is possible as cyberphobics are not generally more anxious than other people, nor do they generally suffer from mathematics anxiety.

It is part of the diagnosis to understand the origin of a problem. But as theories of etiology differ, so do classification systems. Due to learning theory, previous computer experience gained attention as a factor of cyberphobia. Weil, Rosen & Wugalter (1990) investigated retrospectively the influence of media exposure and childhood and adult experiences with mechanical devices and with computers. Cyberphobics were introduced to mechanical experiences by their mothers rather than by their fathers. They recall early computer experiences more negatively than non-phobic computer users. The strongest influence comes from previous computer interactions that were unfavourably evaluated by a significant person. Memory of parental encouragement is the best predictor for current attitudes to the computer. The expectation that time and experience may solve the problem is therefore a myth. Cyberphobics will reinforce their phobia with every exposure to computers once the negative emotional quality of their previous encounters is habituated.

A second way of explaining cyberphobia is to associate it with a personality complex. Certain standard personality traits may distinguish cyberphobics from others. Cyberphobic students tend to be impatient, intolerant, and incapable of persevering in a task until a solution is found (Weil *et al.* 1990). However, attempts to establish a reliable cyberphobic personality profile based on measurements such as femininity, ambiguity intolerance, rigidity, alienation, introversion, persistence or problem-solving style have not produced consistent results (Rosen & Maguire 1990).

Screening and therapy of cyberphobia: how many cyberphobics?

Stories about cyberphobia make sensational news: a study of the American Phobia Society found that 3% of US managers and industrial professionals are cyberphobic: 'computerphobia costs millions' is the title (Tlustos 1985). Other studies declare 33% of the 45 million US information sector workforce cyberphobic (Meier 1985; Nikodym *et al.* 1989). A recruitment advertisement for university graduates in the London *Guardian* (8 January 1994) depicts a PC and a snake crawling over the display unit. Together with the text 'what are you afraid of?' this image represents symbolically the myth of cyberphobia by associating the computer and the snake, the prototype of an object of phobia. Dorothy Nelkin (1987, p. 22) noted how sensational headlines on 'computer phobia' in the press mystify science as a solution to all human problems. News on 'cyberphobia' suggests a problem that calls for a scientific solution so that America can be prepared for international competition.

A standardized measure allows us to assess the prevalence of a condition in the whole population. A meta-analysis of 109 empirical studies provides 'sufficient evidence that cyberphobia exists' (Rosen & Maguire 1990). Gender differences receive much attention, but without clear conclusions. US studies suggest that cyberphobia is stronger among women than among men; international data show the opposite. Also an age effect, with older people being more cyberphobic, cannot be substantiated. Two-thirds of all studies report data from students or teachers from primary, secondary and tertiary education. About a third of the studies report data on adults in professional groups. The authors conclude that between 2.7 and 5% of the US population is 'cyberphobic' and around 20% run a high risk of developing it; 11 to 27% of the population is computer anxious. Hence around a quarter of the US population 'feels less than comfortable with computers'.

In an international study, Weil and Rosen (1991) compare data from 38 universities in 23 countries making use of a network within the Society for Text Anxiety Research. The study reports national results from samples of first-year students with as few as 50 respondents – hardly a representative national sample of students. Absurd results emerge. Indonesian students, with little computer experience, are 100% computerphobic; the figures are 60% for Japanese, and 25% for US students, who have more computer experience.

Cyberphobia is claimed to be costly for business: by avoiding computers, employees and managers are not using their equipment optimally. The college drop-out rate of cyberphobic students is twice as high as that of normal students (Weil *et al.* 1990). Cyberphobia slows the speed of interaction with the computer; it reduces professional choices as people avoid jobs with computers; managers with cyberphobia put their promotion at risk; the job possibilities for the cyberphobic get more and more limited. Cyberphobia reduces peoples' life choices by causing disadvantages at school and in the job market. In the light of all this, the cyberphobic needs treatment for his or her own sake. Any diagnosis is used to allocate resources or to remove resources by selecting recipients. The clinical approach uses diagnosis to allocate additional resources; in the job market diagnostic tools are more likely to be used for purposes of selection.

In the 'Computer Phobia Reduction Program', over a period of three years students and members of an aircraft company were assessed for cyberphobia (Weil *et al.* 1987). After an initial test cyberphobic people are assigned to special training that will enable them to cope with their discomfort when facing a computer screen. The clinical treatment consists of four elements: first, relaxation training; and secondly, systematic desensitization to extinguish the individuals' faulty learning history. Systematic desensitization is a standard method used in behaviour therapy to treat phobias. Thirdly, the 'individual cognitive–behavioral thought-stopping and cover-assertion program' redirects the dysfunctional internal dialogue that interferes during computer interactions. Fourthly, regular small group meetings provide a supportive social environment. The programme lasts five hours spread over five weeks, plus the weekly meetings in the support group. Pre- and post-training measures of cyberphobia assess the effectiveness of the therapy. Ten per cent of all students went for the treatment. The treatment is effective with a 50% reduction of phobia scores on average after the treatment. The most cost-efficient treatment to reduce cyberphobia, and a clearer understanding of the factors of cyberphobia, are problems for future research (Rosen & Weil 1990).

A critique of the concept of 'cyberphobia' and its uses

Having described the rise of cyberphobia since 1980, its network, definition, prevalence, and treatment, I continue examining critically this 'clinical eye' on resistance to computers as a form of medicalizing everyday life.

A simplistic diagnosis leads to absurd results

Generally phobias comprise a pattern of intense anxiety, avoidance behaviour, and physiological reactions to a particular object or situation. The person recognizes that his or her fear is excessive. The stimulus situation is avoided or endured with extreme fear. Zoophobic persons suffer intense anxiety in the presence of animals; they sweat, shake and tremble, feel stomach cramps, and experience a profound sense of dread; their breathing rate increases; they may even vomit, defecate and urinate in fright. This multi-faceted phenomenon is taken into account by the professional standard of multi-modal diagnosis. It urges the use of data on experience, physiology and behaviour, obtained by self-report and by observations, to form a diagnostic judgement (Himadi, Boice & Barlow 1985; Seidenruecker & Baumann 1987). In violation of this standard, the diagnosis of cyberphobia is based on one single type of data: self-report of experience. The Weil and Rosen test instrument distinguishes three measures: computer anxiety, computer cognition, and general computer attitudes. This multi-dimensional assessment falls short of multi-modality. No behaviour observation is obtained, neither are physiological reactions assessed. Psycho-metric self-description is only partial information in the clinical judgement. At times it may be impractical to obtain multi-modal observations because it is incommodious, costly or laborious. One may then rely on incomplete data, be aware of it, and explore the implications of a false diagnosis; simply reducing the standard of practice results in empirical artefacts. Furthermore the reported 25% of potential cyberphobia in the USA is excessively high in the light of a recent survey. The most common phobic experiences have a rate of up to 8% over a single year period; and 13% over a life time.[12] Based on one data type only, the validity of the diagnosis is overstretched and even tautological: 'cyberphobia' is what the scales measure.

Furthermore, the present assessment of cyberphobia leads to absurd empirical results, another indicator of artefact. The international data with a cut-off point validated in California show that 100% of Indonesian students and about 60% of Japanese are 'cyberphobic'. Such a high incidence of cyberphobia is absurd as no phobia has such a high rate in a population, as mentioned before. This result stems largely from the calibration of the test measure in a particular environment, Southern California, which is unlikely to be representative of other populations. We may learn from these results that Indonesian and Japanese students respond to self-report questionnaires differently from Californians, but beyond that caution seems to be appropriate. The authors acknowledge linguistic difficulties with the translation of the questionnaires, and the different factor structure of responses across countries which indicates a cultural factor in the expression of anxiety (Weil & Rosen 1991). For the concept of 'cyberphobia' to be useful it must refer to a significant functional disorder. It is very unlikely that 100% of a student population will experience such conditions. Californian 'cyberphobia' is hardly a problem in Indonesia. To diagnose cyberphobia for a whole population seems premature to say the least. To make distinctions is the very idea of 'dia-gnosis', not between populations, but within populations. To summarize, the diagnosis of 'cyberphobia' with single-modality type data is tautological and leads to absurd results about its prevalence both nationally and internationally.

The average experience and social control

The second problem with 'cyberphobia' is its statistical definition. The deviation from average real-life experience is claimed to indicate the pathological. This implies (1) that the average is an optimal form of life, and (2) that deviance is inherently pathological. Social deviance is neither a necessary nor a sufficient condition for illness. A person may be socially deviant and perfectly healthy and vice versa, be ill but not socially deviant. Illness requires that part functions are impaired (objective criterion) and that the person feels ill (subjective criterion). Social deviance indicates illness only in the presence of additional physical and psychological impairments (Lewis 1953), and the more part functions are impaired, the more ill somebody is likely to feel. Unless cyberphobia is separated from non-conformism, inappropriate social control may result. However, diagnosing functional impairment may still rely on statistics. Canguilhem (1991) has problematized this issue further. A functional norm refers to a specific environment; any norm is only temporarily adequate because environments change. In this view illness is a reduction of variability of functioning, temporary or chronic, and not the level of functioning *per se* (Tiles 1993).

First, 'normal' is best seen to indicate a state of affairs that is conducive to well-being in a specific environment. Different environments result in different (ab)normalities; hence the assessment of norms is relative to a certain environment, biological, psychological and social. It is inadequate to claim that cyberphobia, measured by subjective well-being in California, is a universal psychological problem. All it measures is the deviance from the Californian way of life and the part that computer (non-)interaction may play in it. The exposure to a computerized world may be so intense in California that otherwise normal reluctance appears 'abnormal'. Indonesians have a different way of experiencing and dealing with computers, but are not therefore necessarily in need of special treatment, albeit that according to Californian standards they appear to be. This difference is an interesting result, and we need not pathologize it.

Secondly, the environment is not fixed. To identify deviations from the population average with pathology is inadequate, because this norm may become dysfunctional in a changing environment, and supposedly 'dysfunctional' experience and behaviour becomes a 'functional' resource. Variability is the essential resource for living in an uncertain environment. To cope optimally with unpredictable changes we require flexibility and the social system requires variability. The significance of present deviance for future adaptations cannot be ruled out. Deviance in experience and behaviour is a social resource that needs encouragement rather than extinction. Speculatively, one could expect that by large-scale diagnosis and treatment of cyberphobia a society might regress towards the temporary functional average, succumb to 'mediocracy', and thus reduce its cultural variability as a resource for the future. We would end up with an active population who have no interfering thoughts during computer inter-action; no sensitivity to problematic operations, organizational settings, and societal impacts of that technology; and no fears and discomforts to motivate the critical assessment of the computer design and its implementations in hardware and software. Among 'normalized happy users' we lose the impetus of fear, which renders us watchful of the uses to which new technology is put. I think we need to regard fear and anxiety primarily as functional signals about potential dangers which require our urgent attention and action. The functional analysis precedes the analysis of the dysfunctional.

Thirdly, the diagnosis of cyberphobia sets a certain pattern of human–computer interaction as the norm. A 'norm' expresses a temporary arrangement of a system, be it an organism, an individual or a collective, as part of a feedback system. Every mode of life, even pathological ways of life, are 'normative' in that sense. Pathological conditions and ageing are conditions of reduced variability. A cyberphobic person sweats, gets tense, or may even urinate in anticipation of computer interaction; he or she will not use a teller machine to save time, does all paperwork by typewriter or by hand, and may even change his or her job. This reduces the person's daily flexibility and reduces life choices as they become excluded from more and more activities. The criterion is the person's own standard set in the past. They may experience this as problematic and may seek expert advice; but equally they may not. The crucial criterion must be that the personal experience of discomfort qualifies as pathological. Any imposition of another norm requires legitimation. To offer a cure to cyberphobics is appropriate in the health market, but it may be culturally problematic. Expanding psycho-pathology to computer behaviour is not the best strategy to resolve the problem of resistance. I shall, however, not deny its value in individual cases.

The French historian Canguilhem (1991, p. 275) some time ago suggested replacing the concept of 'disease' by the concept of 'error'. An error is a 'misreading' that generates temporarily inadequate structures of activity. Systemic error is an inherent possibility. Errors are normal events which indicate a mismatch between a system and its environment. In that light cyberphobia is not an individual disorder, but a temporary mismatch between persons and their environment. It indicates the users' misunderstanding of computers, and equally the computer's misunderstanding of the users, due to inadequate design. The technological environment is a creation of designers and innovators. The causal attribution of my problem to either the user or the technology is a matter of priority. This leads to the problem of dispositional attribution.

Dispositional attribution: who needs therapy?

Back in 1981, Jay asserted that immature technology and bad organizational design may be a factor of people's aversion to computers. This crucial idea got lost on the way to the clinical diagnosis of 'cyberphobia'. The clinical approach personalizes the problem. Technology is taken as fixed in the equation. To explain resistance to computers with a disposition succumbs to the 'funda-mental attribution error' (Hewstone & Antaki 1989, p. 126). Common sense favours dispositional over situational explanations in problematic situations. However, far from being universal, this bias expresses a culture of individualism that stresses internality, subjectivity and self-responsibility. We have seen that cyberphobia is mainly a US concern. It is commonly agreed that American culture celebrates individualism more than any other culture. Therefore one would expect a stronger inclination to make dispositional attributions in the

US than in other places in the world, other things, such as computer exposure, being equal. European research, for example, is not beyond dispositional attribution, but shows more interest in 'computer addiction' and the 'hacker personality' than in 'cyberphobia' (Shotton 1989).[13] Furthermore, European researchers seem to prefer the concept of 'stress' to that of 'cyberphobia'. Stress is clearly related to the concept of person–environment mismatch, but with the idea of the 'individual's coping style' it may lead to a similar attributional bias. It seems that the concept of cyberphobia expresses US individualism and a different priority for some applied psychologists, rather than indicating a widespread problem.

The diagnosis of cyberphobia focuses unilaterally on the user. The definition of resistance in clinical terms suggests a personal deficiency and need for treatment, and is part of a secular process of medicalization. Medicalization means 'defining a problem in medical terms, using medical language to describe a problem, adopting a medical framework to understand a problem, or using a medical intervention to treat' (Conrad 1992, p. 211). On the one hand, extending clinical expertise to new technology opens up prospects of sharing the resources of the health sector. In practical life, a clinical diagnosis may allow us to claim insurance coverage or legal protection otherwise not available. On the other hand, in Western culture we experience an increase in the number of labels which declare us in need of special treatment – treatment we may or may not want to undergo. Here may lie a real distinction between, say, spider phobia and cyberphobia. With spider phobia it is the individual who decides to refer him or herself for treatment. However, by refusing the treatment of cyberphobia, in the context of widespread testing, we may incur disadvantages on the job market for 'not acting responsibly'. For this reason it seems increasingly desirable to avoid pathological labels where possible.

Furthermore, to pin the problem of resistance on the user whitewashes designers and producers of technology. It clouds the possibility that user problems point to the deficiencies of technical design. The design is as much a variable as people are, but as alterations cost money, changing people often seems to be the cheaper solution or a way of externalizing costs. Resistant people object to being engineered to the expectations of the designers. Faced with a problem of technical standardization and human heterogeneity, the attribution of the resistance problem is a matter of priority. The professional imperative for applied psychology must be to assist the redesign of technology and jobs, object-psychotechnics rather than subject-psychotechnics, to use the language of the 1920s. Technical design before person training, therapy and selection seems to be the appropriate order of professional priority (Hacker 1984; Ulich 1990). The 'redesign' of persons may be unavoidable, but it should only be complementary and secondary.

The problem of resistance needs to be seen with another eye to allow more long-term benefits. I propose to analyse resistance as a signal of mismatch in the socio-technological process (Bauer 1993). The idea is captured in the metaphor of 'pain in the process': user resistance is an alarm signal with implications for the design process.

The signal function of resistance in software design

If resistance has a signal function, then we start with a different question. Rather than asking what causes people's reactions to new technology, we ask how their reactions influence the development of new technology. Resistance is an independent variable of the design rather than a dependent one. In living systems this distinction is arbitrary; processes are circular and outputs influence inputs; for research purposes only we may focus on one or the other. Resistance is not about changing or not changing, but about the direction and the rate of change. The problem is to accommodate human variability in two dimensions: over time due to learning, and human diversity at any moment in time (Ulich 1990).

Resistance is therefore more a problem of the designer than of the user. But in avoiding the trap of shifting the attribution from user to designer, one needs to see the systemic process between designers and users over time. The virtual user is present in the design process as an implicit representation of present and future needs. This representation is reality tested when confronting the real user. Resistance is 'testbench' information that makes a contribution to the process.

Designers and users are not the only actors; there is capital involved, and so are shareholders who hardly care about how profit is made, as long as it is made; there are buyers and sales people, all following a different logic. Highlighting the contributions of resistance to that 'game' empirically in different contexts serves to enhance the user perspective. Any design process is a spatio-temporal network that produces retrospective accounts and visions for the future. Computer design is part of what has been called the 'computerization movement' (Kling & Iacono 1988). This process is beyond prediction; the future is open.

Directing social processes by self-monitoring[14]
For the purpose of analysis I see a technological project as a living social system (see Bauer, Chapter 19). Living systems face a twofold task: they direct their activity by adapting to both internal and external demands in maintaining a relative autonomy and an identity (Cranach et al. 1986). For the analysis of resistance, I focus on self-monitoring, which is the elaboration of messages about

the systems' internal state of affairs, its various, often conflicting, needs and requirements for future actions. Living systems normally have several such processes, which influence each other: they neutralize, enhance, integrate or suppress each other's contributions. It can be shown that resistance functions as an alarm signal when things go wrong, in functional analogy to 'acute pain': it is more appropriate to understand resistance in relation to its effects than to its stimuli.[15] Resistance influences the course of a project in three ways: by reallocating attention and leading to an increased awareness of organizational dynamics, by evaluating the project, and by urging alterations to the process. The analysis starts with functions and sees dysfunction as decay from normal functioning.

I investigated a software development project in a major Swiss bank between 1983 and 1991; this was to assist retail banking in local banks with centralized processing. Resistance in the design process is defined as:[16]

> A temporary, informal and unanticipated network of communication, that includes contributions from designers and from users who are linked together in conflict. The function of this temporary structure, in form and content, is to alter the project.

This definition stresses the temporary nature of resistance and the importance of communication; its informal nature distinguishes resistance from opposition which is institutionalized; it is unanticipated in the sense that members of the designer task force do not expect it in form and content. Resistance is communication with 'resistance' as its theme. The designers are the team of programmers in the central bank; the users are the local bank employees. Resistance to computerization in my study was not spectacular, in line with what observers of the information 'revolution' have noted in general (see Nelkin, Chapter 18, or Bauer, Chapter 1). Resistance was noted at two stages of the project: at the beginning and at a later, mature stage. Rather than asking about the dispositions of users, I explored the effects of the resistance on the project by analysing project documents and interviews with bank staff. I took the computer architecture of 1991 and reconstructed how it had come about, and what the contribution of resistance had been. The methodology developed for this purpose allows me to show when (timing) and how strongly (intensity) resistance appears as an issue during the project, how the issue is construed (discourse elements), and what effects can be associated with it (functions). The alarm function of resistance focuses the analysis of effects on three specific contributions: attention allocating, evaluating and inducing alterations and reinventions. I will briefly illustrate these functions with evidence from the study.

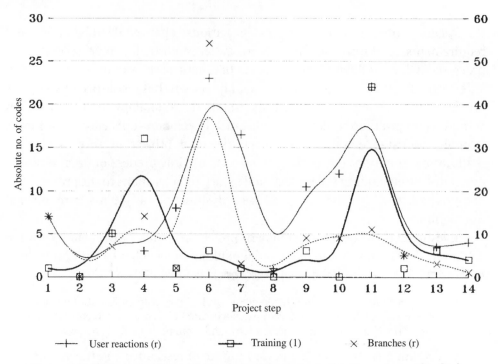

Figure 5.2. Shifts in the focus of attention of the project: the frequency with which three user issues are documented in formal project documents. The 14 project steps extend from 1983 to 1991. User reactions and issues of local bank organization (branches) are on the right scale; issues of user training are on the left scale. Total number of documents analysed: $N = 134$ (Bauer 1993).

Focusing attention on problem areas

The attention function states that resistance shifts the focus of attention internally during a project. It increases awareness of users' needs and functional requirements and the mismatch between these and the offered design solution. In a software project attention is drawn to where it is required. User resistance focuses the attention of the project team to unexpected problem areas, and awareness of the internal dynamics of the project increases.

Figure 5.2 shows the changes in the structure of overall attention between 1983 and 1991 as reported in documents such as memoranda, organizational notes, and regular reports ($N = 134$ over 8 years). The project period was divided into 14 project stages. For each stage the content of documents was analysed; the time-series shows the number of coded units for each step. The curve is smoothed and shows three concerns over time: awareness of specific users (user reactions), issues of user training (training), and work organization in local banks (branches). The graph shows that these issues come and go through the project. Resistance coincides with these shifts in a significant way. User resistance was officially communicated during steps 1, 2 and 4, at the beginning and, in step 11, at a later stage of the project. In steps 4 and 11 we find issues of user training directly coinciding with resistance. The project team's awareness of user reactions and of problems with the work organization in local banks is delayed by about one year, in the first round, and coincides with resistance in the second round. The increased awareness of the users in step 6 is facilitated by an internal conflict in the designer team. Two factions quarrel over the best way of serving the needs of the users. Bringing the user in can be seen as an attempt to solve the conflict by calling for a 'user referendum'. The conflict opens the factions to new information and facilitates the internal flow of communication. Attention allocation does not depend only on the strength of resistance, but on the constellation within the project team. It seems designer conflicts constitute an opportunity structure for resistance to be noticed. Furthermore resistance increases the sensitivity of the system. Sensitized by the earlier resistance the designer team commissioned a study of users' needs and perceptions during the pilot phase of the new software which started in phase 8. This leads to the second function: evaluation.

Evaluating the project

The evaluation function indicates that resistance is a negative evaluation of the on-going project. The evaluation mobilizes values and criteria which may differ from those of the designers. I analysed criticisms voiced by local bankers in a series of in-depth interviews in steps 9 and 10. Resistant bankers are a prolific source of information for the designers. Figure 5.3 shows the average number of points voiced by local bankers and compares those who accepted the project easily (accepting) with those who resisted the project (resistant).[17] In all 250 different points of criticism from a total of 510 points were recorded.

Three areas of project deficiencies have been identified: first, concerning information input, such as the kinds of customer data that are recorded and the structure of the human–computer interface; secondly, concerning information output, its content and its layout for internal use (account lists), and for external use (customer statements); and thirdly, concerning the local organization of work, such as data archiving, security, availability of up-to-date data, costing and finance, training, and, on a meta-level, demands for front-line participation and the need to 'have a say'. Resistant banks delivered on average more criticism than accepting banks; this is marked for input (12.9 vs 8.8), and less so for process (11.3 vs 8.8) and output issues (6.1 vs 4.7). The criticism of resistant banks is richer and yields more different issues. Resistance is voiced differently in the first and second round, moving from the general to the specific. Resistance in the first round basically voices strategic concerns such as project costs and concerns about the autonomy of local banks;[18] in the second round, during the interviews, the criticism concerns the project details in terms of software,

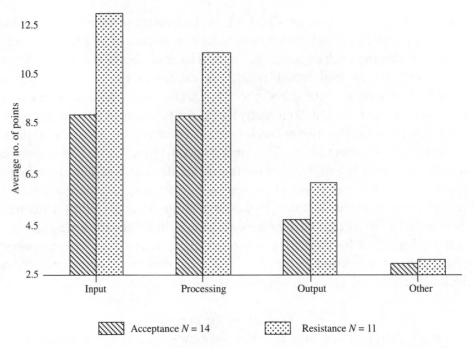

Figure 5.3. Proliferation of project criticism and resistance. Average number of criticisms by bank type (14 accepting banks and 11 resisting banks), and by four categories of criticism: data input, data processing and work organization, data output and 'other'. Total number of points collected in 25 interviews with bank managers: $N = 510$ (Bauer 1993).

hardware and work organization. Acceptance of the project correlates positively with the size of the bank (yearly balance sheet) and negatively with the share in the local market. The more local bankers enjoy their present jobs and work climate (Job Diagnostic Survey measures), and the more they see themselves temporarily protected, the more they are inclined to resist the project. On the other hand, bankers who see their business at risk tend to favour the project as it stands.

Reinventing the project

The alteration function of resistance shows the material effects of communication on the project outcome. Two such effects can be observed in the case of the banks. First, the pace of the project slows down, and elements of the project change (reform of the project). The general expectation about resistance is that it slows the pace of innovation. This often categorizes it *a priori* as a universal nuisance. However, delay does not need to be disadvantageous; delay gives time to reflect, to explore, and to import better solutions. The late comer may have an advantage. The slower pace of change at this bank is shown by comparing the penetration of computers into the work of the bank with other Swiss banks of the time (Luethi,

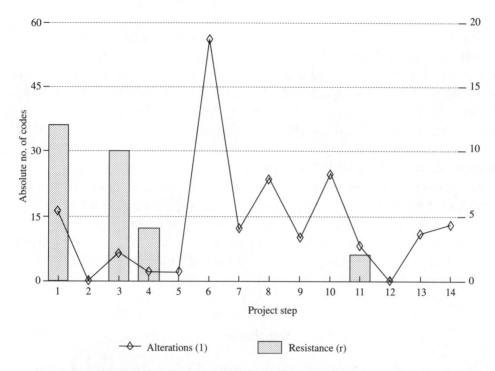

Figure 5.4. Project alterations in relation to resistance: absolute number of codes of documented alterations and resistance over 14 project steps. Project steps extend between 1983 and 1991. Resistance is on the right scale, alterations are on the left scale. Resistance precedes a rise in alterations from 5 to 6 and from step 12 to 13. Total number of documents analysed: $N = 134$ (Bauer 1993).

Schaller & Schweizer 1990). The gap increases from zero in 1982 to 4% in 1990. In 1991, 91% of banks in the study had direct access to data processing as opposed to 94% of other Swiss banks: a small but significant difference.

The second impact of resistance consists of concrete alterations of the project in response to communicated resistance. Figure 5.4 shows the number of documented alterations through the 14 project steps. Reported alterations in step 6 rise to over 50 from two in the previous step, and in step 13 to ten alterations from none in the previous step. Alterations follow on the reported resistance in steps 4 and 11 with a reaction time of two project steps, corresponding to about one year in the first round, and to about five to seven months in the second round. As the project gets more hectic, the reaction time of the design team decreases.

Alterations also occur which are unrelated to resistance. Step 6 is the time of internal conflict in the project team, and many reported changes mark the solution of this conflict. Changes in the project have many other sources, not least the self-criticism of the team itself. However, the coincidence of resistance and alterations strongly suggests that user resistance successfully signalled the need for, and led (with some delay) to alterations in the software project of the bank.

Voicing criticism is one thing, but it needs to be heard by the designer. Designers make selective use of the information that is available to them according to a variety of agendas. By 1991, two years after the user interviews, some 40% of all problems voiced by users had been solved. Designers classify 'valid' and 'invalid' points; invalid ones are rejected on the basis of constraints set by the hardware and by the costs the alteration would involve. Most alterations concerned the organization of work in local banks where the flexibility is greatest. The in-built flexibility of technology is crucial in the process.

In summary, it can be shown that resistance, adequately voiced in concrete criticisms, functions as an alarm signal for a technical project. It indicates that something is going wrong; it relocates temporarily the focus of attention and changes permanently the structure of attention; it evaluates the ongoing activity negatively; and it alters the pace of change and the direction of the project on various levels. The project is reinvented in its course.

Conclusion

I have shown how 'cyberphobia' defines resistance to computers and information technology, and emerges as a research topic: up to 25% of the American public were supposed to be at risk at the end of the 1980s. Technophobia is a clinical discourse that is periodically revitalized to talk about people's reactions to new technology: from siderodromophobia in the nineteenth century to nuclear phobia and cyberphobia in recent years. I criticize this clinical eye on three counts, and present an alternative view to frame different questions. 'Cyberphobia' is an example of how institutional resources give rise to a research topic and a service package that, although efficient in operation, is misleading in solving the problem of resistance. The diagnosis of 'cyberphobia' focuses on the individual dispositions of computer users, reacting to a cultural bias of individualism of the type we may find in the USA. European researchers, in personalizing the problem, seem more interested in the attitudinal opposite: computer addiction and hacking. 'Cyberphobia' is a weak diagnosis by professional standards, based on self-report data only. The psychometric test, calibrated in California, leads to methodological artefacts both nationally and internationally. The statistical definition of normality overlooks the fact that normality and deviance are at best temporary adaptations to a particular environment, for example to the US West Coast in the 1980s. We may speculate that widespread screening and therapy for cyberphobia would lead to a regression towards a cultural 'mediocracy'. Variability of experience is an indispensible resource for an uncertain future. Ready-made solutions are the risks of institutionalized practices. 'Cyberphobia' appeals:

it has news value, it is simple to measure and cheap to 'cure'. A standard diagnostic instrument, Government funding, institutional legitimation, and an efficiency tested therapy give it credibility. However, the clinical approach to resistance blinds us to features of the problem that only a different approach can reveal. It is likely that this 'blind spot' in the clinical eye incurs opportunity costs. A different psychological professionalism gives priority to the design of technical products rather than to the design of persons. Design is a spatio-temporal system which includes designers and users among other players. The metaphor of the 'pain in the process' shifts the interest from individual dispositions to the effects of resistance on the project. Resistance, motivated by anxiety, functions as an alarm signal; it focuses attention and increases the awareness of organizational dynamics, user needs and front-end functional requirements; it evaluates the ongoing activities in this light; and it induces alterations, and often slows the pace of activity. On the whole the focus of the analysis is shifted from resistance as an individual deficit to resistance as a social resource. Resistance is the benchmark of local reality to test new technology.

Notes

1 I am grateful to Larry Rosen, who commented on an earlier version, and to Dan Wright and Sandra Jochelovitch for helpful comments.
2 In experimental behavioural studies with animals the term 'neophobia' has a specific meaning and refers to the animals' naturally adverse reaction to the presentation of unfamiliar smells and odours, which can be conditioned.
3 I owe this historical point to Joachim Radkau.
4 Medical institutions may impose their diagnosis and treatment relatively independently of the condition of the person, as demonstrated in a famous field experiment. A group of otherwise healthy researchers volunteered to be admitted to mental hospitals under the pretext of hearing voices. Each of them was given a temporary diagnosis of schizophrenia and was retained on average for 19 days (Rosenhan 1973).
5 Four data bases were used: ABI/INFO, Social Science Citation Index, Psychological Abstracts, and Dissertation Abstracts. The keywords used were 'computer phobia', 'computer anxiety', 'cyberphobia', 'computer aversion', and 'computer attitudes'. A bibliography of cyberphobia, computer anxiety, technophobia, computer aversion, and negative computer attitudes since 1970 can be obtained from the author.
6 Rosen & Weil (1990) report 500 publications on cyberphobia, of which about 100 were empirical studies. It may be that the difference from my count of 300 lies in the fact that they collected unpublished reports, which do not appear in electronic data bases.
7 This is the only book which takes 'phobia' in the title. Enquiries with the authors revealed that the book was mildly successful and sold a few thousand copies; the target markets were (a) managers and secretaries who had a computer placed on their desks; (b) executives who had never learnt to type; (c) students and teachers who were forced to use computers as learning aids. The publisher has since gone out of business.
8 Studies on cyberphobia are published in journals such as *Anxiety Studies, Computers in Human Behaviour, International Journal of Man-Machine-Studies, Behaviour Research Methods, Instru-*

ments and Computers, Behaviour and Information Technology, Human Factors, Journal of Educational Computing, Educational and Psychological Measurement.

9   The attention of British psychology was recently documented by an article in the BPS journal *The Psychologist*, focusing on the problem of cyberphobia as a disposition of women (Brosnan and Davidson 1994). The authors make use mainly of North American sources.

10 Constructing a reliable and valid measure for other researchers to use is a well-confined task, therefore attractive, and it is a relatively prestigious social science activity. During the 1980s such scales to measure computer phobia and anxiety proliferated: there were 66 according to Rosen & Maguire (1990, p. 179). Scales vary in length, reliability, and the validity of criteria used. Some are validated on expert judgements; others by other measures.

11 Short of an external criterion the statistical standard deviation often provides a formal criterion to cut off 'pathological' cases; two standard deviations from the average define 'phobia'; one defines 'anxiety'. On this criterion 13.5% of the population would be 'anxious', 2.5% would be expected to be 'phobic'; 16% of the population are either pathologically anxious or phobic. One assumes a percentage of the population to be mentally ill. This assumption is later declared to be an empirical finding.

12 D. Goleman in a news article of a recent survey of the prevalence of psychiatric disorders in the American population wrote: 'half of the US population suffers mental disorders', London *Guardian*, 15 January 1994. The survey does not to mention cyberphobias among the prevalent phobias.

13 The alarm concerning computer addiction and the hacker came from the USA (Weizenbaum 1976; Turkle 1984). The evidence of computer addiction recently gained relevance in a British court. A 19-year old hacker was acquitted from conspiracy charges under the British Computer Misuse Act 1990. The defence argued that he was addicted to computing rather than conspiracy when intruding the computer systems of the Ministry of Defence, Scotland Yard and the American Navy: Susan Watts, 'Hacker penetrated MoD' in *The Independent*, 26 February 1993.

14 The theoretical framework is owed to the Bernese Group on Action Psychology: they conceive and study self-active systems at the level of individuals, groups and organizations using functional analogies (Cranach *et al.* 1985, 1986).

15 After Wall 1979; see Bauer 1993 or Bauer, Chapter 19 for an elaboration of the pain analogy and the context of system theory.

16 Resistant banks were identified with the help of attitude measures, differentials in the diffusion of computer material, and in interviews with the task force and potential users. Data were collected in a sample representative of around 600 branches all over Switzerland between 1988 and 1990. The whole project spanned the period from 1983 to 1991. Resistance has been observed indirectly by using content analysis of project documentation and directly by interviewing project staff and users (for details see Bauer 1993).

17 To classify 'accepting' and 'resistant' banks external criteria were used: attitudes to computers in general and to the project in particular (1985 and 1989), evidence from interviews, and impressions from the project team.

18 It is important to note that the bank in this case is legally a cooperative with, in 1990, about 1200 local branches, and with a high degree of regional and local autonomy. The central bank provides a corporate identity and sells various services such as auditing, computing, marketing and training. Autonomy is a major concern in the discussion that goes on in parallel to the software project about a banking strategy for the next century.

References

BAUER, M. (1993).   Resistance to change – a functional analysis of user responses to technical change in a Swiss Bank. PhD thesis, London School of Economics.

BERGER, P. & LUCKMANN, T. (1967).   *The social construction of reality. A treatise in the sociology of knowledge.* Harmondsworth, UK: Penguin.

BROSNAN, M. J. & DAVIDSON, M. J. (1994).   Computerphobia – is it a particularly female phenomenon? *The Psychologist*, February, 73–8.

CANGUILHEM, G. (1991).   *The normal and the pathological.* New York: Zone Books. (French original 1943 and 1966.)

CONRAD, P. (1992).   Medicalization and social control. *Annual Review of Sociology* **18**, 209–32.

CRANACH, M. von & OCHSENBEIN, G. (1985).   Selbstüberwachungssysteme und ihre Funktion in der menschlichen Informationsverarbeitung. *Schweizerische Zeitschrift für Psychologie* **44**, 221–35.

CRANACH, M. von, OCHSENBEIN, G. & VALACH, L. (1986).   The group as a self-active system: outline of a theory of group action. *European Journal of Social Psychology* **16**, 193–229.

DAVISON, G. D. & NEALE, J. M. (1978).   *Abnormal Psychology.* New York: John Wiley & Sons.

DUNCKER, K. (1935).   *Zur Psychologie des produktiven Denkens.* Berlin: Springer.

DSM-III-R (1987).   The American Psychiatric Association.

FISCHER-HOMBERGER, E. (1975).   *Die traumatische Neurose.* Bern: Huber.

GLASS, C. R. & KNIGHT, L. A. (1988).   Cognitive factors in computer anxiety. *Cognitive Therapy and Research* **12**, 351–66.

GOFFI, J. Y. (1988).   *La philosophie de la technique.* Paris: PUF.

GREENLY, M. (1988).   Computerphobia: the fear that keeps people 'off-line'. *The Futurist*, Jan./Feb., 14–18.

HACKER, W. (1984).   *Psychologische Bewertung von Arbeitsgestaltungsprocessen*, Band 1, Berlin: Springer.

HEWSTONE, M. & ANTAKI, C. (1989).   Attribution theory and social explanation. In *Introduction to Social Psychology*, ed. M. Hewstone *et al.*, pp. 111–41. Oxford: Basil Blackwell.

HIMADI, W. G., BOICE, R. & BARLOW, D. H. (1985).   Assessment of agoraphobia: triple response measurement. *Behavioral Research and Therapy* **23**, 311–23.

JAY, T. B. (1981).   Computerphobia: what to do about it. *Educational Technology*, January, 47–8.

JAY, T. (1985).   Defining and measuring 'computerphobia'. In *Trends in Ergonomics/Human Factors II*, ed. R. E. Eberts and C. G. Eberts, pp. 321–6. Holland: Elsevier Science Publishers BV.

KLING, R. & IACONO, I. (1988).   The mobilization of support for computerization: the role of computerization movements. *Social Problems* **35**, 226–43.

LEE R. S. (1970).   Social attitudes and the computer revolution. *Public Opinion Quarterly* **34**, 53–9.

LEWIS, A. (1953).   Health as a social concept. *British Journal of Sociology* **4**, 109–24.

LÜTHI, A., SCHALLER, T. & SCHWEIZER, V. (1990).   *Informatikeinsatz in Schweizer Betrieben, Institut für Automation und Operation Research.* Fribourg: FUP.

LUHMANN, N. (1990).   Der medizinische Code. In *Soziologische Aufklärung, 5. Konstruktivistische Perspektiven,* 183–95.

MEIER, S. T. (1985).   Computer phobia. *Popular Computing* 4(10), 132.

MITCHELL, R. C. (1984).   Rationality and irrationality in the public's perception of nuclear power. In *Public reactions to nuclear power: are there critical masses?* ed. W. R. Freudenberg and E. A. Rosa, pp. 137–79. Boulder, Colo: Westview.

NELKIN, D. (1987).   *Selling science. How the press covers science and technology.* New York: W. H. Freeman.

NIKODYM, N. *et al.* (1989).   Computerphobia. *Personnel Administrator* 34(8), 54–6.

PATTON GARDNER, E., YOUNG, P. & RUTH, S. R. (1989).   Evolution of attitudes toward computers: a retrospective view. *Behaviour and Information Technology* 8, 89–98.

ROSEN, L. D. & MAGUIRE, P. (1990).   Myths and realities of computerphobia: a meta-analysis. *Anxiety Research* 1990(2), 175–91.

ROSEN, L. D. & WEIL, M. M. (1990).   Computer, classroom instructions, and the computerphobic university student. *Collegiate Microcomputer,* November, 275–83.

ROSEN, L. D. & WEIL, M. M. (1992).   *Measuring technophobia.* California State University, June, version 1.1.

ROSENHAN, D. L. (1973).   On being sane in insane places. *Science* 179, 250–8.

SEIDENRUECKER, G. & BAUMANN, U. (1987).   Multimodale Diagnostik als Standard in der klinischen Psychologie. *Diagnostica* 33, 243–58.

SHOTTON, M. (1989).   *Computer addiction? A study of computer dependency.* London: Taylor and Francis.

TILES, M. (1993).   The normal and the pathological: the concept of a scientific medicine. *British Journal of the Philosophy of Science* 44, 729–42.

TLUSTOS, C. (1985).   Computer phobia costs millions. *Citation Index Computer Journals,* AN 1176274 gnc 8603.

TURKLE, S. (1984).   *The second self: the computers and the human spirit.* New York: Simon and Schuster.

ULICH, E. (1990).   *Arbeitspsychologie.* Zürich: Poeschel/vdf.

WALL, P. (1979).   On the relation of injury to pain (J.J. Bonica Lecture). *Pain* 6, 253–64.

WEIL, M. M. & ROSEN, L. D. (1991).   Psychological ramifications of the technological revolution from a global perspective. Manuscript, California State University, Dominguez Hills.

WEIL, M. M., ROSEN, L. D. & SEARS, D. C. (1987).   The Computerphobia Reduction Program: year 1. Program development and preliminary results. *Behaviour Research Methods, Instruments and Computers* 19, 180–4.

WEIL, M. M., ROSEN, L. D. & WUGALTER, S. E. (1990).   The etiology of computerphobia. *Computers in Human Behaviour* 6, 361–79.

WEINBERG, S. B. & FUERST, M. L. (1984).   *Computer phobia. How to slay the dragon of computer fear.* Banbury Books.

WEIZENBAUM, J. (1976).   Computer power and human reason, from judgement to calculation. San Francisco: W. H. Freeman and Co.

PART II
Case studies

6

Patterns of resistance to new technologies in Scandinavia : an historical perspective

KRISTINE BRULAND

Introduction

This chapter deals with historical cases of resistance to new technologies in Norway and Sweden. It is primarily an interpretive essay rather than original research, drawing on existing historical work related to economic and social dimensions of technological change in two of the largest and most important Scandinavian industries, fishing and timber, plus a major energy technology of the twentieth century, nuclear power. It is interpretive in the sense that the objective of the chapter is not simply to describe some historical aspects of resistance to innovation in these industries, but also to raise questions about how such resistance should be understood.[1] The main point which is argued is that resistance to new technologies should not be seen purely in terms of a kind of conservative labour resistance; rather, it often involves complex coalitions of actors, and is perhaps best understood as one of the selection mechanisms through which societies adopt or reject new technological opportunities.

In general, in analysing technological change in Scandinavia, we are discussing processes by which foreign technologies enter the system. Like many of the core technologies of the region, the new techniques in fishing, timber processing and energy had in common the fact that they were originally developed outside Scandinavia, and so what is being considered here are in part conflicts related to international technology transfer.

The Scandinavian economies have always been highly open trading economies, firmly integrated with the trading systems of the Hanseatic League, the Dutch Baltic trade, and then the 'new' Atlantic economy (based on trade between Europe and the American colonies) which emerged in the eighteenth century. It has been estimated that even in the pre-industrial period, the Scandinavian economies were trading up to 25% of national income, and this openness no doubt played a role in the sensitivity to foreign technological developments which was exhibited by Scandinavian entrepreneurs and officials as Britain and then Europe as a whole began to industrialize (Hovland, Nordvik & Tveite 1982). Certainly a characteristic feature of Scandinavia has been rather large-scale technology import, beginning in the early nineteenth century and persisting to this day. Even now, when the Scandinavian economies are advanced research-intensive systems with major technology development activities, it remains the case that their technological foundations are predominantly foreign. This was even more the case during the industrialization phase which began in earnest in the 1840s.

It is often argued, rightly in my view, that the industrial successes of Scandinavia stem in large part from this extreme openness to technological opportunities arising from external sources.[2] But it should not be supposed that even societies which are very open to the international diffusion of technology simply accept what arrives without resistance. On the contrary, there can be significant social opposition, and the Scandinavian experience seems to suggest that the forms and extent of resistance to new technologies depend in large part on how those technologies originate, on which social actors seek to introduce them, and how their likely effects are perceived. A further distinction which can be relevant here is the case in which a specific technology is known domestically, but in which the pressures to use the technology come from abroad; this form of foreign influence was important in the case of the fishing industry, which will be discussed below.

Against this background, this chapter makes three broad points. These are, first, that resistance to new technologies (particularly technologies of external origin) can come from many sources, and should not be seen – as it has been in much of the literature – primarily in terms of a form of Luddite labour resistance to innovation. The perception of negative effects is not something confined to workers; on the contrary, there are many potential points of resistance to innovation in business, public administration and politics, and resistance from all of these quarters played a role in the technological evolution of the Scandinavian economies. Secondly, when resistance is seen in this broad way, it can really only be understood in terms of the interaction between the technology and its social context. It is not simply a matter of specific workplace interests being threatened by new technologies, but a much wider process concerning the ways in which technologies accord or clash with social organizations, cultural values, and so on. Thirdly, resistance is by no means irrational. While resistance is invariably based on particular social concerns and interests, it can also be seen as a component of the more general processes by which society selects among technological options; this point is explored in the following section. Taken together these points raise questions as to what we understand by the term 'resistance' in the face of technological innovation; this issue is discussed in a final section.

Understanding patterns of resistance

How should we understand the nature of 'resistance'? An important point is that, from one perspective, resistance is simply part of the selection process through which new technologies are adopted or rejected. Modern market economies tend to generate many technological opportunities, but relatively few are diffused into widespread application; resistance can be seen as simply one of the ways (failure in the market is another) in which technologies fail to achieve acceptance. Recent theories of technological change tend to follow an evolutionary perspective, in which the dynamics of technological change are based on some process which generates variety and diversity among technologies, and some mechanism which selects among the new varieties.[3] It is important to note that in the evolutionary perspective the selection mechanism is not simply the market: it involves the whole complex of non-market decision processes, and there are many ways in which technology can be accepted or rejected by relevant actors. These non-market decision processes may involve, for example, public or semi-public procurement decisions, or political decisions to regulate, control or promote a technology. When there is continuous generation of technological variety, selection among alternative lines of technical advance is always necessary, and this selection process can easily be pictured as 'resistance'; but we should not see this as resistance to technological advance as such. The necessity of selection is integral to the technological change process as a whole.

When foreign technologies are involved, the selection process has a further complication: resistance may not simply be to a specific technology – it may also be resistance to industrialization more widely, or to the foreign ownership and influences which are involved. However resistance to a technology may also simply be part of the process by which it is adapted to a new setting. It is now rather widely accepted that technology should not be seen simply in terms of artefacts, but is best viewed primarily as a social phenomenon: a technology involves the integration of tools and machines with management practices, working routines, skills, and a wider social context whose values shape the acceptance and operation of a technology. Adopting a foreign technology is above all a process of change and modification, in which the physical structure and properties of a machine or device, and the micro and macro social contexts, are simultaneously transformed (to greater or lesser degrees). Svante Lindqvist has shown, for example, that the failure of the early diffusion of the Newcomen steam technology into Sweden was primarily not a technical matter, but one of adapting the complementary technical and social systems which were required (Lindqvist 1984). Lars Ekdahl has shown similar processes in the development of industrial printing in Sweden, with complex patterns of resistance and adaptation to new British technologies, in particular in his study of the so-called Times 'speed-press', first introduced in the printing industry in 1814 (Ekdahl 1983). Seen in this way, resistance might be understood as part of the more or less complex negotiation process which surrounds the introduction of any new technology; it is not necessarily a bad thing, and it might be rather widespread.

This in turn suggests that, when we see technology as a social phenomenon and in a social context, and when we see the selection environment in more nuanced terms, then we might expect resistance from a very wide range of sources. Changing a technology can shift not only the micro-environment of production (that is, the direct organization of work, and the structure of rewards for tasks), but also wider social and economic distributions of income, wealth and power. New technologies change the relative performance of firms, they often change the income distribution, and they can have major effects on the regional location of production (which historically and to the present day has been an important political issue in Scandinavia). This is basically a matter of the links between large-scale technological change (especially the diffusion of new generic technologies) and the dynamics of industrial structure. Via their impacts on the industrial structure, the diffusion of new technologies can influence voting patterns and political power, as well as having effects on the public administration (for example, by changing the relative importance of ministries or public agencies). If these effects are accurately perceived by socio-economic actors, as they often are (even if only in outline), then there is a wide range of potential sources of opposition to the introduction of new technologies. In the Scandinavian context, it is therefore important not to see resistance simply in terms of labour resistance, or resistance by trades unions, which is how historians have often viewed the question of resistance to new technologies. On the contrary, depending on how the effects of new technologies are perceived, we might expect resistance from entrenched business interests, from regional administrators whose local economic bases are threatened, from politicians concerned with electoral or financial support, from civil servants, from the general public and so on. Resistance by workers is but one component of a wider potential resistance to innovation.

These broad aspects of resistance to new technologies – that it is part of general technological selection, that it involves a wider social dimension, and that a potentially wide range of actors is capable of resistance – can all be found in the case studies which are explored below. However in my view these cases also raise a more general issue, which is whether 'resistance' is the appropriate term for the kinds of oppositional attitude to new techniques which are described here; this issue is taken up in a final section.

Historical patterns of resistance in Scandinavia

Against the background of these points, we turn to various historical forms of resistance in the Scandinavian scene, focusing first on two of the largest Scandinavian industries, namely fishing and timber products. These industries were characterized by prolonged processes of resistance to new technologies from the mid-seventeenth century. This will be followed by discussion of a twentieth-century example of more or less explicit resistance to foreign technologies, namely resistance to the development of nuclear power in Sweden.

The case of fishing

Fishing has long been one of the major industries – especially in terms of employment and exports – for Scandinavia as a whole, but in particular for Norway. In the mid-nineteenth century, roughly half of all Norwegian commodity exports consisted of fish products, and between 12 and 15% of the economically active population worked in fisheries (Hodne 1975, pp. 73–9). The fishing industry of Norway is based on two major migrations of breeding fish populations, namely cod and herring (and closely related species), which enter the warm coastal waters of Norway at more or less regular intervals (historically, the herring fishery has been less predictable in terms of size than the cod fishery, but usually more valuable). The basis of the export trade is onshore processing, originally with both herring and cod being dried, smoked and salted; in the nineteenth century canning was developed, and in the twentieth century, freezing. In the nineteenth century, herring had markets throughout Northern Europe, especially in the Baltic countries, and dried and salt cod sold well – as they still do – in virtually all of the Mediterranean countries.

Although the fishing industry had a certain degree of geographical concentration, with an especially intense and productive cod-fishing season based on the Lofoten Islands, a very important feature of the industry is that it was also carried on from small rural towns or hamlets, located along an extremely long coastline. It was, and remains therefore, an industry with deep connections to Norwegian society as a whole, and also to the regional politics of Norway.

In looking at the evolution of the fishing industry, technological conflict and resistance to specific techniques stand out as a persistent feature. This section will describe some aspects of this persistent struggle, emphasizing its roots in different social interests: the roles of civil servants, or regional politicians, for example, were of particular importance; both had interests which were threatened by new technologies, even though they were not directly involved in the economics of the industry.

The technology of fishing involves two closely related dimensions, namely the specific techniques for catching fish (such as the various types of nets, or lines and hooks), on the one hand, and the size and types of boats on the other. As with many apparently simple technologies, fishing techniques are in fact highly differentiated and complex. However, at the risk of oversimplification, for many centuries Norwegian fishing was in most areas an inshore activity involving small boats and a simple line technique known as 'snøre', which was a single hand-held line and hook.[4] Technological advance in fishing essentially involved combinations of larger and more seaworthy boats (capable of longer voyages and storage of larger catches), and new catching techniques (involving either varieties of nets, or multiple lines and multiple hooks). An important point to note here is that the innovative changes through which these techniques were introduced into Norway were not necessarily new in any technological sense: nets also are a very ancient technique in fishing, as are multiple lines and hooks. What was new in the Norwegian fisheries was usually the extension of these techniques into new contexts, and their combination with new boat types; here, the impulse often came from abroad, from technologically much more dynamic fishing fleets elsewhere in Northern Europe, in particular Britain, Denmark, Germany and The Netherlands. This is a case of the type referred to above, where the international diffusion of the technology involves the application of a known technique into commercial practice, rather than the development of a new item of technological knowledge as such.

Prior to 1816, when nets and multi-hook lines were legalized in the Lofoten Islands, there was a sustained conflict over the use of these techniques in the rich cod-fisheries of the area. In what follows, the term 'lines' means the multiple hook, multiple line technique; among other differences, lines could be set from a boat, as opposed to the single line and hook which was actually held and monitored by a fisherman. The resistance to lines can be traced back at least two centuries, with the earliest known example of opposition being found in the early seventeenth century. From the 1620s the state and regional or local authorities began to promulgate regulations, restrictions or outright prohibitions against line techniques. Posti remarks that

> In 1627 the first prohibitions against using lines came in the Lofoten waters. The prohibition was not respected, and complaints were made directly to the Danish king (Kristian IV). He demanded in 1644 that local sheriffs should enforce the law, and punish those who broke it.   (Posti 1991, p. 26)

The basic issue here was that although multiple lines enhanced productivity, their use was confined to relatively well-off fishermen who could afford to invest in extra equipment and suitable boats. These investment requirements, though modest, were beyond the means of the poor fishermen not only of the Lofoten area but of Norwegian coastal regions as a whole. As a fishery newspaper, *Norsk Fiskeritidende*, (Norwegian Fishing Times), put it many years later,

> For the well-to-do fisherman it [multiple lines] made work easier. Instead of working out at sea from early morning to late at night, he could set a couple of hundred lines which after a few hours would yield a catch comparable to days and even weeks of work with a hand-line. The opponents of lines argued strongly that this caused great damage; it was the reason why fish were no longer coming in such numbers to the old fishing grounds, where people had been used, since time immemorial, to 'throw so many lines'. This led to bad feeling, which God would punish by withdrawing his blessing from the coast.... One of the main reasons for the opposition to this equipment was resentment that the person with multiple lines was certain of a bigger catch than the poor fisherman who had only his hand-held line to earn his living from the sea.[5]

During the eighteenth century opposition extended to new techniques of net-fishing in the cod fisheries, a struggle which peaked between 1750 and 1780 and which involved a large part of the coastline between Stavanger and Lofoten (a distance of over 1500 km). The fishermen themselves made their views heard through persistent letters of complaint to the authorities. In 1756 this conflict erupted into violence, with a pitched battle in Lofoten between users of the various techniques.

This conflict emerged in the Lofoten fishery when nets began to be introduced, so that there were now three competing techniques: hand-held lines, multiple lines and nets, with three corresponding social groups. The authorities supported the weakest but most numerous group, the hand-held line users, with the local governor prohibiting all other techniques in 1753. By the mid-1770s the law was being systematically flouted, even by the local sheriff; in 1774, 145 people were punished and the sheriff imprisoned. Nevertheless the law was not much respected, and following further representations to the Government nets were partially allowed from the 1790s (Posti 1991, p. 27).

The greater investments required for nets certainly led to greater productivity, in terms of catches per fisherman. But arguments that this conferred a public benefit continued to be met by concerns about fish stocks; fish were seen as a quasi-free good, but not in unlimited supply. Of course, in many districts the new fishing equipment was introduced without any battle. Lines were used in eastern Finnmark from as early as 1600, and at Sunnmøre nets for cod were introduced later in the century. Furthermore, in some areas the ordinary fishermen were divided in two groups, using different techniques.

However, tension between users of different types of technology persisted into the late nineteenth century, frequently involving public and political debate about forms of regulation, and occasionally erupting into physical conflict. Even after 1816, the law restricted the use of lines to deep waters in Finnmark, to protect hand-line fishers (Solhaug 1983, p. 268). Later in the century, from the 1880s, there was a build-up of resistance among fishers against the increasing use of new net-types known as *not* (seine, or 'closing nets') in fishing cod, and against the use of *synkenot* (sink seines); both had been used for a long time but increasingly so in the 1880s. Opposition to the use of *stengenøter* (seine nets across fjords) had likewise increased: the state was mobilized to counteraction and in 1893 prohibited the use of *not, synkenot, slepenot, posenot* and trawls during the Lofot-fishing. Provisions were made so that the law could apply to other areas, too, if applied for (Solhaug 1983, p. 678).

This may have been linked to fears about other new equipment as well, such as trawl and *snurpenot* (purse seine). Once again, violence ensued, in the shape of the so-called 'Trollfjord battle' in 1890.[6] The Trollfjord clash was a pitched battle between hand-line fishermen and trawler crews, which was followed by a formal demand by the hand-liners to the Parliament in Christiania 'demanding a ban on the use of all types of purse-nets in the cod fisheries... The Storting met their wishes in a fortnight. A new law, the Lofoten Act of 1897, codified the new situation.' (Hodne 1975, p. 84). The result was perhaps surprising, 'a victory for the Luddites', with the Parliament acceding to all of the demands of the hand-line fishermen. At the same time, a law of 1890 banning new tackle in the herring fisheries in the north of Norway was upheld in a new regulation of 1897 (Hodne 1975, p. 95).

This active political role by the Lofoten fishermen continued. In 1900 they 'raised complaints against the Norwegian whaling activities, which they argued decimated the whale stock in those waters. Believing that whales were responsible for driving the fish to the coast, they wrecked a whaling station the following year in 1902. The action predictably induced the Parliament to place a ban on whaling in territorial waters for a test period of ten years in northern Norway' (Hodne 1975, p. 84). The resistance extended into boats: at Sunnmøre there was a struggle against Swedes using a new type of boat (*skøyter* – a new type of fishing smack) in the 1860s and 1870s. They were accused of damaging equipment. In 1868 fishers sent a complaint to the Interior Ministry demanding a prohibition against *skøyter*. Regulations followed: a 'compromise law', which made it illegal to use *skøyter* inside the territorial boundary, and during February–March (Solhaug 1983, p. 228). Later, 'when the new petrol engines were introduced, voices rose in protest. Noise, it was said, would frighten the cod away. Here the advantage to the small fisherman was so obvious, however, that protests soon died away' (Hodne 1975, p. 84).

As Hodne has suggested, 'there developed a permanent tension from the 1890s between users of low-productivity tackle and users of high-yielding equipment', echoes of which can still be found in Norwegian society to this day (they were an element in the recent debates over EU membership, for example). As suggested here, however, this conflict has rather longer historical roots.

What was the basis of this long-running technological conflict? The key point in understanding resistance to new techniques in fishing is that technological change in fishing had implications not simply for relative incomes in the industry, but for the entire social structure of coastal Norway. It is this wider dimension of fishing technology which explains why those who resisted the trajectory of technological change in fishing were able to mobilize a constituency of administrators and politicians who were seeking to enforce what we might call technological stability on the industry. Commercial fishing was a seasonal activity, in which a very large proportion of the coastal population took part over a relatively short period; outside the fully commercial activity, fishing was a component of a semi-subsistence economy, which was combined with relatively small-scale peasant agriculture (and in some areas with forestry). Although fishing was a marginal activity in terms of time, it was nonetheless critical to the viability of the coastal economy as a whole. A key point about the new techniques was that they changed the entry conditions for the industry: the traditional techniques involved virtually free entry, because of the extremely simple equipment requirements. Newer techniques challenged the coastal communities in a number of ways. On the one hand the investment requirements associated with the new techniques, though not large in any absolute sense, closed off participation by the relatively poor peasant-fishermen who made up a large part of the industry. But they also threatened the possibility of combining fishing with agriculture at all, because they tended to require a longer fishing season, and therefore a more specialized division of labour. The new techniques thus threatened not simply a particular category of fishermen, but what we might call the 'economic ecology' of the coastal region as a whole: the balance of activities which supported the overall pattern of settlement, of community organization and – above all – political structure. It was the latter which secured the continuous commitment of both civil servants and regional politicians to control of the new technologies: the connection between technological change in fishery, and their own bases of power was a relatively direct one. This tension between technological opportunity and social structure was never resolved once and for all; Norwegian politics to this day is dominated by regional policies which have as their objective the viability of small coastal communities, and the technological conflicts which characterized the fisheries in the nineteenth century are being repeated as the prospect of entry into the European Union opens up the possibility of new entrants and new techniques in the Norwegian fishing grounds.

Resistance to steam power in the timber industry

Timber and timber products (such as furniture and paper) are among the biggest industries in Scandinavia. Norway, Sweden and Finland have massive forests of fir and pine, and have sustained a long-standing export trade; in Norway and Finland timber products still make up approximately 15 % of total manufactured output today. As with fisheries, the export trade is a very long-standing one, but with three important historical phases. Timber exports expanded significantly during the eighteenth century, primarily to Britain but also to Western Europe as a whole. The export trade diminished sharply in the early nineteenth century, partly as a result of blockade associated with the Napoleonic Wars, and subsequently as a result of trade barriers. Finally, after the removal of trade barriers in the early 1840s (both on timber directly, but also as a result of the repeal of the Navigation Acts in 1842) output and exports of timber and timber products, particularly to Britain, grew rapidly throughout the nineteenth century. During this period between 25 and 30 % of industrial workers in Norway worked in some part of the timber industry.

The technological evolution of the timber and processing industries was shaped by two main technical problems. The first was that of transport of logs from the forest hinterlands to the ports; in practice this was solved principally by the use of river systems and canals for floating logs to their destinations. In this respect, southern Norway had major competitive advantages, since Swedish and Finnish producers lacked access to ice-free ports and rivers, and consequently had a much shorter production cycle; the only land-based transport system, using horses and sleds, was inefficient and expensive, and viable only over short distances. The second problem was power resources: processing of timber involved sawing and planing, both of which were power intensive (the other main use for timber, which emerged during the nineteenth century, was pulp production for paper; this also involved power-intensive shredding and crushing). This second problem was also solved via use of Scandinavia's abundant water resources: sawmills were water-powered, and located on or near the rivers used for transport.

This technological system was challenged by the emergence of the Watt steam engine, and its various applications and modifications, which began quite extensive international diffusion from the early nineteenth century (Robinson 1974; Tann & Brechin 1978). On the one hand, there was the possibility of railway transport of logs, which changed the economic significance of large areas of forest. Norway's first railway was in effect developed as a joint venture, with

50% of the capital coming from Britain, and the remainder from the Norwegian state and a group of four large forest and sawmill owners in southern Norway (Sejersted 1979a). Perhaps more significant for the evolution of the industry as a whole, however, was the development of the steam-powered sawmill, which had the potential to liberate the industry from dependence on suitable water sites, and thereby to change – more or less radically – the competitive position within the industry.

Steam-powered saw technology became available to Scandinavian producers from around the 1840s. But it was at least two decades before it began to diffuse in any significant way in the timber industry, and during that period there was sustained resistance to its spread, taking the concrete form of official control and regulation of the application of steam-powered saws. During the 1840s, for example, official permits were required to use steam saws, and two major applications – in 1845 and 1847 – were refused. Government commissions and representatives of the Finance Ministry discussed the issues associated with steam saws repeatedly throughout the 1840s and 1850s, before a deregulating measure finally became effective in 1860.

Some specific details of the technological debate and conflict will be discussed below, but these should be seen in the general context of the organization of the timber industry at that time, and the debates concerning its regulation (Sejersted 1979a, b). From the eighteenth century, the timber industry was heavily regulated by the state, through a set of Mercantilist licences which directly controlled who could own and work with forest timber. In return for a payment to the state, a relatively small number of timber producers in effect had local monopolies; apart from these producers, individual farmers were permitted to cut timber from their own land. What emerged was an asymmetric industry structure, with most production being in the hands of a relatively small oligopoly, of big timber producers and sawmill operators who controlled the key waterfall sites for water-powered sawmills, with the state deciding who could participate and also – in effect – what technology they could use.

In this context, any decision to allow steam-powered saws was in effect a decision to allow new entrants to the industry. Naturally those who held the existing production privileges opposed any derestriction of the new technology, and so questions about using steam saws became part of a much wider debate concerning liberalization and the removal of Mercantilist privileges. Francis Sejersted has argued that in understanding the persistence of Mercantilist privileges the environmental aspect of the timber sawmill privileges should be emphasized: the privileges were regarded as important for the protection of the forests. Fear of deforestation is a persistent theme in the history of debate on the organization of the timber industry, analogous with the arguments used in the fishing industry concerning exhaustion of fish stocks. He argues that regulation on essentially ecological grounds remained persuasive as liberalism developed, which explains in part why the older privileges lasted so long. Secondly, only parts of the privileges remained intact as long as until 1860. Gradual dismantling had begun as early as 1795 when regulations governing production quantity were removed and export of logs was deregulated. Further liberalization came in 1818, when it was decided that farmers could freely saw from their own forest for exports.

The period 1818–30 was one of stagnation for the industry, which reduced pressures both for technological advance and for changes in the privilege system. But after 1830, demand and growth in the industry increased, and this coincided with important liberalizing measures, notably liberalization of internal trade and artisan laws. Sawmill privileges survived this period, probably because farmers' needs were mostly met by the 1818 law, and because the system of privileges was weakened by the way in which it was used. The state had simply granted many new privileges: 77% of applications were granted. The Interior Ministry used this possibility of granting new privileges in their ambivalent argumentation for keeping the system, in 1851, but also added:

> they want it expressed, that it would be in discord with the spirit of the Law or Fairness, were this possibility used as a means to make the system of existing saw privileges, that had developed over time, in reality ineffectual.
>
> (Sejersted 1979b, p. 95)

The reason for both continuation and opposition to the Mercantilist privileges was that new grants were usually adapted to the interests of existing owners; for example, existing sawmill owners were used as referees for new applicants, and the argument that existing interests should be protected was frequently used in rejecting applications.

However, opposition to the Mercantilist system increased with the growth of the industry from the 1840s, leading to repeal of the system in 1860, and hence to the free diffusion of the new technology. But these two decades were a period of explicit and sustained resistance to the new technology. In the 1840s a committee of the Industry Ministry had already argued that the privilege system was not in accordance 'with our understanding of the state's role'; why, then, did it take so long? Sejersted argued that there were two central elements in the process. First, the availability of the new technology posed a threat which led to increased resistance and more determined defence of the existing system. Secondly, the process of repealing the system uncovered conflicts within the liberal system, which had no practical knowledge of how to deal with this kind of thing – that is, there was no clear view about how the privileges were to be dismantled.

The long formal struggle began in early 1849, when the Government appointed a commission with the task of investigating whether there ought to be restrictions on how forest owners used the forest, whether the privileges ought to be repealed, and if so, how to regulate access to rivers (for log transport), and how to manage communal forests. However an important subtext of the debate was the role and regulation of the new steamsaw technology.

In the Government proposal to set up the commission the connection between privileges and forest conservation was finally rejected:

> To the extent these privileges in fact may have some temporary restrictive influence against damage to the forests, even so, this raises a question about whether the positive effect they might have in that respect, would be possible to achieve by other means. (Sejersted 1979b, p. 97)

The commission emphasized the virtues of free competition, with some reference to the benefits of free trade. Against this, however, they expressed concern that liberalization might open up the industry for unskilled producers with little capital (who could not survive recessions). Overall the commission 'recommends liberalization of the sawmill industry'. When it came to the technological dimensions of this, the commission saw a repeal of little practical importance, the view being that the established sawmill owners had such market power that they would remain the strongest members of the industry if the privileges went. The commission pointed out the unavailability of new waterfall sites, and noted that the opportunity to set up new sawmills after a repeal would be 'more an illusion than reality' (Sejersted 1979b, p. 98). Moreover, established interests were helped: a ten-year period should elapse before repeal was put into operation. Of course, the commission foresaw some competition between water-powered sawmills and the steamsaw technology; the latter had already arrived in the 1840s, and as noted above the Finance Ministry had decided in 1843 that a permit would be needed to use them.

The arguments for permitting steam saws, proposed by a member of the Interior Ministry, rested on benefits of free competition. But the steam saws were seen as costly to use, and although they might entail 'perhaps also loss of some Men' they were no real threat to privilege holders. Thus when the commission proposed abolishing privileges they argued *inter alia* that there was no 'dangerous rivalry from Steamsaws' (ibid., p. 99).

However, as Sejersted has argued, 'the privilege system stiffened when faced by the steamsaws' (ibid., p. 100). The major forest owner Løvenskiold argued that they had radically altered the conditions for existing rules and laws, principally because for steam saws no natural limitations existed as for water saws. Consequently, new laws were needed. At this stage, the debate was inconclusive, and no real change was made in the system, although the pressures remained both for conservatism and change. Two years later, in 1853, the Government established a second commission, including some of the same members, to reconsider the whole issue. This time, the commission took a different view, arguing that although the privilege system should be dismantled, implementation should be deferred until 1860; at the same time the Commission took a much more negative view of the role of steam saws, clearly as a result of intensified argument by those who benefited from the existing system and the water-powered technology.

Three commissioners took the view that the steam saw was in fact highly competitive with water power. Its advantages

> are so decisive that they will soon become established, and that they then will squeeze
> out and eradicate the Water saws.   (ibid., p. 102)

Many applications for steam saws had been sent in within the past two years. It was now perceived very much as a threat:

> the main reason why the three members of the Commission now are of a different
> opinion than they were as members of the Forest and Sawmill commission of 1849,
> is the Fear of the Steamsaw's influence ... which has formed ... in connection with the
> particularly strong Development, which the use of steampower in the later time has
> shown also here in our country.   (ibid., p. 102)

Sejersted has identified four major arguments which were advanced against the steam saw. They were:

> first, that the nation would sustain a direct loss of the substantial invested capital in
> water-powered sawmills;
> second, that opening the doors to steam-powered mills would mean entry into the
> industry by 'foreign capitalists', and hence a loss of the benefits which otherwise
> could 'flow to the State's own Sons';
> third, that the 2400 workers in watermills would face unemployment; and
> fourth, that the move would damage the wealth of existing privilege holders.

The Interior Ministry rejected these arguments on the grounds that 'some exaggeration is present [in the commission's report] in the Fear that Steam saws would squeeze out Watersaws', and that it would be wrong if the possibilities for steam saws were 'shut because of fear for its productive and profitable force compared to the production methods used hitherto' (ibid., p. 104).

A compromise was reached; some compensation was provided, and privileges were continued until 1860. In the longer run the established interests lost. Although the established owners had the capital and opportunity, market knowledge and so on, which might have been used to support their competitive position, they failed to do so. Their arguments were that repeal would kill the old industry. It did.

What happened between the two Commissions, of 1849 and 1853, to change the situation such that the Government felt confident in repealing the regulatory system which had for so long inhibited the use of steam power in the timber industry? Of course there were straightforward economic realities: the industry was growing, and there were increasing constraints on sawmill capacity which could only be resolved via the new technology. But there may have been wider cultural factors. I noted above the development of Norway's first railway, which was partly financed with timber industry money, and used to overcome the problems of the river-based log transport system. Francis Sejersted has argued that this railway had a social and cultural impact which went far beyond its transport functions:

> ... the railway provided the solution to what had for so long been the bottleneck in the Christiania timber trade, an efficient answer to an urgent need. More clearly than anything else, it also demonstrated to the man in the street the great potential of the new technology. It pointed the way ahead to a new and better society. It may have contributed more than anything else to the feeling of being at the threshold of a new era.   (Sejersted 1993, p. 80)

It seems reasonable to suppose that this may have played a major part in the changing policy line of the government at that time, in which case the spread of the steam technology was a result of social conditions much wider than those within the industry itself. The key point here, of course, is that both resistance to the steam saw, and the ultimate overcoming of that resistance, can only really be explained in terms of the specific constellations of social forces and interests within which both new and old technologies operated. These social forces and interests involved potential new entrepreneurs, old-established capitalists, an emerging modernizing bureaucracy, and so on; the pattern of resistance to and support for the new technology was socially very complex.

Resistance to nuclear power in Sweden

Nuclear power began early in Scandinavia; Norway, somewhat surprisingly, was the third country in the Western world to build a nuclear reactor, after Britain and the United States, and planning for a Swedish nuclear supply system began in the early post-war period.[7] The Swedish programme was much more ambitious and long-lasting than the Norwegian, however. Unlike other Scandinavian countries (Norway in particular, although Denmark also has hydrocarbon resources), Sweden has no major domestic power sources, and this led to a national programme for the construction of a power supply system based on nuclear technology; in terms of per capita expenditure, Sweden had the world's largest nuclear programme by the early 1970s (Jasper 1988, p. 363). The nuclear system did not come under serious question until the mid-1970s, when the

Centre Party (the largest of the non-socialist or social democratic parties, mainly representing rural interests) began to oppose the nuclear power programme, and succeeded in drawing increased support in the election of 1976. A period of increasingly intense debate and media interest began at that time, and the government attempted to assuage public concern through a policy which aimed at completing the existing reactor programme, and switching to other energy sources for any further energy needs (Sahr 1985).

This policy, which was probably not viable given the increasing degree of public concern, finally came to grief as a result of the Three Mile Island accident in March 1979. A key development was a changed policy on the part of the Social Democratic Party, in which the party agreed to support a referendum on the future of nuclear power in early 1980. The precise meaning of the referendum result is difficult to interpret, given the wording on the ballot (there were three questions, two giving a variety of approval, and one expressing rejection). However the all-party policy consensus which emerged as a result of the referendum was that use of nuclear energy sources should be abolished by the year 2010; this will involve a major programme of decommissioning, plus alternative infrastructure construction, and there appears to be some scepticism on the part of the Swedish population as to whether it will be achieved.

The background to the Swedish nuclear programme was a decision to adopt a 'Swedish line' policy on industrial development in the mid-1950s; this policy essentially involved massive public and private investment in new technologies, especially related to the development of the engineering sector of the Swedish economy. The strategy immediately raised major questions concerning energy supply: a key assumption at the time was that oil was not necessarily a reliable energy source, and that Sweden should – as a component of its industrial policy – aim for energy self-sufficiency.

However, the shift to a nuclear energy policy was an element in a perhaps more fundamental shift in policy thinking; from an economy and society based on heavy industry, to one based more on science-based technologies and knowledge-intensive industries. Nuclear power was seen not only in terms of energy supply, but also in terms of its backward linkages to the heavy engineering sector of the Swedish economy. The view was that the programme should draw on internationally recognized techniques, but should use Swedish equipment suppliers, and thus initiate a process of learning and competence development which would enhance the long-run capabilities of Swedish heavy industry. Two firms in particular, ABAtomenergi and ASEA,[8] had access to Government. Thus the engineering firm ASEA became the main supplier to both the Marviken and Ågesten plants, and later delivered nine of twelve light water reactors to the Swedish nuclear energy industry.

The main debates on nuclear policy in the 1950s and 1960s, in the first phase of construction, concerned not the programme itself, but the specific technological choices which had to be made within it. These particularly concerned reactor types, and the choice between domestically developed heavy water reactors or US-developed light water reactors. The specific choices are not so important here, but basically it was thought in the 1960s that developing countries were potential customers of heavy water reactors, and so the main development project at Marviken used this technique, although ASEA kept open both heavy and light water options. More important is the fact that nuclear decision making at this time was regarded as a matter of technological expertise, in which decisions were taken in an interplay between politicians and experts. So experts become indirect decision makers, but without the responsibility for decisions made; the nuclear policy was in effect de-politicized, and certainly not open to public participation (Schagerholm 1983, p. 3).

The rise of public debate on nuclear energy in Sweden took the discussions well beyond the technical issues which had been the province of nuclear engineering experts and their political masters. But according to Schagerholm the origins of the public debate lie in something wider even than the nuclear policy. From the late 1960s there emerged increasing scepticism and concern about the whole industrial growth policy of the government; concerns about environmental and resource issues, about the 'limits to growth', and about welfare issues all coalesced into concerns about the general direction of Swedish society.

During the 1970s these concerns focused much more closely on environmental issues, as a specifically 'green' movement emerged. With the exception of the rurally based Centre Party this was essentially an extra-parliamentary movement; the parliamentary majority at the beginning of the 1970s still saw nuclear energy as a clean system, at least in comparison to the alternatives. Of course the development of the environmental movement brought to the fore a concern with the general ecological impact of nuclear and other industrial technologies. But it also brought the question of risk onto the agenda: there was a much sharper focus on the uncertainties of nuclear power and the risks of accident. Until that time the probabilities and consequences of nuclear accidents had been enclosed within the 'circle of expertise' in which nuclear decisions had been taken, and it seems likely that this was not regarded as a key issue. Not the least effect of the environmental movement and the politicizing of nuclear power was a dramatic enhancement of the perceptions of risks and possibilities of failure, and a publicization of risk assessment.

One element in this changed perception of risk, in Sweden, was the failure of the Marviken project, which was halted as a technical failure in 1970. On the one hand, there was considerable criticism of the Social Democratic government on technical grounds, in that they ought to have changed from heavy to light water technique earlier. But more significant was that the overall technocratic confidence in nuclear power was shaken, and the idea that this was some kind of complex but neutral technology never recovered. This in a sense 're-technicized' the debate, because much of the issue of risk required technical assessments based on engineering expertise; but by now the issue was out in the open, and the legitimacy of nuclear engineers was by no means automatically accepted. The question of who could assess risks, what methods they should use, and what were acceptable techniques of assessment were now part of a wider debate on this complex technology.

The risk debate centred on two main issues (Brante 1984, Ch. 6). These were, first, the risks of major reactor accidents and the possibilities for core meltdown, and secondly, the long-term problems of managing nuclear wastes. The broad positions which were taken in the debate are well known: pro-nuclear advocates argued that the probability of major reactor accident was extremely low, and that consequences were manageable (this extended into quite technical analyses of prevailing winds in Sweden, and the distribution of radiation fall-out), while opponents argued that even if risks were small the negative pay-offs, as it were, were extremely high. On waste management, opponents argued that there was simply no known technology for acceptably managing wastes over the long term (which, given the half-life of fission products, is very long indeed).

By the mid-1970s the nuclear issue was not only a major debate in Sweden, it was politicized to a degree that threatened the viability of the political parties; that is, they were internally split on the issue, and there was no question of any party taking a clear line on the issue without threatening its own stability; at the same time, the parties promoted debate, and actively encouraged studies and discussion. The attempt by the Conservative Party in the early 1970s to resolve the issue by reference back to experts definitively failed, and it was ultimately this which led to the referendum idea as a means of resolving the national conflict.

In international perspective, the Swedish case was unique in three ways: in the speedy politicizing of nuclear power, in the responsiveness of the political system to public concern, and perhaps in the clarity of the public decision that was taken (the latter conclusion needs to be qualified due to the obscurities of the referendum questions). A feature which differentiated Sweden was that it was often the political parties which had the initiative and which formulated the political alternatives; they did not seek to avoid the public concern. We can distinguish three broad phases in the evolution of nuclear power in Sweden: first, a debate enclosed with the circle of politicians and experts, in the 1950s and 1960s; secondly a debate concerned with broad issues of morality and ecology, in the late 1960s to the mid-1970s; and, finally, a 'retechnicized' debate on specific technological features of nuclear power from the mid-1970s through to the national referendum in 1980. But regardless of how one assesses the respective arguments in the nuclear debate in Sweden, this is an almost unique case of a major industrial society making a major technology selection decision via public debate.

Conclusion

One of the more prevalent myths about Scandinavia is that it is a region characterized by high levels of consensus, in which social conflicts are more or less rapidly resolved through compromise. While most Scandinavian citizens would be reluctant to accept this view of the Nordic area, many of them probably would take the view that their societies are untroubled by some of the conflicts – in particular labour union conflicts – which are associated in many countries with the introduction and operation of new technologies. What I have tried to suggest in this chapter is that this is not necessarily the case. Even a cursory examination of some of the biggest and most prevalent technologies in the Nordic region uncovers persistent and sometimes very long-lasting conflicts, with many different social actors involved. This should not be surprising, for two reasons. First, when technologies enter a system from outside, their potential social impacts are a very sensitive issue. Secondly, the recognition of technology as a social phenomenon should lead us to see that the sources of potential opposition and resistance to innovation are as complex and multifarious as society itself; moreover the motives for such opposition, both implicit and explicit, can be highly varied also. In the cases I have cited, resistance came from religious sources, from peasants, from regional administrators, from politicians, from the general public, and even from large capitalists; motives ranged from direct economic interests, to concerns for the social structure, to egalitarian values, to a concern for the environment. The large-scale cases I have suggested could, in my view, be extended. For example, despite the fact that Scandinavia as a whole is a region apparently well suited to the use of private cars (given the extreme difficulties of running an effective collective transport infrastructure over a very large area with a relatively small population) there has been a persistent opposition to this technology and to the infrastructure which it requires. This resistance has a popular aspect, but is also expressed through a wide range of government measures: for a significant part of the post-war period (until 1960, in fact) owning a car in Norway actually required a government permit, and even today the diffusion and use of cars in Scandinavia generally is controlled via a stringent tax policy (Østby 1991b, p. 4). This has its basis partly in environmental concerns, but also in widespread social resistance to the social implications of car ownership.

For example, an official Norwegian government report on *The car, environment and society* in 1975 recognized the various advantages offered by cars, but went on to remark that

> However the many negative aspects of personal car use are significantly more clear, such as direct environmental damage; but there are many indirect undesired developments in social life as a result of greatly increased personal car ownership.
>
> (Østby 1991a, p. 21)

I have emphasized the social complexity of these types of historical patterns of resistance to new technologies. A final point which I should like to make is that we should not necessarily regard such resistance in a bad light; we might even want to reconsider the perjorative aspects of the term 'resistance', and question whether it is the right term for discussing such phenomena (see Bauer, Chapter 1). If the only standard for judging a technology was productivity, then the resistance which I have described to new fishing or timber techniques, or even to nuclear power, might attract condemnation. But societies have social as well as economic aims, and economic perspectives are just one element of social and political values. This makes judgements on resistance to certain technologies a much more nuanced matter; if a particular culture wishes to retain low-productivity rural fishing communities, or if it wishes to forego the risks associated with nuclear power plants, then economic criteria are no more relevant than any other standard of judgement. This kind of 'resistance' could thus be seen as a positive part of a social selection process, not an obstacle to the inevitable march of technological progress.

Notes

1 I would particularly like to thank Francis Sejersted, Keith Smith, Stein Tveite, Per Østby and an anonymous referee for comments and assistance in the writing of this chapter.
2 See for example, my own arguments and the various papers in Bruland (1991).
3 See Richard Nelson (1987) for a succinct overview of the evolutionary approach.
4 In fact the hand-held line goes back to the Stone Age when it consisted of a spun thread – of plant fibres, leather or tendons – and bone hooks and stone sinkers (the hook and sinker changed to iron and the later sinker to tin or lead). Technical change in this apparently simple technology has continued up to the present day, primarily through the use of new materials.
5 *Norsk Fiskeritidende*, 1891, cited in Posti (1991), p. 25.
6 This complex event is described in Posti (1991).
7 Apart from the works cited below, this section draws on A. Schagerholm (1983).
8 Now ASEA-Brown-Boveri, one of the largest engineering companies in the world.

References

BRANTE, T. (1984). *Vetenskapens Sociala Grunder – en studie av konflikter i forskarvälden.* Chapter 6 ('Kärnkraftsdiskursen'). Stockholm: Raben and Sjögren.

BRULAND, K. (ed.) (1991). *Technology Transfer and Scandinavian Industrialisation*. New York: Berg Press.

EKDAHL, L. (1983). *Arbete Mot Kapital. Typografer och ny teknik – studier av Stockholms tryckeriindustri under det industriella genombrottet*. Lund: ARKIV Avhandlingsserie.

HODNE, F. (1975). *An Economic History of Norway, 1815–1970*. Trondheim: Tapir.

HOVLAND, E., NORDVIK, H. & TVEITE, S. (1982). Proto-industrialization in Norway, 1750–1850: fact or fiction? *Scandinavian Economic History Review* **30**, 45–56.

JASPER, J. (1988). The political life cycle of technological controversies. *Social Forces* **67**, 363.

LINDQVIST, S. (1984). *Technology on Trial. The introduction of steam power technology into Sweden, 1715–1736*. Uppsala Studies in History of Science No 1. Stockholm: Almquist and Wicksell.

NELSON, R. (1987). *Understanding Technological Change as an Evolutionary Process*. Amsterdam: Elsevier.

ØSTBY, P. (1991a). *Mobilitet eller Miljø*. Konflikter knyttet til bil og bilisme 1950–1990. Working Paper 3/91. Trondheim: Centre for Technology and Society.

ØSTBY, P. (1991b). *A Steel Phoenix? The social construction of a modern car*. Working Paper 4/91. Trondheim: Centre for Technology and Society.

POSTI, P. (1991). *Trollfjord Slaget. Myter og virkelighet* [The Trollfjord Battle. Myths and Reality]. Tromsø: Cassiopeia Forlag.

ROBINSON, E. (1974). The early diffusion of steam power. *Journal of Economic History* **24**, 91–107.

SAHR, R. (1985). *The Politics of Energy Policy Change in Sweden*. Minneapolis: University of Minnesota Press.

SCHAGERHOLM, A. (1983). *För Het Att Hantera. Kärnkraftfrågan i Svensk Politik 1945–1980*. Uppsatser Från Historiska Institutionen i Göteborg No 3.

SEJERSTED, F. (1979a). *Fra Linderud til Eidsvold Værk*, Vol. III, pp. 156–179. Oslo: Dreyers Forlag.

SEJERSTED, F. (1979b). Da sagskuren bli frigitt. In *Makt og Motiv: et festskrift til Jens Arup Seip*, ed. O. Dahl *et al.* pp. 94–109. Oslo: Gyldendal Norsk Forlag.

SEJERSTED, F. (1993). *Demokratisk Kapitalisme*. Oslo: Universitetsforlaget.

SOLHAUG, T. (1983). *De Norske Fiskeriers Historie 1815–1980*. Bergen/Oslo: Universitetsforlaget.

TANN, J. & BRECHIN, M. (1978). The international diffusion of the Watt engine, 1775–1825. *Economic History Review* **31**, 541–64.

7

Henry Ford's relationship to 'Fordism': ambiguity as a modality of technological resistance

JOHN STAUDENMAIER

Historiographical observations about resistance to technology

Emphasizing resistance to new technologies, as this conference at the Science Museum has, highlights technological creativity, the process of defining a goal and then trying to achieve it. For some centuries now such projects have fascinated, even mesmerized, Western storytellers and social theorists alike. And it is a great story: that a person, or team, or institution would cast an imaginative eye out onto the broad field of the existing order and conceive a plan to insert something new into that field, that such an agent would have the intelligence and power to gather resources and shape them according to the imagined plan so that, one day, the agent could take a deep breath and exult, 'It works!'

It is a powerful idea to be sure, and legitimately so. Technological creativity has been central to most of the creative work in the history of technology for decades. It is over-simple but still helpful to distinguish historians of technology as tending toward one or the other of two paradigmatic descriptions of technology.[1] Traditionally, technology has been understood in terms of rational achievement. Recently, however, other scholars have begun to describe technology in terms of conflict, as an arena wherein one group wins by succeeding in designing a technology representing its values and vested interests, while other groups lose that same struggle. Those who see technology as primarily rational see the unknown and the uncertain as the main form of resistance. Those who see technology as conflictual emphasize competing world views and vested interests instead.

Uncertainty in design and vested interests

Several books exemplify uncertainty as resistance to technological achievement. Walter G. Vincenti's *What Engineers Know and How They Know It* uses case studies from aeronautical engineering to explore workaday engineering thinking (i.e.

where political contests are muted by the dynamics of routine problem solving within already-defined project parameters). Vincenti's engineers frequently encounter uncertainty as a challenge that requires painstaking data gathering and testing before it can be rationalized into theories. Speaking of the quarter-century evolution of flying quality theory he observes:

> ...the learning process over the intervening years provides an example of how an engineering community translates an **ill-defined** problem, containing in this case a large subjective element, into an **objective, well-defined** problem for the designer.
>
> (Vincenti 1990, p. 51)

Similarly, David Hounshell's study of mass production tracks a century of efforts to overcome areas of ignorance or uncertainty by gradually improving processes of manufacturing objects assembled from pre-formed metal parts. Ordinarily, historians taking this approach recognize the retarding action of those whose vested interests are threatened by new technologies but they tend not to emphasize conflicts so much as the creative struggle to achieve the new technology (Hounshell 1984).

Competing world views

Other historians of technology see technological resistance in terms of competing world views as in David Noble's studies of an early twentieth century alliance between large corporations, government, and universities against small business and skilled workers or his later study of conflict over machine tool design in the post World War II era. Likewise, Bryan Pfaffenberger studies the tension between British colonial ideology and Sri Lankan religion as each influenced the design of irrigation systems. Occasionally, rational and conflictual approaches are combined, as in Merritt Roe Smith's study of the Harper's Ferry Armory. Smith treats two themes as central: the first, conflict between the skilled-craft ideal of the Harpers Ferry community and the US Ordnance Department's ideal of uniformity; the second, the technical challenge and eventual achievement of interchangeable parts technology (Smith 1977; Noble 1979, 1984; Pfaffenberger 1990).

Sociologists who use the Social Construction of Technology approach treat conflict not so much in terms of the content of the competing ideologies, but as a universal process wherein some group of potential technological practitioners is confronted by a recalcitrant, even hostile, context which must be overcome and reorganized by the protagonists of the new technology (e.g. Bijker, Hughes & Pinch 1987).

The symbolic and ambiguity

In the main, these several approaches to technological resistance privilege rational over non-rational consciousness. Most students of technological change pay much more attention to purposefulness, that is, to defining goals and trying to reach them, than they do to affective responses which do not operate according to pre-defined strategies. When technological behaviour is portrayed as exclusively purposeful, however, it can appear to operate in an abstract world, disconnected from the awe, play, fear, and delight, that human beings repeatedly report as a powerful dimension of their conscious experience of specific technologies.[2]

When technological cognition is construed exclusively as intentional behaviour, scholars concentrate on technological actors – the inventors, engineers, entrepreneurs or business leaders who design, market or manage technologies – and workers, who either adapt or resist the technology in question. The attention is certainly warranted. Technological actors who design and execute strategies make the modern world go round and the history of technology makes a major social contribution by continually rethinking the complexities that attend their behaviour. Nevertheless, important as these approaches obviously are, they need to be complemented by a disciplined awareness of other conscious styles, both to understand the blend of rational and non-rational in technological actors themselves, and to make sense of the visceral, even numinous, power that characterizes the symbolic presence of very successful technologies in twentieth century society.

Despite the obvious symbolic influence a successful technology exerts in its host culture, that symbolic character is difficult to pin down. How much, for example, does symbolic meaning accrue, intentionally or not, from technical design and how much is crafted by the marketing strategies of those who design and preside over its operation? How again are such meanings revised by those who benefit from, or suffer because of, the technology? These questions have begun to evoke interest in the field.[3] They are still so new, however, that no coherent set of themes or interpretative vocabulary exists. The present chapter is understood as part of this emerging area of interpretation. I will discuss some of the many stories that have clustered about Henry Ford to explore technological resistance, not as explicit strategy but as emotional response. First I will consider the multi-faceted symbol set – amorphously composed of 'Mr Ford', 'The T', 'Fordism' – which showed such remarkable staying power over several turbulent decades of the man's life. I will call attention to the durability of the symbol set especially in light of its internal ambiguities. Mr Ford's personal behaviour, particularly during the 1920s, reveals his own troubled ambiguity about the very symbolic forces over which he presided.

By 'technological ambiguity', I mean a seemingly contradictory but remarkably stable mix of emotions: exuberant delight and purposeful energy entangled with discomfort and solipsistic retreat. I see this bonding of exultation with anxiety, personal power with personal impotence, as a kind of affective escape mechanism. It operates with a back and forth rocking movement, from exultation to intimidation, that inculcates in ordinary people a feeling of power and betterment even as it warns them not to get in the way of Progress. This kind of technological ambiguity, so common after the turn of the twentieth century, may help explain the remarkable passivity of Americans when it comes to debate over technological policy.[4] This study, then, is less about resistance to technology than about its obverse, namely the staying power that successful technologies of the twentieth century so often demonstrate in the face of the various resistant forces at work in its societal context.

'Henry Ford' and 'Fordism'

It could be argued that the immensely powerful twin symbols, of 'Henry Ford', the man, and of 'Fordism', the movement, were born in the revolutionary technological events of January 1914: the completion of the moving assembly line for making Model Ts at the Highland Park factory and the announcement of the 'Five Dollar Day'. Before 1914, the public salience of the Ford name had grown only gradually.[5] Ford was known for car racing, for a rather ordinary mix of business success and turmoil, and for standing up to the Selden Patent claims when other manufacturers knuckled under. More important than any of these, however, the symbolic events of 1914 depend on growing public awareness of what would be called 'Fordism', a term that came to include the Model T as a car for ordinary people and the massive 'Crystal Palace' factory at Highland Park as the cutting edge of world class engineering. Highland Park opened for operation in 1910 and became the site where Ford engineers, in one of the most remarkable sustained bursts of technical creativity in history, designed one subsystem after another to fit into what would in 1914 become the moving assembly line.

The enormous financial resources required for the Highland Park expansion depended, of course, on the amazing popularity of the Model T almost from the day of its announcement in March 1908. Few if any technological artefacts have evoked such affection as the 'T', arguably the best match between technical design and market conditions in US history. Early automobiles were notoriously unreliable, required considerable wealth to own and operate, and had to negotiate appalling roads. Nonetheless, evidence suggests that during the early automobile years, American response to the horseless carriage transcended wonder at a new form of hi-tech and took on the idealized character of a love affair. For years after the Model T appeared in 1908 the social vacuum created by this mix of inadequate technology and increasing desire sucked up every vehicle Ford could push out of the door.[6] Durability of the vanadium steel construction and a high wheelbase made the T much more reliable on bad roads than competing autos; the car was designed so simply that many owners, especially mechanically inclined farmers who purchased the 'T' by the hundreds of thousands, could repair it themselves; its bare bones style fostered a host of small companies which provided accessories. Ease of maintenance and the habit of customizing through add-on accessories had much to do with the affection which surrounded Ford ownership. Songs and poems celebrating 'Tin Lizzie' and 'The Flivver', which appeared as early as 1909, marked the first mature stage of American car culture. Thus, from one of the early favourites of 1914,

> Oh the little old ford just rambled right along
> Oh the little old ford just rambled right along
> The gas burned down in the big machine
> but the darn little Ford don't need gasoline ...
>
> smash up the body and rip out a gear
> smash up the front and smash up the rear
> smash up the fender and rip off the tires
> smash up the lights and cut out the wires
> throw in the clutch and don't forget the juice
> and the little old ford will go to beat the deuce ...
>
> patch it up with a piece of string
> spearmint gum or any old thing
> when the power gets sick just hit it with a brick
> and the little ford will ramble right along.[7]

In January 1914 the nearly complete line was put into operation at Highland Park and, vastly more important from a media perspective, the company announced its labour reform programme, popularly known as 'The Five Dollar Day'. Ford doubled the daily wage to five dollars while cutting the work day from nine to eight hours and hiring a third shift of workers. The triple announcement astonished the world, infuriated his competitors, and won Ford an almost indestructible reputation as the working man's friend. His homey explanations to the press (e.g. 'Profits should be shared between capital and labor, and labor ought to get most of the profits, because labor does most of the work that creates the wealth') further enhanced his stature with ordinary people.[8]

Americans in 1914 had myriad reasons to rejoice at this. As far back as the infamous Molly Maguire trial of 1876, and 1877's rash of violent strikes, through the pitched labour battles of the 1890s at Pullman, Homestead and elsewhere, mainstream and elite Americans alike worried about the nation's increasingly violent urban context. Labour strife and mean streets were exacerbated by the flood of peasant immigrants from Eastern and Southern Europe, Jews and Catholics in the main, which numbered over one million per year by 1910. A host of proposals appeared in academic publications as well as the popular and business press prescribing social order through the application of professional expertise. Engineering heroes such as Edison, Steinmetz, F. W. Taylor, and Ford became technocratic role models for a troubled nation.[9]

Ford's reforms took bold action that promised to provide new employment, solve the problem of factory strife, and even Americanize unwashed immigrants through Ford's inspection of workers' personal lives and the English Language School. At graduation exercises for the Language School graduates came on stage dressed in Old World clothing and jumped into a large melting pot. While they changed clothes off-stage, their instructors mounted a catwalk behind the pot and stirred it with large wooden spoons. Then the graduates emerged from the pot wearing new suits, straw hats, and waving tiny American flags.[10]

Thus, the combination of a phenomenally popular product and irrefutable technological prowess were fused at white heat by the astonishing labour announcements of January 1914. These elements stabilized in a complex symbolic nexus composed of multiple overlapping and mutually reinforcing images. Most enduring and most central were the deeply ambiguous dyad: 'Mr Ford' and 'Fordism'. 'Mr Ford' came to be seen as the homey friend of the working man, shrewd about what a plain man needs in order to make it in the modern world. Thousands of letters from ordinary people to Mr Ford remain at the Ford Archives as witness to Mr Ford's accessibility for working and lower middle class Americans. In its most technical sense, 'Fordism' delineated the Ford way of doing business, namely: mass production via the moving assembly line; producing a high volume of products sold at low profit margins; distribution through a network of franchised dealers. As a world-wide symbol, however, 'Fordism' ('Fordismus' on the European continent) came to represent a new technocratic order made possible by engineering and science, the very cutting edge of technological progress. By the 1920s, as we shall observe below, its most visible shrine had shifted from the Model T and the Highland Park factory to 'The Rouge', Ford's incomprehensibly enormous world of production.

The durability of this nest of symbols shows up most vividly when we consider its staying power against the background of a number of well-publicized events which might have seriously eroded Ford's popularity. Consider a few of the most notable. Late in 1915 Ford chartered the ocean liner, *Oscar II*, and announced his intention to sail to Europe and settle the Great War. 'We are going over there to see if we can do any good. We are going to see if we can't get the men out of the trenches on Christmas Day' [*New York Times*]. The Peace Ship was portrayed as a ridiculous fiasco almost from the start and Ford as a naïve fool (e.g. 'The world has no objections to private citizens making fools of themselves in an impossible effort to establish an inopportune peace' [*Memphis Scimitar*]).

In 1919 Ford sued the *Chicago Tribune* for libel. The press portrayed Ford on the witness stand as a bumpkin with embarrassingly childish notions about the larger world. Although he won the case, much was made of the size of his award, $0.06 in damages. Editorializing after the trial, *The Nation* saw Ford as a diminished public figure:

> Now the mystery is finally dispelled. Henry Ford is a Yankee mechanic, pure and simple; quite uneducated, with a mind unable to 'bite' into any proposition outside his automobile and tractor business, but with naturally good instincts and some sagacity ... He has achieved wealth but not greatness; he cannot rise above the defects of education, at least as to public matters. So the unveiling of Mr. Ford has much of the pitiful about it, if not of the tragic. We would rather have had the curtain drawn, the popular ideal unshattered.[11]

In the mid-1920s sales of the Model T began to plummet. Ford refused to change until he faced economic disaster in 1926 and shut the entire plant down for six months in 1927. The extraordinary difficulty, and well-publicized expense, of Ford's transition from the Model T to the Model A revealed a company's troubles and its owner's increasing rigidity. Five years later, during the depths of the Great Depression, some 3000 jobless marched on the River Rouge plant where they were met with police gunfire. Accounts vary but the 'hunger march' yielded around four killed and 50 wounded. Ford received even more lurid coverage in 1937 when company thugs severely beat UAW leaders Reuther and Frankenstein on the overpass leading into the River Rouge plant; the event was covered in great detail by photographers and news reporters generally.[12]

These highly publicized negative incidents appear to have done relatively little damage to 'Ford'/'Fordism' which retained a powerful hold on public consciousness. A few of the more notable indications of the remarkably durable symbolic power follow. Very much like Ross Perot in 1992, Henry Ford was the most popular candidate for president (unannounced) in 1923. His presidential popularity suggests that the Peace Ship fiasco and libel trial had relatively small influence on his popular appeal. A decade later, Ford stayed out of the Chicago 'Century of Progress' Exposition when it opened in 1933 but rethought the matter and mounted an exhibit in 1934. The Ford Rotunda portrayed assembly line manufacturing through a host of production-specific exhibits and George Ebling's wall-sized photo murals of the Rouge factory. The Rotunda stole the show during the 1934 season. It was moved, after the Exposition, to a site near the Rouge and became Ford's main tourist centre. In 1940 there were 951,558 visitors to the Rotunda and another 166,519 toured the Rouge. Clearly, 'Fordism' retained its star quality through the years of the Depression.[13]

Much of the mail that poured into Dearborn addressed to 'Mr. Ford' was astonishingly intimate. Letters, sometimes written in pencil on dime store tables, told touching stories of hard luck, gave homey advice, asked favours, expressed support. The authors, simple people from across the land, wrote to a man who seemed believably like themselves, an accessible friend. Most striking of all, perhaps, are the several hundred letters that specifically refer to the battle of the overpass. The vast majority favour Mr Ford and encourage him to keep up the good fight. Space prohibits extensive citations but a single letter from a woman in Ohio captures the frequent combination of affirmation for Ford's stance toward labour and an intimate request for advice:

> I may have misunderstood the radio the other day, as there was so much static, but I thought I heard someone say that you were paying $6.00 a day for an 8 hr day, but the labor unions said they were going to make you pay $8.00 for a six hr day. I couldn't resist the impulse to tell you that I hope you don't let the unions run your business.

And a few paragraphs later:

> Would you be too busy, Mr. Ford, to take time to suggest what we could raise to make a living, that is something I could do, if my husband should have to give up his shop work again? I'm going to get some strawberry plants – for one thing. Would soy beans be practical in a small way? I had thought of a patch of ginseng, but don't know where I could get a start of the roots. I am starting some little chickens here in town. I never raised any before and have lost three ...[14]

Finally, from 7 to 10 April 1947 a massive outpouring of grief and respect marked Ford's lying in state in the lobby of the Lovitt Building at the Henry Ford Museum and Greenfield Village complex. The thousands who stood waiting to pass by his bier offer still another indication of the long-lived power of his hold on the American imagination.

River Rouge: a new world order in the making

The sketches just presented of events in the symbolic life of Ford and Fordism do not do justice to the complexity of public feeling for a major technology and technological hero. They only hint at the myriad stories, many of them small articles used by publishers everywhere as tried and true filler material, playing on what quickly became the basic Ford themes. The stories were held together by the dynamic tension between 'Fordism' and 'Mr Ford'. 'Mr Ford' expresses, as we have seen, an accessible common man, trusted with family tragedies, hopes and secrets by so many of the nation's little people. In sharp contrast, 'Fordism' meant sophisticated technology on a scale never before imagined. The Highland Park plant and later the massive River Rouge complex belched plumes of smoke and streams of motorized vehicles from the largest manufacturing operation on earth. More than anything else, more even than affordable cars for a mass market,

'Fordism' came to mean overwhelming technological power creating a new world order as you watched.

And watch they did! Somewhere between 100,000 and 500,000 visitors toured the Rouge each year in the late 1920s making it perhaps the world's largest working exhibit of technical progress.[15] Some of the tourists recorded their impressions in vivid language indicating, sometimes in the same sentence, the ambivalence we have been observing above. German engineer Otto Moog recorded his impressions in a language combining intimidation with an exultant sense of liberation in almost schizoid fashion.

> No symphony, no *Eroica*, compares in depth, content, and power to the music that **threatened and hammered away** at us as we wandered through Ford's workplaces, **wanderers overwhelmed** by a **daring expression of the human spirit.**[16]

Mexican painter Diego Rivera echoes the same feelings. Ruminating on the power of the Rouge, which he was soon to immortalize in his Detroit Institute of Art murals, he juxtaposes worker achievements with a near preternatural power that overwhelms human agency.

> I thought of the millions of different men by whose combined labor and thought automobiles were produced, from the miners who dug the iron ore out of the earth to the railroad men and teamsters who brought the finished machines to the consumer, so that **man**, space, and time might be conquered, and **ever-expanding victories be won against death** ...[17]

Rivera's rhetoric ('so that man, space and time might be conquered') salts his triumphant tone with the ominous hint that humans serve as victims rather than masters of progress.

Ford's escapism: Greenfield village and the museum

About two miles from the bustling Rouge, around the corner from Ford's secluded Fairlane estate, Mr Ford began accumulating technological relics into what became the largest private collection on earth. In October 1929 Ford opened the only partially complete Edison Institute and Greenfield Village with a ceremony of extravagant homage to his revered octogenarian friend, Thomas Edison. 'Light's Golden Anniversary' marked the Edison team's successful testing of an incandescent light bulb 50 years before to the day. Ford invited hundreds of the wealthiest scions of the industrial era to a banquet in the foyer of the partially completed museum building. The gala began next door in Greenfield Village's reconstructed Menlo Park complex. With President Hoover in attendance, Edison switched on a replica of the first lamp while the nation listened over a live NBC radio hookup.[18]

Despite the massive publicity of its inaugural event, the Museum and Village did not open to the public for four more years. The Museum and Village were conceived of primarily as an educational endeavour: students in the small trade school were to learn among the machines and historic buildings. Other visitors were admitted only on a random basis until early 1933 when Ford was finally convinced by thousands of requests to build a gatehouse and begin a paid admission policy for the general public. The haphazard process by which the Museum and Greenfield Village became a place of public technological display contrasts strikingly with the Rouge visitor programme which, as we have noted above, routinely handled more than 100,000 tourists annually for most of the 1920s.[19]

Ford's ambivalence about the public identity of the Museum and Village, to which we will return below, reflects an uncertainty about his public identity as the world's primary symbol of technological progress. Another disturbing symptom was his growing obsession with control over every aspect of his domain. Thus, the 1914 labour reform required that workers qualify for it by passing a home visit inspection. Members of the newly created Sociological Department checked for cleanliness, debt, drinking habits, and a list of other equally intimate matters. A bad grade meant that the profit sharing bonus was put in escrow until the worker mended his ways. Failure to comply eventually meant firing. The design of the Rouge itself – its access to Great Lakes shipping lanes and intersecting rail lines, its dramatically expanded capacities for primary input manufacture – sought to integrate vertically a wholly-owned, mine-mouth-to-dealership production system.[20]

Ford's dream of total control showed itself in other areas during the years that the Rouge was being completed. After major stockholders had sued (and won) because of Ford's practice of ploughing profits back into company expansion, Ford designed an elaborate and deceptive strategy for buying them out and completed the transactions in July of 1919. Almost simultaneously, three of 'his ablest lieutenants' (C. Harold Wills, John R. Lee, Norval Hawkins) resigned under pressure.[21]

Thomas Hughes has suggested that Ford's increasingly autocratic managerial style reflects his inability to make the transition from the intimate scale of operations at the 1906 Piquette Street plant to the colossus on the Rouge River. The work-force at Piquette Street had been so small, and highly innovative operations so informal, that Ford could relate with everyone in folksy comfort.

> When he and his team were creating the Model T and the Ford system of production, there were no lines of authority, routine procedures, or experts. Theirs was a resourceful, ingenious, hunt-and-try probing into the unknown future... [Ford] continued to advocate a leadership style suited for times of invention and great

change long after the Ford company had become an extremely large and a relatively stable managerial and technical system with high inertia. Ford would not, or could not, make the transition in leadership style from the inventive stage to the managerial.

(Hughes 1989, pp. 214, 216)

The employees assigned to work at the Museum and Village project look a great deal more like Piquette Street than the Rouge or Highland Park. Tales abound of familiar, though hardly peer, interactions between Ford and his boys.

> There was one fellow named Long Brown. He was good with the broadaxe. Mr. Ford came up to Long one day and said, 'Long, you must have learned that in the north woods, didn't you?' Long said, 'No sir. I learned that in Greenfield Village.'[22]

Edward Cutler, who oversaw the construction of the Village, recalled years later his intuition that Mr Ford sought out the Village as a retreat from the Corporation.

> I believe the pressures of Mr. Ford's work were relieved by the work in Greenfield Village and my office. The office and his work in the Village were safety valves for the pressure and strain of the Ford Motor Company. He spent so much time around the Village. Several times he would make a crack, 'Well, I guess I'll have to leave you now, and go and make some more money for us to spend down here.'
> It was a relief for him to get down here. For years he wouldn't let me have a telephone. When I would ask him about it, and I had a lot of running around to do, he would say, 'Oh, forget that stuff. I come down here to get away from that gang.' He didn't want any way for them to get a hold of him.[23]

Given Ford's penchant for control and his solipsistic tendencies, it is not surprising that when he turned from private hideaway to public technological display, his aesthetic should take the form of sanitized machinery portrayed in abstraction from its working context. The Museum and Village embody this motif by combining a powerful strain of romantic nostalgia with an equally powerful commitment to contemporary technocratic motifs. Greenfield Village ignored twentieth century technologies almost completely and, indeed, made an expensive hash of any historical chronology. Ford bought what he liked and installed it. Shrines to American heroes dotted the landscape: Abraham Lincoln's early court house, homes of Noah Webster, Ford himself, the Wright brothers, and the jewel of the collection, a worshipful reincarnation of the laboratory complex at Menlo Park where Thomas Edison had invented the electric light half a century before.

The adjacent museum aimed more at aesthetically pleasing arrays of artefacts (most notably: steam engines, automobiles, locomotives, agricultural equipment, machine tools and domestic appliances), each series arranged in chronologically ascending order to demonstrate the march of inventive progress. Thus the historically specific working context of each machine was replaced by a progressivist context in which design changes follow one another with seeming inevitability, where no design change appears to result from the political, cultural or ideological forces at work in the artefact's context of origin. Wherever he could,

Ford displayed large machines as sensuous, quasi-sacred, icons. In the Village, he ordered the walls opening into the steam engine rooms of the Loringer grist mill and adjacent Armington and Sims machine shop changed from the original wood to glass so that he could watch the machines operate as he passed by. The powerhouse at the Highland Park plant was exhibited to Woodward Avenue passers-by through walls of showroom glass; inside, the dynamos were enthroned amid gleaming brass and immaculate tile. Not a few observers have noted that Ford's secluded 'Fairlane' estate (c. 1916) could boast of only one really elegant building, the hydro-power plant, a shrine to dynamos very like Highland Park's.[24]

One of the Rouge's design departures from its predecessor at Highland Park would later become a world famous symbol of Ford's rejection of debate and dissent. Just as Ford pursued integrated control of inputs through his network of transportation lines converging on the Rouge, so he sought even more control over workers than the house-to-house inspections and in-factory spy networks of the 1914 labour reforms had offered. Highland Park opened directly on Woodward Avenue, giving Ford management no say about who mingled with workers on the public streets fronting the factory. Plant accessibility meant unwanted interference. In dramatic contrast, the new Rouge compound was insulated by a fully fenced perimeter and tightly guarded gates. The Miller Road Gate No. 4 became world infamous when photographers caught Ford thugs beating United Auto Worker activists Walter Reuther and Richard Frankenstein in 1937's 'battle of the overpass'. Independent minded workers seeking to organize a union were anathema to Ford; they apotheosized the intrusion of pluralism into the ideologically standardized interior of the Ford universe. Ford's aesthetic, in short, showed itself not only in design choices affecting the display of sensuous and elegant machines but also in the seemingly pragmatic design decisions that determined architecture as well. In this regard, the layout of the new Ford museum would consistently echo the Rouge's dedication to control over workers. Architect Robert O. Derrick was hired to design the Museum building. In his oral history interviews he recounts his conversation with Ford about the Museum's floor plan.

> He said we would have to have a model [of the proposed Ford Museum building] made, so we had a model made and it showed the balconies, naturally, and the basement, and he said, 'What is this up here?' I said, 'That is a balcony for exhibit.' He said, '**I wouldn't have that; there would be people up there, I could come in and they wouldn't be working. I wouldn't have it.**' He said, '**I have to see everybody.**' Then he said, 'What's this?' I said, 'That is the basement down there, which is necessary to maintain these exhibits and to keep things which you want to rotate, etc.' He said, '**I wouldn't have that; I couldn't see those men down there when I came in.** You have to do the whole thing over again and put it on one floor with no balconies and no basements.' I said, 'Okay.'[25]

Henry Ford is sometimes portrayed as an eccentric oddity driven by deep-rooted compulsions and with good reason.[26] It is helpful, nevertheless, to balance the notion of Ford as eccentric with its opposite. In many ways Ford's position as the owner of his own empire, a man with extraordinary resources in public reputation as well as money, permitted him to act out affective responses to the power of modern technology that reveal dimensions of public consciousness generally. We see, in Ford, not the concern of a handful of elite critics like Thoreau and the Transcendentalists in the nineteenth century. Ford's ambiguity resides in the mainstream of the first half of the twentieth century. In fact, ambiguity pervades the art and rhetoric between the two World Wars. To cite only two instances, Charles Sheeler was hired by the advertising firm of N. W. Ayer to photograph the Rouge for Ford publicity use. The photographs became models for a series of paintings, Sheeler's industrial landscapes, which stand among the best known celebrations of factory-scape aesthetics in the period. Still, as Karen Lucic's recent study indicates, Sheeler's paintings also reveal an ominous, brooding tone: 'In his beautifully crafted and masterfully composed images of a machine-dominated world ... he seems to mourn the death of the creative self's autonomy in a mood of anxious melancholy' (Lucic 1991, p. 139).

Perhaps the most striking example of the ambivalence of Progress would appear in the main foyer of the Hall of Science at the 1934 Chicago 'Century of Progress' Exposition, itself an astonishing romanticization of life's promise during the depth of the Depression. Visitors to the Hall of Science were met in the foyer with Louise Lentz Woodruff's three-piece sculpture, 'Science Advancing Mankind'. Two life-sized figures, male and female, faced forward with arms uplifted. Both were dwarfed by the massive figure of a metallic robot twice their size. In the words of Lenox Lohr, general manager of the exposition, the robot typified 'the exactitude, force and onward movement of science, with its hands at the backs of the figures of a man and a woman, urging them on to the fuller life' (Lohr 1952, p. 96). The sculpture's iconographic ideology was reinforced by the official Guidebook's stunning, bold-faced thematic motto:

'Science finds, industry applies, man conforms'[27]

How does this pervasive ambiguity – technology as liberating progress and technology as intimidating master – relate to the question of resistance to technology? The ambiguity we have sketched so briefly here does not operate like organized and intentional strategies in conflict nor is it like the uncertainty with which engineers wrestle in pre-theoretical areas of technological practice. We see, instead, that the modernist vision of inevitable technical triumph reaches deeply into the bowels of twentieth century consciousness. That so successful and powerful a player as Henry Ford would manifest such deep-seated anxiety about his own triumphs helps us understand the latent insecurity that surfaces in so many guises after 1900. Ford is helpful, too, because he cannot be neatly contained under the heading of anxiety alone. Although he appears to have retreated from the Rouge to the Museum and Greenfield Village, he clearly did not intend them to be a condemnation of progress. The world of elegant machines and of the inventors and heroes who brought them forth are enshrined there with a depth of wonder and enthusiasm that continues to move visitors to the present day.

Could we find the same mix of fear and delight in bureaucrats who do not own their own companies? Could it be that hired managers, like Alfred Sloan at General Motors, were insulated from their own ambiguity, leaving personal feelings at home under the illusion that they are only responsible for the efficient functioning of a system that is not theirs? Perhaps, then, Ford's unusual combination of proprietary capitalism at a massive corporation, ownership with managerial control, might reveal more than has heretofore been thought, that society at large has yet to learn how to integrate the heady enthusiasm that attends possession of amazing technological systems with the burdens those same systems carry with them.

Notes

1   These remarks are limited, for the most part, to scholarship in the United States. As my reading of the literature from other parts of the world is in its early stages, I do not feel competent to comment in those areas.
2   Examples abound. See John Kasson's account of public responses to the mighty Corliss steam engine that physically and symbolically dominated the 1976 Philadelphia World's Fair (Kasson 1976, pp. 161–5) or Leo Marx's classic (Marx 1964) in its entirety. A monograph by David Nye treats the topic as central (Nye 1994).
3   A sampler of recent work shows that sensitivity to affective response has begun to appear in a wide range of topical areas: see Rosalind Williams (1990); Christopher Hamlin (1990); David E. Nye (1990); Robert L. Frost (1991); David Rosner and Gerald Markowitz (1991); Claude S. Fischer (1992); Peter Fritzsche (1992).
4   The argument – that twentieth century Americans view Progress as liberating and intimidating simultaneously – is complex. For more detailed treatment see Jackson Lears (1983, pp. 1–38); Staudenmaier (1989, pp. 268–97).
5   In addition to the many biographies of Ford – the best being Nevins (Nevins & Hill 1954, 1957, 1962), Sward (1948), Lacey (1986) and Lewis (1976), I am indebted for much of the detailed archival research on which these observations are based, to Mr Thomas Bohn, a graduate student research assistant who spent many fruitful hours in the Henry Ford Museum Archives tracking media references to Henry Ford in a host of categories over the period 1904–23. For Ford's public image, before and after the 1914 explosion of publicity, see especially David L. Lewis (1976, Ch. 2–5).

6 See James Flink (1970, *passim*) and Lewis (1976, Ch. 3 and 8). For a thorough account of the 1910–14 production innovations at Highland Park, see Hounshell (1984, Ch. 6).

7 Selected from 'The Little Ford Rambled Right Along', words by C. R. Foster and Byron Gay, music by Byron Gay (Los Angeles, Calif.: C. R. Foster Publishing Co. 1914). Taken from recorded version on file at the Ford Museum Archives. I am indebted to Mary Seilhorst of the Ford Museum for access to the music.

8 W. H. Alburn, 'Profit Sharing to End Wars Between Capital and Labor, Says Henry Ford', Omaha, Nebr. *News* 14 January 1914; typed version in Box 87 Henry Ford Profit Sharing, Ford Archives.

9 Some of the voluminous literature about social anxieties during the Progressive Era include John Higham (1975) and Daniel T. Rogers (1974). On Utopian and dystopian novels of the period and literary references to technology generally, see Cecelia Tichi (1987) and Howard P. Segal (1985). For a recent study of a technocratic hero of the period, see Ronald R. Kline (1992).

10 Stephen Meyer III (1981), see especially Ch. 7, 'Assembly-Line Americanization'.

11 'Unveiling of Henry Ford', *Nation* CIX (26 July 1919), p. 102 cited in David L. Lewis, 'Henry Ford Meets the Press', *Michigan History* XIII/3 (1969), pp. 283–5.

12 Hounshell's account of the transition from the Model T to the Model A offers the most thoroughly researched account of production changes (1984, Ch. 7). On the Hunger March of 1932 see Steven Babson (1986). In addition to standard accounts of the Battle of the Overpass see Acc 292, Box 46 for letters to Mr Ford in the wake of the event.

13 On Ford's Presidential popularity see 'Ford Threatens the Old Parties', *Current Opinion* July 1923, pp. 9–11. For letters to Ford urging his candidacy see Acc 62, Box 119, File 1. On Rotunda and Rouge visitor figures for 1940 see Acc 629 A-1, Public Relations General 1936–40, Box 1.

14 Margaret Meacham/Ashland Ohio/28 May 1937 [selections from 5 pages typed copy of handwritten]. Acc 292, Box 46, Ford Archives. Lewis cites a May 1937 opinion survey by the Curtis Publishing Company indicating that 59.1% of Americans believed the Ford Company treated workers better than any other firm. *Public Image*, p. 75.

15 Some scholars follow a *New York Times* figure of 500,000 per year. However, I have yet to find records of more than 135,000 in the Archives (Acc 162, Box 2: 'Secretary's Office-General-Visitors').

    In his current work in progress, *Creating the corporate soul: the rise of corporate public relations and institutional advertising*, Roland Marchand observes the assets and liabilities of plant tours for business advertising. Designing the tours to present the company's best face to the public and to educate visitors about the technical processes involved were balanced against the relatively small number of visitors who were able to travel to the plant site. One of the most effective at such advertising was the H. J. Heinz company where it was estimated that an average of 20,000 people visited the Pittsburgh plant yearly by 1900. Heinz hit upon a dramatic marketing ploy when, in 1898, he bought the Iron Pier at Atlantic City, rebuilt it on the grand scale, and created a demonstration site that replicated the Pittsburgh tour. The company estimated that in the 1920s as many as 15,000 people came onto the pier during a single day during the summer season. In contrast to the Heinz Atlantic City pier, however, the Rouge attracted visitors in very large numbers and from around the world to the actual production site. (see Ch. 10, Corporate images on parade: Atlantic City, the 'Silver Fleet' and the 'Parade of Progress'.) Cited with permission.

16 Otto Moog, German Engineer, in Thomas P. Hughes (1989, p. 291; author's translation of Otto Moog, *Druben steht Amerika: Gedanken nach einer Ingenieurreise durch die Vereinigten Staaten* [Braunschweig: G. Westermann, 1927], p. 72 [my emphases]). Hughes cites another German

engineer, Franze Westermann, saying: 'the most powerful and memorable experience of my life came from the visit to the Ford plants...' p. 99.

17 *My Art, My Life*, pp. 187–8 (my emphases).

18 Veteran Graham McNamee of NBC was the announcer. See E. I. #52, Boxes 1–4 and Vertical Files 1–5 for archival materials including a transcript of the radio coverage. For a helpful summary see Geoffrey C. Upward (1979).

19 The museum opened on 22 June 1933 (Upward, 1979, p. 76). For some of the fragmentary archival records of Rouge visitors see Acc 162, Box 2. On the Edison Institute conceived as a school see E. I. 145 Box 1 Folder: Guide Training 1934 and 'The Edison Institute: Its Scope and Purpose', *Herald* ('Published by the Children of the Edison Institute'), Vol. I, No. 1, 11 February 1934.

20 Nevins and Hill summarize the new Ford vertical integration as follows: '...by 1926 the entire productive activity of the company had been impressively developed. Raw materials were now flowing from the iron mines and lumber mills of the Upper Peninsula, from Ford coal mines in Kentucky and West Virginia, and from Ford glass plants in Pennsylvania and Minnesota, much of the product traveling on Ford ships or over Ford-owned rails. Ford manufacture of parts had been expanded – starter and generators, batteries, tires, artificial leather, cloth, and wire had been manufactured by the company in increasing quantities' (Nevins & Hill 1957, p. 257), cited in Hughes 1989, pp. 209–10. On the five dollar day inspection system see Meyer 1981, Ch. 6.

21 On the stockholder buyout, see Nevins & Hill (1957, pp. 105–11). Nevins interprets the three resignations as follows: 'Ford... looked aback with distaste on the period of Couzen's activity in company affairs, when he had been unable to move freely. The Dodge suit had of course intensified his desire for absolute authority. He was therefore irritated by the presence of anyone in the company who might not work with him in complete harmony.' (Ibid. p. 145).

22 In Edward Cutler, *Reminiscences* (Ford Museum Archives), p. 119 or p. 47.

23 Cutler, *Reminiscences*, Vol. 2, p. 138.

24 Lacey, for example: '...here was only one beautiful room in the entire building: the powerhouse. This was a spare, clean chamber which Henry had designed himself... and he created a very Ritz of power stations, all marble and gleaming brass dials and pipes. Around the floor were set out little generators, raised on plinths like so many modern sculptures...' pp. 149–50. See also Collier & Horowitz (1989, p. 71); Nevins & Hill (1957, pp. 20–1).

The clean, uncluttered, 'Ford' style that Charles Sheeler would make famous with his late 1920s photographs and paintings represent the *continuation of*, and not a completely fresh *artistic reflection on*, the Ford style. See Mary Jane Jacob, 'The Rouge in 1927: Photographs and Paintings by Charles Sheeler', in *The Rouge: The Image of Industry in the Art of Charles Sheeler and Diego Rivera* (funded by the Ford Motor Company Fund and Founders Society Detroit Institute of Arts).

25 Robert O. Derrick, Oral Reminiscences, p. 50 [emphases mine]. Cited in Upward (1979, p. 50). For a discussion of the striking parallels between Ford's need to 'see everybody' and utilitarian philosopher Jeremy Bentham's principle of constant scrutiny for the 'Panopticon', see Staudenmaier (1990).

26 See especially Anne Jardim's (1970) psychological interpretation.

27 Chicago Century of Progress International Exposition, *Official Book of the Fair*, (Chicago: A Century of Progress, Inc., 1932), p. 11. I am indebted to Lowell Tozer (1952, pp. 78–81) for first calling my attention to the Exposition and to Cynthia Read-Miller, curator of photographs and prints in the archives of the Henry Ford Museum and Greenfield Village, for copies of the Official Book and photos of the iconography referred to here (emphasis mine).

References

BABSON, S. (1986). *Working Detroit.* Detroit: Wayne State University Press.

BIJKER, W. E., HUGHES, T. P. & PINCH, T. (ed.) (1987). *The social construction of technological systems.* Cambridge, Mass: MIT Press.

COLLIER, P. & HOROWITZ, D. (1989). *The Fords: an American epic.* New York: Summit Books.

FISCHER, C. S. (1992). *America calling: a social history of the telephone to 1940.* Berkeley: University of California Press.

FLINK, J. (1970). *America adopts the automobile 1895–1910.* Cambridge, Mass: MIT Press.

FRITZSCHE, P. (1992). *A nation of fliers: German aviation and the popular imagination.* Cambridge, Mass: Harvard University Press.

FROST, R. L. (1991). *Alternating currents: nationalized power in France, 1946–1970.* Ithaca, NY: Cornell University Press.

HAMLIN, C. (1990). *A science of impurity: water analysis in nineteenth century Britain.* Berkeley, University of California Press.

HIGHAM, J. (1975). *Send these to me: Jews and other immigrants in urban America.* New York: Atheneum.

HOUNSHELL, D. A. (1984). *From the American System to Mass Production, 1800–1932: The Development of Manufacturing Technology in the United States.* Baltimore: Johns Hopkins Press (Dexter Prize 1987).

HUGHES, T. P. (1989). *American genesis: a century of invention and technological enthusiasm, 1870–1970.* New York: Viking.

JACKSON LEARS, T. J. (1983).   From salvation to self-realization: advertising and the therapeutic roots of the consumer culture, 1880–1930. In *The culture of consumption,* ed. R. W. Fox and T. J. Lears. New York: Pantheon.

JARDIM, A. (1970). *The first Henry Ford: a study in personality and business leadership.* Cambridge, Mass: MIT Press.

KASSON, J. (1976). *Civilizing the machine: technology and republican values in America, 1776–1900.* New York: Penguin Books.

KLINE, R. R. (1992). *Steinmetz: engineer and socialist.* Baltimore: Johns Hopkins Press.

LACEY, R. (1986). *Ford: the men and the machine.* Boston: Brown, Little & Co.

LEWIS, D. L. (1976). *The public image of Henry Ford: An American folk hero and his company.* Detroit: Wayne State University Press.

LOHR, L. (1952). *Fair management: the story of a century of progress exposition.* Chicago: The Cuneo Press, Inc.

LUCIC, K. (1991). *Charles Sheeler and the Cult of the Machine.* Cambridge, Mass: Harvard University Press.

MARX, L. (1964). *The Machine in the Garden: technology and the pastoral ideal in America.* New York: Oxford University Press.

MEYER, S. III (1981). *The five dollar day: labor management and social control in the Ford Motor Company, 1908–1921.* Albany: State University of New York Press.

NEVINS, A. & HILL, F. E. (1954). *Ford: the times, the man, the company.* New York: Scribner's.

NEVINS, A. & HILL, F. E. (1957). *Ford: expansion and challenge: 1915–1933*. New York: Scribner's.

NEVINS, A. & HILL, F. E. (1962–3). *Ford: decline and rebirth, 1933–1962*. New York: Scribner's.

NOBLE, D. F. (1979). *America by design: science, technology, and the rise of corporate capitalism*. New York: Knopf.

NOBLE, D. F. (1984). *Forces of production: a social history of industrial automation*. New York: Knopf.

NYE, D. E. (1990). *Electrifying America: social meanings of a new technology*. Cambridge, Mass: MIT Press.

NYE, D. E. (1994). *The technological sublime*. Cambridge, Mass: MIT Press.

PFAFFENBERGER, B. L. (1990). The harsh facts of hydraulics: technology and society in Sri Lanka's colonization schemes. *Technology and Culture* **31**, 361–97.

ROGERS, D. T. (1974). *The work ethic in industrial America, 1850–1920*. Chicago: University of Chicago Press.

ROSNER, D. & MARKOWITZ, G. (1991). *Deadly dust: silicosis and the politics of occupational disease in twentieth-century America*. Princeton, NJ: Princeton University Press.

SEGAL, H. P. (1985). *Technological Utopianism in American culture*. Chicago: University of Chicago Press.

SMITH, M. R. (1977). *Harpers Ferry Armory and the new technology: the challenge of change*. Ithaca, NY: Cornell University Press (Frederick Jackson Turner Prize, 1977).

STAUDENMAIER, J. M. (1989). Perils of progress talk: some historical considerations. In *Science Technology and Social Progress*, ed. S. L. Goldman, pp. 268–97. Bethlehem, Pa: Lehigh University Press.

STAUDENMAIER, J. M. (1990). Science and technology: who gets a say? at Conference: Technological development and science in the 19th and 20th centuries, Technische Universiteit Eindhoven, The Netherlands, November.

SWARD, K. T. (1948). *The legend of Henry Ford*. New York: Reinhart & Co.

TICHI, C. (1987). *Shifting gears: technology, literature, culture in modernist America*. Chapel Hill: University of North Carolina Press.

TOZER, L. (1952). A century of progress, 1833–1933: technology's triumph over man. *American Quarterly* **4**, 78–81.

UPWARD, G. C. (1979). *A Home for Our Heritage: The Building and Growth of Greenfield Village and Henry Ford Museum, 1929–1979*. Dearborn, Mich: The Henry Ford Museum Press.

VINCENTI, W. G. (1990). *What Engineers Know and How they Know It*. Baltimore: Johns Hopkins Press.

WILLIAMS, R. (1990). *Notes on the Underground: An Essay on Technology, Society and the Imagination*. Cambridge, Mass: MIT Press.

8

Resistance to nuclear technology: optimists, opportunists and opposition in Australian nuclear history

ROY MACLEOD

Introduction[1]

Alain Touraine (Chapter 2) situates the history of social resistance to new technologies in the 'ambiguous project of modernity, and in the long history of ideas associated with democratic control of instruments of power'. In this context, the nuclear history of Australia presents both a paradox and a possibility. Today, Australia has no civil nuclear power and no nuclear weapons. Its government participates actively in the programme of the International Atomic Energy Agency, has protested against French tests in the South Pacific, and is a keen advocate of the Nuclear Non-Proliferation Treaty. Yet, the last 40 years have seen a long history of nuclear optimism and opportunism, played out among conflicting political and economic interests, domestic and overseas. Has, then, 'resistance' to nuclear technology been successful? If so, what has made it so?

To answer this question requires a grasp of the salient features of Australia's affair with the atom. That relationship, which once embraced nuclear energy as a quintessential element in Australia's post-war development, also embraced characteristic features of resistance, operating upon a landscape well known for its contours of intellectual conservatism and radical dissent. Australia's nuclear destiny has depended, in part, upon the prospect of achieving economic nuclear power for domestic use, and advantageous terms for the sale of uranium in overseas markets. But its nuclear history has also come to reflect changing perceptions of the role that Australians wish to see their country play in the nuclear arena. That prospect and that possibility are the subject of this chapter.

The nuclear present

At present, something over 1600 spent fuel elements – stored behind the security fences of Lucas Heights, the research establishment of the Australian Nuclear Science and Technology Organization (ANSTO), 35 km southwest of Sydney – represents the most enduring product of Australia's 40 years of nuclear history. To this may be added perhaps 800 tons of waste from the twelve British tests in South Australia between 1953 and 1963, and an uncertain number of 44-gallon drums of yellowcake lost in mining operations between 1970 and the present. Today, ANSTO, which has two small research reactors, is proposing to decommission its largest and oldest – HIFAR (High Flux Australian Reactor) – at 35 years of age, now nearing the end of its useful life; and to replace it with a reactor that is twice as large and potentially six times more expensive.[2]

In early 1993, a committee of review, appointed by the Federal Government to test the argument and weigh its implications, excited comment from sources both at home and overseas. What might in an earlier day have been considered a routine (if expensive) technical decision, to be taken in Canberra by the appropriate departmental interests, suddenly became a matter of general interest to the community as a whole. 'Resistance' to a proposal for a new research instrument was re-configured into a deep-seated scepticism about the entire nuclear project. In the process, postures of resistance once viewed as 'radical' by the national press were re-constructed into legitimate responses by concerned citizens, commanding the heights of a new rationality. The review has appeared, recommending the postponement of a decision for five years; but in the meantime, we may forsee the beginning of a new discourse, in which 'proponents' and 'resistors' enter a relationship much more complex than any before seen in Australia.

With the end of the Cold War, the nuclear debate was momentarily relegated to the back pages. It was just possible to put aside thoughts of nuclear disasters, potential and real, in Eastern Europe, the perilous expansion of the nuclear industry in Japan, and the advent of nuclear power in neighbouring Indonesia. But suddenly, within the last year, nuclear technology has again become a domestic issue. And in its train, in an era sensitive to calls for republicanism and increased respect for Aboriginal land rights, have awakened slumbering memories of alien presences – of British testing, American bases, nuclear ship visits, and failed attempts to establish a nuclear power industry with reactors developed overseas. Coupled with these memories has come more recent news of proposals for fresh uranium mining on tribal lands and in areas of outstanding natural beauty. Such developments are encouraging Australians to review their understanding of the circumstances that brought their country into the atomic age, and that still shape debate surrounding its participation in the nuclear fuel cycle (Moyal 1975; Cawte 1992).

Bauer (Chapter 1) has proposed an integrating framework that embraces the language and forms of resistance, its causes and effects, and its significance in demonstrating the limits of technocracy and planning. In Australia, as elsewhere, the history of nuclear technology has been inextricably bound up with questions of population, geography and industrial development. But in Australia, that history has also been structured by the particular nature of the country's colonial past and its traditional role as a primary producer (see Bolton 1990).

That history began in 1944, with the wartime needs of Britain and the United States, and continued through the work of the Australian Atomic Energy Commission (AAEC), created in 1953.[3] For the next two decades, Britain and the US saw Australia as a provider of testing space and raw materials – a natural laboratory for the nuclear weapons industry. For the next 30 years, the prospect of a nuclear future held a special place in the hymnology of modernism, in a country that has historically had a high regard for practical skills and new technologies. Today that situation has changed, and with it, the status of nuclear issues. We can best understand the ways in which Australians now view these issues by recapturing the history of optimism, opportunism and opposition they involve, and by examining the watershed which the recent reactor enquiry may signify.

Optimism and Opportunism, 1945–1975

To some, nuclear technology has always held the promise of wealth, international influence, and national independence; to others, it has always been a certain source of fear, sickness, and dependence on the preferences of superpowers. It is little wonder that, as a social technology, it has been read as an instrument both of well-being and of destruction, its advocates and adversaries divided between those wanting its benefits at any cost, and those preferring to question its value at all stages.

It is possible to reconstruct Australia's nuclear history in three phases: from the late 1940s to the mid-1970s, from the mid-1970s until the early 1980s, and from the 1980s onwards. It is convenient to see the first phase in terms of the optimism and opportunism of a small group of political leaders, mining interests, scientific administrators and nuclear physicists. The latter included Sir Leslie Martin, Sir Philip Baxter, Sir Ernest Titterton, and Sir Mark Oliphant – the 'nuclear knights', as they have been called.[4] Thanks in large part to their energies, nuclear science for a short period absorbed vast sums of public money, briefly eclipsing Federal research funds for biology and chemistry, mining and agriculture. Thanks in large part to their forceful presence, the public image of science became significantly shaped by physical science and big technology.

To understand their hold on Australian thinking, it is necessary to remember that Australia emerged from the Second World War with a small base of manufacturing wealth and industry, but with large ambitions to sustain (or recapture) the standard of living its people had enjoyed at the turn of the century. That was to be achieved through national development – with expanded provision for science-based agriculture, medical research, industrial growth, migration, education and investment. Into this picture of urban and rural development came the need for vast quantities of cheap electrical power. This would be achieved partly through an enormous hydroelectric scheme in the Snowy Mountains, and partly through the use of coal-fired stations on the coast (Johnston 1969; Collis 1989). Alongside these, for the country's inaccessible and energy-poor regions, Australia, blessed with natural uranium, could look to the bright future promised by the advocates of nuclear power.

Aside from brief and unrewarding efforts to locate uranium for the American bomb project, Australia had taken no part in the events leading to Hiroshima and Nagasaki. In the immediate post-war world, however, there were attempts to share in the international presence that nuclear knowledge bestowed. Overseas, Australia participated actively in the short-lived attempts of the newly formed United Nations to establish a 'new nuclear order' in the world; while at home, the Menzies Liberal government created the Australian Atomic Energy Commission (AAEC), linked by its structure with the defence and supply departments, and given a mandate to promote the discovery, mining, treatment and sale of uranium; the construction and operation of plants for the 'liberation and conversion of atomic energy'; and programmes for the training of research workers. The overall intention was twofold: to bring, in the optimistic words of Oliphant, 'prosperity and fruitfulness such as few nations have known', and to ensure Australia a place in the international 'nuclear club'.[5]

In Australia, resistance against the nuclear project would ultimately take several forms – bureaucratic, political, technical, and popular. But for nearly twenty years, there was little to speak of. On the one hand, mining exploration was widely encouraged, and notably successful at the isolated Rum Jungle in the Northern Territory and at Mary Kathleen in northwest Queensland. By 1959, the mining of uranium at Rum Jungle amounted to a quarter of all economic activity in the Territory (Cawte 1992). On the other hand, however, little information about the implications of Australia's nuclear project was shared with the general public, as nuclear facilities were on a security par with US satellite and intelligence sites – in themselves, nuclear targets. The Atomic Energy Act of 1953 contained provisions by which anyone 'reasonably suspected' of conveying restricted information was liable to twenty years' imprisonment. The Act also made applicable the Approved Defence Projects Protection Act of 1947, which was designed to protect the procurement of strategic materials and the safety of the rocket ranges at Woomera (Cawte 1992).

Under these circumstances, critics of Australia's involvement in the nuclear fuel cycle were obliged to defer to the commitment made by successive Australian governments, for the purpose of achieving a trusted place among 'great and powerful friends' in the non-Communist world. However, the effect of these strictures was to stifle criticism, particularly in the press, and to induce a sense of collective apprehension – both among those living near the site, and among those who were simply curious. Security especially surrounded the choice of observers at the British tests. Sir Ernest Titterton, an Englishman whom Chadwick had recruited to the Manhattan Project in 1943, was recruited in 1952 by Oliphant to join the new Australian National University in Canberra. In England, Titterton had worked at Cockcroft's newly established Atomic Energy Research Establishment (AERE) at Harwell, and had close contact with the Atomic Weapons Research Establishment (AWRE) at Aldermaston. In 1952 and 1953, he became the 'Australian' representative at the British tests at Monte Bello and Emu Fields, and in 1955, joined Martin on the highly secret, three-man Atomic Weapons Test Safety Committee. Titterton, unlike Oliphant, was accepted by the British in part because he was British. Few Australians in government and scientific circles were filled in on the facts he knew and it would be much later before the Australian public knew (Royal Commission 1985; Arnold 1987, pp. 28–9).

Titterton's views were largely shared by (Sir) Philip Baxter, a Welsh-born ICI industrial chemical engineer and research manager who, thanks again to Chadwick, had gained extensive wartime experience of the Manhattan Project's uranium hexafluoride and U-235 separation plant at Oak Ridge. Leaving ICI for Australia in 1950, Baxter rose quickly in the small circle of academic managers, and from 1956 until 1972, became the first – and by far the most influential – chairman of the AAEC.[6] If, in a kingdom of the blind, the one-eyed man is king, Baxter's presbyopic sentiments permeated the nuclear vision. Australia was, in Baxter's words, 'the last big continent which the white man has to develop and populate' (Baxter 1957, p. 75), and nuclear power was to be his instrument. His vision, iterated from AAEC headquarters overlooking the beach at Coogee, was manifested behind barbed wire at the research establishment at Lucas Heights, centred around HIFAR, where research training, international collaboration, and a plan for uranium enrichment began in the mid-1960s (Baxter 1963).

The Australian nuclear research and uranium programme thus emerged under the guidance of men profoundly influenced by their experience of Britain's war, whose vision of the future was shaped by their perception of white Australia's isolation and vulnerability. Earlier, between 1952 and 1958, all Australia's uranium went to the United States or Britain for defence purposes; after 1958, most went under contract to the UK for power reactors; but by the 1970s, new markets opened in Japan and France. Uranium was not only a domestic issue, but also a matter of foreign policy. With the oil embargo and the energy crisis of the early 1970s, Baxter and Titterton had easy remedies to the potential danger that, without uranium mines and nuclear energy, the world would disintegrate into civil war, as nation fought nation for the prize of fossil fuels. It was in Australia's vital defence and economic interest to secure nuclear power, enrich uranium, and sell it to the world.

With the optimism of the AAEC and its champions went a kind of corporate opportunism. Behind its security clearances, during its first 25 years, the Commission's budget grew by a factor of thirty. By 1972, it had cost $170 million, almost twice the overrun budget of Sydney's Opera House, and was employing 1200 staff, including 310 scientists. While it was neither the nation's largest nor richest scientific organization (CSIRO was larger and Defence was more expensive), it was the biggest single civil research programme ever undertaken in Australia. It was also a great adventure, raising the hopes of energy-poor South Australians and mineral-rich West Australians in anticipation of a future of development and trade. Indeed, it was in this hopeful spirit that Commonwealth and State governments enthusiastically endorsed the Operation Plowshare proposals of Glen Seaborg and Edward Teller; these, in the name of the 'peaceful atom', were to blast out port facilities at Cape Keraudren and to mine iron ore in the Hammersley Ranges (Findlay 1990). Criticism, let alone resistance, directed against any part of this optimistic programme was politically and academically difficult; legally, it might be viewed as commensurate with subversion, and even possibly treason. This situation did not wholly suppress criticism,[7] but it undoubtedly inhibited an informed resistance.

Resistance was an anathema to Baxter and Titterton, in particular, who defended the proposition that Australia should cover the entire 'nuclear continuum', from uranium mining to the production of nuclear power, with the possibility of nuclear weapons kept as a reserve option ('X' [Baxter] 1969, p. 31). Secrecy also pervaded discussion of the Nuclear Non-Proliferation Treaty, which the Commission – against the advice of the Department of External Affairs – pressed the government not to sign (Richardson 1968; Encel & McKnight 1970, p. 20). There was speculation that, at a time when Australia enjoyed relatively low-cost power from conventional fossil fuel sources, a nuclear power station based on natural uranium could be justified only as a producer of weapons grade plutonium for an Australian bomb (Baxter 1968, pp. 57–63; Richardson 1969,

p. 70; Temple 1970, p. 11; Edwards 1971b).[8] Given the post-war history of Communist movements in Malaya and Indonesia, the renewal of Soviet testing, and China's unclear intentions, Titterton, who frequently appeared as the Commission's 'front-man', argued that

> ... a nuclear free zone in South East Asia is neither possible nor desirable ... It is already too late to talk about making South-East Asia into a nuclear free zone ... Much as we may dislike the idea of nuclear weapons, we have to learn to live with them. There is no possibility of wishing them away and, indeed, if our policies are right, these weapons can benefit all mankind by ensuring that there will never be another world war.   (Titterton 1965a)

To safeguard the region, Titterton argued, the UK, the USA, or preferably both, could be invited to stockpile nuclear weapons in Australia, so as to 'tie us even more firmly to our major allies ... increase Australian security at minimum cost, and ... allow us to continue to devote our resources to developing the continent as rapidly as we can' (Titterton 1965b).

His message struck a chord within conservative Democratic Labour Party and Liberal party circles and among uranium interests, who relegated export bans to the rubbish tip:

> To suggest that uranium should be withheld because of possible military developments is like suggesting that we should cease to sell iron ore lest it go into tanks or fighting ships, or stop selling wheat, beef and wool because these could be used to feed and clothe soldiers ... The peoples of the world, poor and wealthy, want more and more energy to improve their standards of living; the demand comes from the customers – as is shown by the rapid increase in consumption of electrical energy all over the world ... Those nations, like Australia, who have far more uranium than is required for their own energy needs have a duty to sell the excess to those less fortunate.   (Titterton 1977)

Nuclear critics: resistance past

During the early 1970s, a sequence of events began to disturb the believers' dream. Opposition to the nuclear enterprise grew and took shape in several forms. First, the Liberal Government, which in earlier incarnations had shared the nuclear vision, now began to take fright at its cost. Despite pressure from Baxter, the Commission's plans to build a 500 MW power station on the unspoilt harbour at Jervis Bay, 200 km south of Sydney, were suspended in 1971, and effectively cancelled by the return of the Labour government the following year.[9] The cost of the proposed reactor was estimated variously at between $150 million (Temple 1970, p. 13) and $1.3 billion (Angyal 1991, p. 193). Secondly, a smouldering core of resentment among suburban residents in the bushland around Lucas Heights (by 1971, 14,000 people were living within a 5 km radius of the reactor) flashed into flame, as representatives in the New South Wales parliament demanded enquiries into the Commission's record on leakages and waste disposal, and forced a debate on environmental jurisdictions between state and federal governments (Watson 1971). Thirdly, the public service bureaucracy in Canberra, long sceptical of the Commission's cavalier methods and limited accountability, decided to call in its marker. When Baxter retired, he was replaced not by Titterton, as predicted, but by a former Canberra public servant, who knew better the routines of government (Edwards 1971b).

This was not, however, the end of nuclear optimism; it was only the beginning of the end. If the Jervis Bay decision spelled an end to nuclear power, it also signalled a renewal of the Commission's interest in a 'major enrichment facility to enhance the export value of Australia's uranium' (ANSTO 1993). It was well known that a full-scale enrichment plant was out of reach; Stuart Butler, professor of physics at Sydney University, and later Director of the Research Establishment at the AAEC, estimated in 1977 that such a plant would cost $3 billion (Butler *et al.* 1977, p. 103). But the AAEC persevered, blessed by accidental news from the north. Here events favoured the deed. In September 1970, Queensland Mines announced the discovery of the 'world's richest uranium strike' on an Aboriginal reserve in western Arnhem Land (Cawte 1992, Ch. 5). It was called Nabarlek (a local aboriginal name for a small kangaroo). Almost simultaneously came huge finds at the Ranger fields, 50 km to the southwest. These were followed by discoveries at Koongarra and Jabiluka, north and south of Ranger. By the early 1970s, the whole Alligator Rivers region of the Northern Territory had become a uranium province of global importance, and Australia had become the fourth largest non-communist source of uranium ore (Cawte 1992).

With these finds, the scene of debate returned to primary production, instead of nuclear power *per se*. In politics, it has been observed, some issues come and go; some are resolved and forgotten, but others linger on like festering sores (Smith 1979, p. 32). Such has been the fate of the uranium mining issue since about 1974. A complex of issues – whether, where, and to what extent to mine, process and export; and how to deal with large but historically unspecified Aboriginal land rights – transfixed Australian politics. By their nature, these issues involved all major political parties, the trade unions, state governments, multinational companies, and at least ten different departments of the Federal Government. They had implications not only for domestic politics, but also for Australia's relationship with the United States, Britain, the European Community, and the Asia–Pacific region.

The return of Labour in 1972 offered opponents of uranium mining the hope that the new government would impose export bans, and effectively neutralize the nuclear issue. However, by 1975/6 the Australian Labour Party, under pressure from the Australian Council of Trade Unions (ACTU) and its chairman, Bob Hawke, led the government to backtrack on its formal policy. Instead, Labour announced that the AAEC would exploit the finds in the Northern Territory for the benefit of private industry. As expected, the proposal sowed dragons' teeth. In 1975, only four months before his dismissal, responding to pressure from his left wing, Labour Prime Minister Gough Whitlam appointed an inquiry into the proposed development of the Ranger mines. The inquiry, under Mr Justice R. W. Fox, sat for eighteen months and produced two major reports, in 1976 and 1977. Commissioned to review the environmental and financial aspects of the development, it widened its scope to cover the entire spectrum of nuclear power, uranium exports, safeguards, and the probity of the AAEC – a supposedly objective government agency, acting as an interested party (Ranger Enquiry 1976, 1977). This proved to open another phase in a Great Debate that has no end in sight.

In 1975, the return of the Liberal government encouraged a new wave of exploration, particularly in the Northern Territory, where Aboriginal groups made financial settlements. Division within the Labour movement robbed the party of political unity on this issue, but between 1975 and 1977 there seems to have been a decided shift in public opinion. In 1975, polls suggested that 60% of Australians supported uranium mining and export; by August 1977, although the issues had become highly confused, less than 50% indicated their approval (Smith 1979, p. 40). Pro-uranium sentiments clustered around seven interests: the mining industry, Titterton and Baxter, some mineral resource economists, some government spokesmen for energy policy, the interests of certain states (notably South Australia, but also Western Australia and Queensland), the AAEC, and the more conservative trade unions.

Against the establishment perception of 'Australia's national interest' stood small and poorly funded bodies of resistance, including the Australian Conservation Foundation, founded in 1975; Friends of the Earth (FOE), which came to Australia in 1975; and the Movement against Uranium Mining (MAUM), founded the following year. With these came newsletters, beginning in 1977 (FOE's newsletter began in 1981) and journals, including *Chain Reaction* (established in 1975), *Independent Australia* (founded 1975), and the anarchist *Action*. Several smaller organisations sprang up, including Campaign Against Nuclear Energy (CANE) formed in South Australia in 1976. Gradually, informed by the Ranger enquiry, anti-uranium sentiments drew increasing general press coverage.

In early 1977, in the midst of all this, and between the two reports of the Fox Commission, emerged scientists anxious to present an objective picture. These were aided – or not, according to taste – by the concurrent appearance in Britain of the Flowers report (Royal Commission 1976). In 1977, Stuart Butler, Charles Watson-Munro, who designed the HIFAR reactor, and Robert Raymond, an environmentalist film maker, produced a book that suggested a range of legitimate differences within the scientific community on Australia's realistic expectations of uranium mining and nuclear research. Even so, their conclusions appeared to vindicate the AAEC's policy. If, as President Carter had decided, breeder reactors and plutonium production were to stop, uranium mining seemed necessary – at best as a contribution to the world's energy needs, and at worst as a stop-gap until the arrival of fusion and solar energy. No one explained why research money was not going in these directions but, in the meantime, there seemed no alternative but to proceed with enrichment, guarded by a close watch on safety and waste disposal (Butler *et al.* 1977).

By this time, faced by so many conflicting agendas, the Australian layman was understandably confused. When published in 1976 and 1977, the Ranger enquiry satisfied no one. For their part, commercial interests rejoiced. Ten days after the second report appeared, the Fraser government granted permission to mine. Yet, the public had on record a long list of environmental arguments which the mining interests, including the AAEC, had failed to resolve. Between 1977 and 1982, an enormous fury possessed the land. For the first time, sustained resistance was mounted against the nuclear establishment and its allied interests in industry and defence. Aided by a handful of academics at La Trobe and ANU, opponents from Sutherland Shire to South Australia raised public consciousness at a time when the international environmental groups were fast gaining ground. These organized seminars, public demonstrations, 'bicycle rides against uranium', and street theatre. While the *Sydney Morning Herald* and the Melbourne *Age* remained broadly favourable towards the nuclear enterprise, the *Australian Financial Review* cast doubt on the economic legitimacy of its calculations. Letters to the *Canberra Times* raised the temperature of Titterton, and reached an influential reading public. By the late 1970s, according to one veteran observer, 'it was difficult to find any major newspaper not reporting something radioactive almost every day' (Smith 1979, p. 32).

In New South Wales, where the immediate likelihood of nuclear power was no longer an issue, resistance shifted to the environmental history of the reactor at Lucas Heights. The AAEC matched a growing problem of waste storage with perennially poor public relations. In 1972, Sutherland Shire Council, whose jurisdiction adjoined the federal land on which Lucas Heights was built, had begun a regular programme of monitoring alleged leakages and safety procedures. Titterton reported himself 'astonished' by these local attitudes, but more was to come. In 1976, in response to the announcement by the Fraser government of permission to export 9000 tonnes of uranium, Friends of the Earth set up an

'Atom-Free Embassy' outside the gates of Lucas Heights. It remained until a mysterious fire destroyed its storage shed in mid-1977.

Titterton and Baxter, now outside the Commission, continued to preach the virtues of nuclear energy, and defied all attempts at compromise (Titterton 1977; Baxter 1976). 'Plutonium is only twenty times as toxic as the caffeine in our coffee', Titterton wrote in 1977; and alleged that those who opposed nuclear power and uranium mining were either misguided or treasonous: 'The Australian anti-nuclear conspiracy is a political thing with links to international communism and the general motive of reducing the economic and military strength of the West' (Baxter 1977, 1979). In 1979, just as the Harrisburg accident seized the headlines of the world's press, Titterton was publishing a series of articles in the *Canberra Times* singing the virtues of nuclear reactors. In the circumstances of the late 1970's, however, such sentiments seemed hopelessly out of touch. Titterton's appeal fell on sceptical ears. None of the protest campaigns was successful in stopping Australian mining. Trade union blacklisting never stopped uranium exports. But it seemed that a critical debate was gathering momentum, inspired to some extent by the argument that Australia belonged to all Australians, and that Australians had an obligation to protect it. This spoke to states rights, and Aboriginal rights, as before – but also to a new sense of national sovereignty, resentful of foreign ownership, and angered by continued, uncritical deference to a small circle of self-appointed nuclear experts.

Thus began the 1980s and the age of economic rationalism. Academic critics took up opposition to the 'deadly connection' between uranium mining, nuclear reactors and weapons, linking Australia's nuclear policy to nuclear proliferation and the 'plutonium economy' (Falk 1983). The destiny of a 'nuclear Australia', enlivened by the prolific writings of Desmond Ball – most notably, *A Suitable Piece of Real Estate* – exposed Australians' limited knowledge of the assumptions on which American facilities at Pine Gap, Narrungar and North West Cape – all prospective nuclear targets – were set up and run. Articles on potential nuclear targets appeared next to apparent nuclear accidents. In 1984, the AAEC was accused of covering up accidents and leukemia incidences: both allegations were hotly denied, which made their eventual disclosure seem all the more ominous. An effluent leakage in 1985, and at least one fire, although brought carefully and quickly under control, fuelled hyperbolic press titles.[10] Meanwhile, uranium mining resistance continued, led by MAUM, and aided by Greenpeace, with its accompanying concern for a nuclear free Pacific. CANE reached its peak in 1984, when it ran a bookshop, café and full-time office, but its finest hours were in 1982, when it organized the first occupation of a uranium mine at Honeymoon, and – later commemorated in the folklore of opposition – the first and second blockades at Roxby Downs in 1983 and 1984.

By the late 1980s, as the price of uranium fell and the costs of nuclear power rose, opponents seemed to have won their case. The overall result of Ranger and its aftermath was public uncertainty. Uncertainty, like secrecy, is always a political weapon, but in this context it became a resistor's weapon as well. At the same time, it became a justification for government in delaying decisions, and an argument for 'safe policy' which, like safe sex, might well mean no policy at all. As a result, the uranium debate was left in suspension, ultimately to become prey to new policy interests dealing with the environment, company policy, export controls, and land rights.

Nuclear resistors had to face the political reality that many decisions affecting uranium were taken in secret, and by competing bureaucratic structures. Speaking on uranium policy during the 1970s were no fewer than six different ministers (representing the departments of Foreign Affairs, Trade, Energy, the AAEC, Prime Minister and Cabinet, and Treasury). A similar situation continued in the 1980s, as portfolios shifted responsibilities for uranium and nuclear research between Foreign Affairs, Primary Industries, and Industry, Technology and Commerce. These overlapping 'dependencies' linked international relations, nuclear proliferation and safeguards to the commercial production of medical isotopes. It was difficult for critics of the nuclear enterprise or uranium mining to gain purchase on such a slippery adversary, particularly as a central part of its political justification – notably, an Australian presence in the negotiation of international safeguards and the advancement of nuclear non-proliferation – could hardly fail to be approved by the deepest critic of Australian investment in the nuclear fuel cycle. Not least for this reason, party politics rarely cohered on nuclear or mining policy. Anti-nuclear sentiment remained a populist cause, but not a people's cause.

By the mid-1980s, the Great Debate fizzled to a political end. After Labour's return to power in 1983, a compromise was struck in the 'three mines policy', whereby uranium mining would be limited to Roxby Downs (Olympic Dam) in South Australia, and to the mines at Ranger and Nabarlek, both conveniently under the Federally-administered Northern Territory. By the mid-1980s, it was also agreed that nuclear energy would be dropped from the AAEC's portfolio; and that, while other fields of energy research would continue, there was no particular reason for the AAEC to be responsible for them.

In retrospect, at least two other factors contributed at this time to a major reversal in government attitudes towards the nuclear enterprise. First, by the late 1980s, thanks to 'big technologies' such as the Boeing 747 and telecommunications satellites, Australians were no longer isolated from the rest of the world. Their governments could no longer be passive where Western defence interests were in question. Neither, for that matter, were they merely bystanders, so long as their government sold uranium, nor could they be innocent of the uses to which that uranium was put. This 'becoming part of the world' gave the nuclear issue a renewed currency. Anti-mining groups resolved into networks for information and public discussion. A coalition for a Nuclear Free Australia combined the forces of Greenpeace, FOE, MAUM and CANE (until CANE disbanded itself in 1988). With this development came a new tactical orientation, as demonstrations gave way to displays and publications – some, ultimately, receiving small grants from the Federal government.

Secondly, in 1984 and 1985, across a heavy canvas of public darkness, flashed the searchlight of a Royal Commission on the British tests. The McClelland Commission opened the eyes of many Australians to decisions taken on their behalf. Australia, it seemed, had been sleeping with the enemy. There was no 'peaceful atom', and there never had been. The public and government looked deeply at the AAEC, and even more critically at the mystique of the 'nuclear knights'. Eminent scientists had been seen to disagree violently with each other 'in public [and] often in a vindictive and arrogant manner'. As compromised servants, 'The Australian scientist and expert [appeared] as a biased promoter of particular values and interests' (Smith 1979, p. 46). Even where the Commission's case was well put, resident Action Groups began to complain vocally about 'work carried out under an unwarrantable and suspicious veil of secrecy'. The Commission vigorously, and accurately, denied wrong-doing, and attempted to improve its public relations, issuing two new popular journals – *Nuclear Australia* (1984) and *Nuclear Spectrum* (1985) – in what proved to be a misguided belief that more publicity would disperse criticism.

In 1971, John Edwards described 'Australia's Atomic Fantasy' as 'an image that would amaze even Jules Verne'. Seventeen years later, it became widely evident that, in his words, 'no single group had spent so much and achieved so little' as the AAEC (Edwards 1971a). Its destiny was no longer to be avoided. In 1987, after four years' discussion and a major review, the AAEC was abolished and replaced by the Australian Nuclear Science and Technology Organization (ANSTO). Out went the old security requirements, and the link between nuclear research and nuclear power. In came a Nuclear Safety Bureau and a new commercial orientation, as the organization attempted to make the most of its capacity to produce radioisotopes, and to profit from its legacy of research in nuclear technology and waste containment. These events set the scene for the events of the first half of 1993.

The new reactor: resistance present

In August 1992, as part of the ministerial budget submission to the Government's White Paper on Science and Technology, Ross Free, Minister for Science, announced an enquiry into the proposal by ANSTO to replace the ageing 35-year-old HIFAR reactor with a new 15–30 MW thermal reactor, to be ordered from overseas at a cost in the region of $200 million. This followed the endorsement of the project by the Australian Science and Technology Council (ASTEC) as one of Australia's six most important developments in science and technology in the next decade. A new reactor was necessary, according to ANSTO, to serve Australia's scientific, commercial, industrial, and national interests – as a major national training facility, as a source for the commercial production of medical radioisotopes, and as a source of neutron beams for metallurgy and crystallography. The current reactor presently occupies about 10% of ANSTO's staff, and between 10 and 16% of its annual budget of $62 million.

For more than half a decade since Chernobyl, nuclear news – always good copy in the foreign affairs pages, when it concerned North Korea or Iraq, Japan or Indonesia - had slipped off the domestic pages of Australian dailies. Meanwhile, in science, fairly large expenditures – $50 million on an Australian telescope in 1988 – had gone ahead, largely undiscussed by a public which is largely uninterested in debates about scientific instruments. Then, all of a sudden, amidst a fine southern summer, the question of whether Australia could afford, and Sydney should have, a new, big and to some, threatening technology on its southern perimeter rippled through the media.

The time was ripe for contest. An election was in the offing. The Liberal opposition promised to favour proposals for uranium if they were returned. Given that the Government had spent considerable sums maintaining the present reactor, the Labour Government had given 'in principle' consent to the idea of a replacement. Why examine the question further? The question was superfluous. The nuclear genie had been disturbed, and left its bottle. A tribunal of three was constituted under the chairmanship of Professor Ken McKinnon. Within a few short weeks, what might have been easily construed as a narrow, technical, almost routine departmental issue, had instead sent up a barrage of unanswered (and perhaps unanswerable) questions:

1. Why does a country which had foresworn nuclear power need a reactor?
2. What were ANSTO's guarantees of cost containment, given familiar experience of cost over-runs in big technologies?

3. What were the implications of siting the replacement at Lucas Heights, or anywhere else? And what arrangements would be made for waste management?

4. What about the old reactor? Decommissioning was estimated to cost $15 million, but this might soar to $32 million if it were decided to entomb the old reactor; and up to $192 million would be needed for returning the location to a greenfield site.

5. What would happen if the reactor were not replaced? Could Australia not make radioisotopes in existing hospital cyclotrons? Or import what it could not?

6. ANSTO put a commercial case for the reactor's operations. But how effective were these arguments? Were they justified in an age of economic rationalism, when even universities had become 'cost centres', encouraging outside contracting?

7. What point was there in investing in a technology which richer countries had, but which could easily become 'frozen' in time? What value was there in having a reactor at all, in the absence of a power reactor programme, or any publicly agreed strategy for uranium conversion or enrichment?

8. Finally, what were the opportunity costs? In 1989, the last year for which there are published figures, the Federal science budget was about $1.5 billion. Federal support to Commonwealth scientific agencies was $862 million. The proposed reactor would cut a swathe through this figure. Possibly, 'research more in tune with Australian industry would be neglected to pay for the reactor and its maintenance' (Panter 1992).

To many of these questions, ANSTO had no quick answers. At its early meetings with local residents in November 1992, it merely stated its intention of proceeding with the reactor. Furthermore, it seemed that ANSTO had declined to submit plans for building approval, as any householder must do for the most routine domestic extension. Would the new reactor be built on the same site, in an increasingly densely populated Sydney suburb? Yes. How would ANSTO deal with its existing build-up of fuel rods? No reply. How was the new reactor to be decommissioned, in its turn? Again, no reply.

With a flair for self-damaging public relations it had apparently inherited from the AAEC, ANSTO then announced dates for a public inquiry, giving interested but under-resourced parties just six weeks over the Christmas and January holidays to make submissions. When they began to arrive, however, they dismembered ANSTO's arguments one by one. What establishes need? Australia, ANSTO argued, needed the technology to remain 'state of the art', and to stem a further 'brain drain'; it needed a reactor presence to justify Australia's seat on the Board of Governors of the IAEA in Vienna; and it needed a reactor because Indonesia was committing more resources to nuclear energy than Australia. Fundamentally, it also needed a reactor for basic research involving neutron-scattering.

Managing big technologies in a world of scarce resources requires sophisticated justifications and clear pay-offs. ANSTO's proposal might have easily been uncontroversial in the 1950s or 1960s; it was not to be so in 1993. In the new context, critics of the nuclear enterprise commanded the high ground. 'Scientific, commercial, industrial and national interests' had to be demonstrated, to be believed. What followed, was a re-representation of the 'object'.

In Canberra, departmental lines were quickly drawn. The science minister was lukewarm about the proposal, as the sum requested could buy all six other technologies his department had prioritized for the 1990s. ANSTO therefore looked outside its immediate circle, and to the Departments of Defence and Foreign Affairs, for support. But this raised other questions. At one extreme, how far would the Government be prepared to go in alienating public opinion, in a pre-election period much concerned with land rights, and when the Minister for Aboriginal Affairs held the Federal electorate in which Sutherland Shire was located?

Analysis yielded the conclusion that nuclear research was not the sole object of concern. Indeed, it appeared that Australia's nuclear future had become problematic. In the present state of the uranium economy, Australia could not be more than a minor player in the nuclear business – and that only as a supplier of yellowcake. While Australia does hold a respected place at the IAEA, and in the international nuclear research community, it is not clear that this would be compromised simply by Australia not having a research reactor of its own, as long as it shared facilities with others, as other countries do. In any case, even in Japan, the United States and Europe, nuclear expertise is too costly to be obtained and maintained except by sharing facilities. Above all, if it needed saying, the argument for keeping up a nuclear presence in the region was two-edged; a nuclear race between Indonesia and Australia would, in the phrase of a report from Parliamentary Library, be 'most unfortunate' (Panter 1992, p. 21).

Within the following weeks, a new debate began. Some 355 submissions were received by the tribunal, including 162 from research organizations (including many overseas), 65 from business, and nineteen from Commonwealth, State and local government agencies. Of these, 39% originated in NSW and 28% in Victoria. The advocates were predictable – justifying the desired outcome by arguments that used what must necessarily be rather vague projections, backed up by commercial cost-benefit analyses that begged more questions than they answered. Overall, 197 submissions supported the reactor, 128 opposed it, and 22 took neither side. The press divided on editorial lines. 'Most in Favour of Nuclear

Reactor', was the title given by the *Sun-Herald* to Peter Pockley's report; yet, 'Nuclear Reactor under Threat', said the *Campus Review* to the same article.[11] In May, the chairman of the review visited a number of nuclear facilities overseas, and in June, the review held a final public meeting at which it heard replies and final submissions from ANSTO, Greenpeace and the Sutherland Shire Council. The Reactor Review cost $2.5 million. Its outcome is still far from clear. What is clear are the new terms that appear to govern Australian debate in the nuclear field.

In this contest, the image of nuclear enterprise remained largely unchanged. The arguments were intended to stand for themselves. Against them, what form did the new resistance take? First, there were no longer merely 'T-shirt' activists. Instead, opposition to the reactor was dominated by councillors and advocates, highly articulate, educated spokesmen, representing local authorities, community groups and environmental bodies employing academic and financial consultants.[12] Secondly, with what information they could get – not always easy, as ANSTO controlled the information on which it based its commercial case – the nuclear opponents transformed an issue of technical interest into an issue of community importance.

In relation to waste management, for example, opponents focused public attention on the little known ANSTO Amendment Act of 1992, which the Federal parliament passed over some local opposition, and which gave ANSTO, as a Federal body, power to override state law, including environmental legislation. By this Act, ANSTO is not required to file Environmental Impact Statements. And while the Government has apparently agreed that Lucas Heights is not necessarily to be the national nuclear waste storage facility, it has not said where that is to be; while the states are vying hard against each other to reject the compliment.

Fourthly, the opposition called into question the independence of the Nuclear Safety Bureau, which was lifted by the Act of 1992 symbolically outside the fence at Lucas Heights.[13] Its acting director and several of its technical staff were drawn from ANSTO. While this does not in itself suggest that their views are compromised, the Bureau necessarily relies largely upon the information ANSTO gives it, and there can be legitimate questions about the extent to which it can be truly independent.

The reactor opposition has been helped by contextual factors. Following its election miracle, the new Labour Government was not inclined to reopen the uranium issue, and it seems likely that no additional mining developments will be undertaken in the life of the present parliament. Nor, given the Government's agenda for Aboriginal land rights, is it either timely or promising to raise divisive matters of a nuclear nature. The press have raised reminders of traditions of nuclear secrecy, all too familiar to those who lived through the debates of the

1970s. When, in March 1993, South Africa claimed it had made nuclear weapons in the late 1970s, apparently undetected by the rest of the world, at least one popular Australian journalist asked whether Australia might not also have gone, unannounced to the public, 'beyond the elementary stages of planning the possible development of nuclear capacity' (Robinson 1993). The very suggestion prompted memories of secret covenants, and probably served the opposition's case.

In August, 1993, the Reactor Review enquiry issued its first (and so far, only) report (*Future Reaction*, 1993). After all the evidence it had received, the Review concluded, it was 'utterly wrong to decide on a new reactor before progress is made on identification of a high level waste respository site'. It further acknowledged that HIFAR is decreasingly competitive as a research tool. It suggested that, as no such research facility can ever be completely commercial, any decision concerning its replacement 'must rest primarily on the assessed benefits to science and Australia's national interests'. Under the circumstances, the Review recommended that HIFAR be kept going, that additional funds be given to international neutron scattering facilities, and that a final decision on a new reactor should be deferred for five years, when the relative arguments 'might be clearer'. It asked that work begin immediately on a high level waste site; and suggested that any future decision would be influenced by arguments relating to the economic production of isotopes and the performance of reactor research in the context of Australian science. While not dismissing the case for a new reactor, the Review left the matter open – not least, because of the political benefits the fact of having a reactor may bring Australia in the world of international nuclear diplomacy.

After months of vigorous debate, some commentators viewed this outcome as weak and unsatisfactory; others, no better or worse than the circumstances allowed. As this chapter goes to press, the future contours of the debate are yet to be seen. But it is clear that a chapter in the history of Australia's nuclear expectations has closed, and that another is beginning. If the era of nuclear optimism is past, and the era of nuclear opportunism is fraught with difficulties, we may be entering a new era of nuclear scepticism, with at least two defining characteristics. First, the process of review, while adversarial, has become increasingly accommodating to citizens' interests. It follows that government agencies cannot hope to win cases by slighting local government or citizens groups. Where parliamentary processes are not appropriate, participatory democracy is finding more direct ways of dealing with issues. Secondly, nuclear power stations were once debated and defeated on economic grounds alone. Now, nuclear research must meet criteria which are at the same time environmental, commercial and scientific. Reactor opponents have lifted the discussion above the specific local consequences of a reactor, and have forced the entire question onto a national plane, where nuclear technology must compete among other proposals for scarce resources (Lowe 1993). This circumstance has the effect of putting single-issue experts on the defensive, requiring them to show not only that what they propose is valuable, but also that there are not other, wiser uses for the same resources.

Conclusion

These characteristics of this 'new' nuclear scepticism and ordered resistance return us to the question that Martin Bauer (Chapter 1) has posed: what can we learn from resistance to new technologies that will increase social benefits and lower social costs? Australian experience offers three tentative answers. First, it appears clear that Government investment in new technologies, particularly where issues of property, health and environmental damage may be involved, requires a clearly stated and argued case, accompanied by a full disclosure of information. If they wish to be persuasive, government departments, with their history of secrecy, must embrace a culture of accountability. Secondly, what constitutes a 'social benefit' must be defined in terms of competing opportunity costs. Thirdly, while it is perhaps inevitable that political decisions are finally made on grounds other than those advanced by either advocates or opponents, any tendency on the part of those advocating new technologies to dismiss those who resist them as uninformed or emotional is counterproductive to advocates' interests. As any teacher knows, clear questions help make clearer arguments. As and when errors are found, they must be accepted, not denied.

The social costs of nuclear testing, uranium mining and nuclear technology have been high. The social benefits – apart from mining profits, accruing partly to overseas interests – are at best, difficult to measure. There has been a clear voice against nuclear testing. And while there is unlikely ever to be a consensus on mining, there may well be agreement on what levels of mining will be tolerated by a country increasingly concerned about the fragility of its environment and the long-term interests of its indigenous population. Finally, it may also become clear within the coming months what kinds of nuclear expertise are appropriate for a country with Australia's geography, population and economic position. There are grounds for a new optimism, replacing the opportunism of the post-war period. In the last ten years, the old nuclear establishment has given way to a new nuclear enterprise, which is learning not to overstate its claims. Australian nuclear science and technology have established a role internationally (O'Reilly 1993). Their advocates must now build a firmer partnership with the Australian people, a relationship that accepts dissent as a normal part of the political process

(Nelkin & Pollack 1977; Hirsh & Nowotny 1977). This was not the case in the past. At present, and in the future, the nuclear enterprise will succeed in Australia only if and when it can persuade the public of the legitimacy of its case.

Notes

1 This essay forms part of a study of the Australian Atomic Energy Commission and its place in Australian nuclear history, supported by the Australian Research Council and encouraged by the ANSTO. I am indebted for the assistance of Daniel Hanna, Leanne Piggott and John Geering. My participation at the conference on 'Resistance to New Technology – Past and Present' was made possible by a travel grant from the Department of Technology, Industry and Commerce, Canberra.

2 HIFAR (the High Flux Australian Reactor) is one of six DIDO-class, 10 MW, heavy water moderated, high neutron flux, materials testing reactors, which was begun in 1955 and achieved criticality on Australia Day, 1958. It was based upon the DIDO reactor at Harwell, which first went critical in November 1956. A third, the Dounreay Materials Testing Reactor in Scotland, was closed in the 1980s; two others survive – in Denmark (Risø) and in West Germany (Jülich). HIFAR cost £4 million to build, and uses 60% enriched U235. For further details see 'HIFAR ANSTO's Major Nuclear Facility', *Nuclear Spectrum* **3** (2), (1987), pp. 10–13.

3 In 1944, with the cooperation of the Australian Government, British geologists explored for uranium in South Australia. The first results were poor, but prospects improved with the new discoveries at Radium Hill, and later in the Northern Territory. For the British background, see Margaret Gowing (1964, 1974).

4 Brian Martin (1980); Caro & Martin (1987); Newton (1992); Cockburn & Ellyard (1981). Martin was knighted in 1957, Oliphant in 1959, Baxter in 1965, and Titterton in 1970. No full biography of Baxter has been written, but see Angyal (1991), Willis (1983) and Dillon & Bowman (1987). For a discussion of their attitudes and activities see Roy MacLeod (1995).

5 Mark Oliphant, Industrializing Australia: Needed: Power and People, *Nation*, 18 June 1955; Peaceful Uses of the Atom Bomb, *Professional Officers' Association*, 183 (February 1963), 3–6.

6 For an appreciation, see Alder (1990) and Angyal (1991, pp. 191–4). Between 1955 and 1969, Baxter was also Vice-Chancellor of the University of New South Wales, and so was only part-time chairman of the AAEC until his retirement in 1969; he then became full-time chairman until his resignation in 1972.

7 See, for example, criticism of the 'patronising attitude' of the Commission by R. B. Temple (1970).

8 John Edwards, quoting Baxter's address to the Rotary Club in March 1968.

9 Fourteen tenders were received from American, German, British and Canadian firms; four were shortlisted, and after much discussion of a Canadian (CANDU) reactor that would have used Australian natural uranium, the British enriched uranium steam generated heavy water (SGHWR) option was instead tipped as winner. However, Academic economists estimated that on the best calculations, the likely unit cost of nuclear-produced electricity would be 60% greater than the cost of electricity produced by any one of Australia's several conventional coal fired stations. Bruce Ross, quoted in Jervis Bay Reactor Costs 'Not Justified', *Australian Financial Review*, 1 June 1970. For criticism of Baxter's proposal, see Ian Moffitt, 'Baxter's Vision Splendid', The Australian, 19 August 1970, p. 11.

10 Cf. Paul Baily, 'Lucas Heights Forum Blows Up', *Sydney Morning Herald*, 28 June 1985; for more recent allegations, with a typical editorial gloss in the title, see 'Acid Rain Fuels Atomic Centre Fears', *Sydney Morning Herald*, 1 June 1993.

11 Peter Pockley, 'Most in favour of Nuclear Reactor', *Sun-Herald*, 7 March 1993; 'Nuclear Reactor under Threat', and 'Highest Price Tag in Science puts ANSTO Reactor under Searching Public Microscope', *Campus Review*, 18–24 March 1993.
12 Cf. submissions by the Australian Conservation Foundation, Greenpeace, and the Sutherland Shire Council to the Reactor Review, February and May 1993.
13 The first report of the Nuclear Safety Bureau, established on 30 June 1992, appears in the *Annual Report of the Safety Review Committee*, 1991–2 (Canberra: AGPS, 1993).

References

ALDER, K. F. (1990). Sir Philip Baxter – Founder of Lucas Heights. *Nuclear Australia* 7(1), 1–4.

ANGYAL, S. J. (1991). Sir Philip Baxter. *Historical Records of Australian Science* 8, 183–97.

AUSTRALIAN NUCLEAR SCIENCE AND TECHNOLOGY ORGANIZATION (1993). *Submission to the Research Reactor Review*. February, p. 3. ANSTO.

ARNOLD, L. (1987). *A Very Special Relationship: British Atomic Weapon Trials in Australia*. London: HMSO.

BAXTER, J. B. (1963). The first ten years, 1953–1963. *Atomic Energy*, April, 3–12.

BAXTER, P. (1957). *Atomic Energy and Australia*. Supplement to the Royal Australian Chemical Institute Proceedings, November.

BAXTER, P. (1968). Australia's nuclear power. *Quadrant*, XII, 57–63.

'X'[BAXTER, P.] (1969). Australian doubts on the Treaty. *Quadrant*, XII (3), (1968), 31.

BAXTER, P. (1976). The case for nuclear power. *Current Affairs Bulletin* **53**, August, 18–23.

BAXTER, P. (1977). Radioactivity and radiation: a guide for the layman. *Canberra Times*, 19 October, 20–1.

BAXTER, P. (1979). Is the Anti-Nuclear Campaign an international conspiracy? *Quadrant*, June, 10–12.

BOLTON, G. (1990). *The Oxford History of Australia*, Vol. 5: *1942–1988*. Melbourne: Oxford University Press.

BUTLER S. T. *et al.* (1977). *Uranium on Trial*. Sydney: Horwitz Group Books.

CARO, D. E. & MARTIN, R. L. (1987). Leslie Harold Martin, 1900–1983. *Historical Records of Australian Science* 7(1), 97–107.

CAWTE, A. (1992). *Atomic Australia, 1949–1990*. Sydney: University of New South Wales Press.

COCKBURN, S. & ELLYARD, D. (1981). *Oliphant*. Adelaide: Axiom Books.

COLLIS, B. (1989). *Snowy: The Making of Modern Australia*. Sydney: Hodder and Stoughton.

DILLON, L. & BOWMAN, L. J. (1987). *John Philip Baxter: an interview*. Sydney, University of New South Wales University Interviews Project.

EDWARDS, J. (1971a). Australia's Atomic Fantasy. *Australian Financial Review*, 23 June.

EDWARDS, J. (1971b). Atomic energy in search of a new role. *Australian Financial Review*, 23 June.

ENCEL, S. & McKNIGHT, A. (1970).   Bombs, power stations and proliferation. *Australian Quarterly* **42**(1), 20.

FALK, J. (1983).   *Taking Australia off the Map: Facing the Threat of Nuclear War.* Ringwood, Victoria: Penguin.

FINDLAY, T. (1990).   *Nuclear dynamite: the peaceful nuclear explosions fiasco.* Sydney: Brassey's Australia.

GOWING, M. (1964).   *Britain and Atomic Energy, 1939–45.* London: Macmillan.

GOWING, M. (1974).   *Independence and Deterrence.* London: Macmillan.

HIRSH, H. & NOWOTNY, H. (1977).   Information and opposition in Australian energy policy. *Minerva* **15** (1977), 316–34.

JOHNSTON, G. (1969).   *Clean straw for nothing.* London: Collins.

LOWE, I. (1993).   *Science policy implications of the ANSTO proposal for a new reactor.* First Report, February 1993, Second Report, May 1993.

MACLEOD, R. (1995).   Nuclear Knights vs Nuclear Nightmares: Experts as Advocates and Emissaries in Australian Nuclear Affairs. *Public Understanding of Science* (in press).

MARTIN, B. (1980).   *Nuclear Knights.* Canberra: Rupert Public Interest Movement.

MOFFITT, I. (1970).   Baxter's Vision Splendid. *The Australian*, 19 August, p. 11.

MOYAL, A. M. (1975).   The Australian Atomic Energy Commission: a case study in Australian science and government. *Search* **6**, 365–84.

NELKIN, D. & POLLACK, M. (1977).   The politics of participation and the nuclear debate in Sweden, The Netherlands and Austria. *Public Policy* **25**, 335.

NEWTON, J. O. (1992).   Ernest William Titterton, 1916–1990. *Historical Records of Australian Science* **9**(2), 167–87.

OLIPHANT, M. (1955).   Industrializing Australia: needed: power and people. *Nation*, 18 June.

O'REILLY, D. (1993).   Atomic bonds. *The Bulletin*, 22 June, pp. 22–3.

PANTER, R. (1992).   Should Australia build a new reactor at Lucas Heights? *Parliamentary Research Service*, 22 December, p. 21.

RANGER ENQUIRY (1976).   *Ranger Uranium Environmental Enquiry.* First Report. Canberra: AGPS.

RANGER ENQUIRY (1977).   *Ranger Uranium Environmental Enquiry.* Second Report. Canberra: AGPS.

RESEARCH REACTOR REVIEW (1993).   *Future Reaction.* Canberra; August.

RICHARDSON, J. (1968).   *Australia and the non-proliferation treaty.* Canberra Papers on Strategy and Defence, No. 3. Canberra: Australian National University Press.

RICHARDSON, J. (1969).   Nuclear follies. *Quadrant*, XIII, 70.

ROBINSON, P. (1993).   'Nuke Club' Conspiracy. *Sun Herald*, 28 March.

ROYAL COMMISSION (1976).   *Nuclear power and the environment.* Sixth Report of the Royal Commission on Environmental Pollution. London: HMSO.

ROYAL COMMISSION (1985).   *Report of the Royal Commission into British nuclear tests in Australia.* Canberra: AGPS.

SMITH, T. (1979).   Forming a uranium policy: why the controversy? *Australian Quarterly* **51**, p. 32.

TEMPLE, R. B. (1970).   Tiger by the tail: nuclear power in Australia. *Outlook*, August, p. 10.

TITTERTON, E. W. (1965a).   An Australian bomb? *The Bulletin*, 19 June, pp. 20–4.

TITTERTON, E. W. (1965b).   Australia's nuclear weapon dilemma. *Australian International News Review* **1**, 7 December, pp. 28–30.

TITTERTON, E. W. (1977).   World energy requirements and resources. *Australian Quarterly* **49**, June, pp. 18–36.

WATSON, G. M. (1971).   *The environmental monitoring programme at the AAEC research establishment, Lucas Heights*. September 1971.

WILLIS, A. H. (1983).   *The University of New South Wales: the Baxter years*. Sydney: University of New South Wales Press.

New technology in Fleet Street, 1975–80

RODERICK MARTIN

Introduction

Since 1970 technology has transformed the mass media of communication. This transformation is obvious in the cinema, television and radio, where new products (videocassette recorder, satellite television, the Walkman) and new cultural forms have proliferated. The products of the newspaper industry have remained relatively stable: newspapers have not changed their role, and in many cases formats remain traditional, although ownership patterns have changed. However, the production system of newspapers has been transformed; in particular, methods of text production have changed radically. The technological transformation has had major implications for work organization, both in terms of the job tasks themselves and social relations, as well as in employment levels, pay and union organization. Although several other changes occurred simultaneously, the transformation has been primarily associated with the introduction of computerized photocomposition. This chapter examines developments in Fleet Street, the major centre of national newspaper production in Britain until the 1980s, linked with this process. The focus is upon industrial relations, institutions and processes, and the extent to which they provided an effective framework within which the conflicting interests associated with the introduction of new systems could be reconciled. In doing so, the purpose is to illustrate four major themes concerning technological change, social organization at work and resistance to new technology.

Contradictions in management and unions

The first theme is the need to recognize the contradictory strands in the approaches of both management and unions. Neither group is united: it is rarely a case of management pursuit of a coherent strategy of technological innovation, faced by total employee resistance. Amongst managements the commitment to technological innovation is likely to be contingent, depending upon specific balances of interests. Established middle management is especially likely to be sceptical of the benefits of new technology, since it faces the problems of disturbances in routine whilst being unlikely to reap the rewards of success – which are likely to be earned by senior management and by product champions. Similarly, there are contradictory strands amongst employees. Survey evidence indicates general support for technological innovation amongst employees, whilst formal union policies usually support investment in new technology, subject to safeguards: unions have criticized managements for lack of investment in new technology rather than for too much (Daniel 1987; Dodgson & Martin 1987). All the unions in the printing industry were formally committed to supporting the Trades Union Congress (TUC) policy of encouraging investment in new technology (Trades Union Congress 1979). Yet there were obvious differences in approach between unions in the industry, especially between the National Graphical Association (NGA) and the Society of the Graphical and Allied Trades (SOGAT), reflecting the particular circumstances of craft and non-craft unions. Within individual unions leaders were subject to conflicting pressures, especially between members directly affected by technological changes and others.

Institutional structures and processes

The second theme is the importance of institutional structures and processes. This chapter outlines the system of industrial relations in the national newspaper industry, especially the dual system of industry and company level negotiations. In normal circumstances industry level negotiations between the employers association and the unions individually established a framework agreement, which was filled out at individual newspapers: for example, the London Scale of Prices for compositors' work had been negotiated centrally, but individual managements negotiated over special allowances. The Newspaper Publishers' Association (NPA) had played little role in technological development historically: responsibility for production processes, and the distribution of labour, was the sole responsibility of individual house level management. The introduction of computerized photocomposition was recognized as a major change by both managements and unions. Both sides recognized that a special approach was needed. However, it was difficult for individual newspapers to delegate responsibilities which had been carried out traditionally on a company basis to a collective organization – to share responsibility for organizing responses to major new developments with their commercial competitors. On the union side, it was equally difficult for competing unions to establish common policies: the institution for developing common policies – the Printing Industries Committee of the TUC – was even less firmly grounded than the NPA machinery. The institutions for articulating – and even more for aggregating – interests on both sides were ill-developed. Limited institutional development has a major impact on the ability of managements – and unions – to deal successfully with the social consequences of technological change. Similar problems in reconciling collective industry level initiatives with individual company strategies occur in other industries.

Interest aggregation

The third theme follows directly from the second: it is the significance of different levels of interest aggregation for the ability of groups to resist the introduction to new technology. 'Encompassing' organizations are more effective in aggregating interests than particularistic organizations. Modifying Olsen, it is argued here that the lower the level of interest aggregation, the greater the likelihood and ability of groups to oppose technological innovation (Olsen 1982). Decentralized systems of interest aggregation narrow the focus on the interests of the specific group, as well as cumulatively providing for greater numbers of opportunities for resistance. Hence trade union structures which push the major decision-making responsibilities to lower levels of the organization – as chapels in the printing industry unions – and in this way are in one sense more democratic, may be more resistant to technological innovation or more capable of securing a higher price for its acceptance. Unions which aggregate interests at a higher level in the organization are less likely to resist new technology since they necessarily represent a broader range of interests and provide scope for greater flexibility. It is for this reason that issues of the institutional structure in general, and trade union structure in particular, are important for understanding the process of implementing technological innovation.

Resistance: voice in need of an audience

The fourth theme is the overall significance of resistance to new technology for social processes. In his classic work, *Exit, Voice and Loyalty*, Hirschmann distinguished three responses to change; exit, loyalty and voice (Hirschmann 1970). This analysis has been subsequently extended to the overall role of trade unionism: trade unions are a means of providing 'voice' for employees affected by changes (Freeman & Medoff 1984). Resistance to new technology is a means of expressing 'voice'. Such voice may be reasoning or unreasoning, individual or shared, particularistic or general: the voice directs attention to issues which might otherwise be neglected. This chapter is predicated on the assumption that attention to the form, content and functions of resistance to new technology represents a form of listening to the voices of employees affected by technological change.

This contribution outlines the structure of the industry's market and industrial relations up to the 1970s, before summarizing the early development of computerized photocomposition in the United States. The chapter then discusses management objectives and union policies on new technology, and indicates the range of positions held. Union policies could be summarized generally as conditional acceptance rather than direct resistance, although in some cases the conditions were so stringent as to amount to rejection, and some groups rejected new technology outright. The chapter then outlines two alternative approaches towards introducing new technology in highly unionized environments, where management preferences for relying on consultation alone cannot be realised (Daniel 1987). The first involves negotiations via peak level industry associations. The second involves separate company by company negotiations. It is argued that for industries structured like the national newspaper industry an industry wide approach is necessary to control the potentially disruptive effects of new technology. In the event competitive pressures, both between newspapers and between union members, prevented the realization of such an approach. Individual newspapers followed independent strategies, which in some news-papers (e.g. *The Times*) were to cause major industrial relations problems. The eventual outcome of the introduction of computerized photocomposition in national newspapers was the end of Fleet Street as a production centre, the transfer of editorial, origination and printing to new sites elsewhere in London and renewed impetus for regional printing. The end of Fleet Street as the centre of newspaper production was not caused by the development of computerized photocomposition. Technological conditions combined with other pressures to bring this result about – it was not the result of new technology or resistance to new technology alone. But technological change and its often maladroit handling by employers and unions made the end of Fleet Street more likely.

Management and unions believed that they had both complementary and conflicting interests over the introduction of computerized photocomposition. Managements perceived it as an opportunity to reduce production costs as well as to enhance competitive responses (through 'editionizing') to developments in the competitive environment in the industry; in the mid-1970s, the industry was greatly concerned about competition from television as an advertising medium as well as a source of information, and with the threat from freesheets as a local advertising medium. The desire to reduce production costs involved conflict with union members' interests, the second concern with competitive responses did not. Some unions saw new technology as a means of improving the opportunities available to their members, if necessary at the expense of members of other unions, whilst other unions were concerned primarily to defend existing practices, including the division of labour, as far as possible. The role of industrial relations structures and procedures in technology and innovation is to enhance the complementarity of interests, whilst providing means for the resolution of conflicting interests. The industry's institutions failed to achieve this. The disrupted history of the attempt to introduce computerized photocomposition in Fleet Street in the 1970s is not simply the story of resistance to new technology: it is the story of institutions failing to provide means for handling conflicts of interests in a turbulent environment.

Industry structure

The newspaper industry comprises three groupings, with different interests – and economic fortunes: the broadsheet national papers (*The Times*, *The Financial Times*, *The Guardian*, *The Daily Telegraph* and now *The Independent*); the tabloid national press; and the provincial press, both weekly and daily, paid for and free; this paper is concerned only with the national press – regional and local newspapers followed a different pattern of technological development. Revenue is derived from two sources – direct sales to the public and the sale of advertising space. The importance of the sources of revenue differs between sectors: the sale of advertising space is especially important for broadsheet national newspapers and for the provincial press – a higher proportion of revenue is derived from cover price sales for the national tabloid press. Broadsheet newspapers carried more column centimetres of advertising than tabloid newspapers, but because they carried more column centimetres in total the proportion was no greater. The proportion of space devoted to advertisements varied surprisingly widely between newspapers, even within the same category. In the three sample weeks covered by the Economic Intelligence Unit (EIU) investigation in 1965 and 1966, the average space devoted to advertisements varied between 49% (*Daily Telegraph*), and 26.7% (*Financial Times*) amongst the broadsheets and between 37.5% (*Daily Express*) and 18.2% (*Daily Sketch*) amongst the tabloids (EIU 1966; no similar analysis was carried out by the Royal Commission on the Press in 1976). Although income from cover price sales may be of lesser importance for broadsheets, advertising revenue is sensitive to the level of circulation especially amongst wealthier social groups: premium prices may be charged for advertising seen by high proportions of social groups A, B and C1. The relative importance of different sources of revenue directly affected the focus of technological innovation. For example, newspapers which relied heavily upon small advertisements (e.g. *The Daily Telegraph*), frequently received over the telephone, were concerned with innovations which facilitated direct input by telephone sales girls, whilst other newspapers were less concerned with such innovations (e.g. *The Financial Times*). The level of ownership concentration in the sector has always been high: in 1975 there were only nine companies in Fleet Street. Despite the high level of concentration, competition between newspapers for circulation was strong, especially within each of the sectors.

Fleet Street has played a major role in the demonology of British industrial relations; stories of 'Spanish customs', 'phantom shifts', payslips made out to 'Mickey Mouse' have enlivened industrial relations textbooks at least since the 1960s (accounts of the industry are found in Sisson 1975; Royal Commission on the Press 1976; Martin 1981; Cockburn 1983). It has been viewed as an industry with major industrial relations problems, examined in detail in the report of the Royal Commission on the Press 1974–6 chaired by Lord Macgregor. Conflict between management and unions was seen as endemic, with strong trade unions forcing employers to pay employees higher wages than their levels of skill or effort justified. Union fragmentation and rigid demarcation lines led to inflexibility and low productivity. Major efforts at technological innovation were thwarted. The return on investment was low. Trade union power was seen as resulting from the nature of the product, trade union control of the labour supply, and owners' apparent lack of concern with unit production costs. Newspapers are highly perishable products, having an effective 'shelf-life' of only 12 hours – yesterday's newspaper is waste paper – with high levels of substitutability: other sources are available for information, entertainment and advertising. Trade unions had secured control of the labour supply in the late nineteenth century, recruitment for both skilled and unskilled labour being managed by the unions. Historically, owners had appeared to be relatively indifferent to unit production costs because the motivations for newspaper ownership were as much political, social and cultural as economic. Newspaper proprietorship was regarded as 'special', conferring influence and social prestige which did not accrue to proprietors in other industries – *The Times'* claim to be a national cultural institution was generally accepted. For proprietors, newspapers appeared to be regarded as a form of conspicuous consumption rather than a means of generating profits. This overall analysis has limitations; levels of strike action throughout Fleet Street's history have been low, and the industry's introduction of the linotype machine was regarded as highly successful. However, the evidence is clear that throughout the industry's history until the 1970s, management's major preoccupation was with avoiding disruptions to production, even at the expense of high unit labour costs and restrictions on technological innovation.

On the management side, industrial relations were organized on the two-tier basis characteristic of manufacturing industry, with an industry level employers' association, the Newspaper Publishers Association (until 1968 the Newspaper Proprietors Association) and industrial relations officers at individual newspapers. In comparison with the engineering industry, or even provincial newspapers, the industry provided a favourable environment for an effective industry level employers' association: there were only nine separate companies, the senior management teams in each house were familiar to each other, the companies shared a common distribution system organized by the NPA on behalf of the whole industry, the companies were geographically concentrated (and managers ate in the same restaurants), and the production systems were broadly similar. (Mirror Group Newspapers (MGN) was not a member of the NPA in 1975, but shared in the distribution system and coordinated its industrial relations policies with the Association.) But the NPA never developed as a strong central organization; it was less effective as a negotiating body than the Newspaper Society which negotiated on behalf of owners in the provincial newspaper industry. Indicative of the respective roles is the relation between centrally negotiated minimum wage rates and average earnings; for all groups of production workers average earnings were more than double the industry rate, the gap growing wider throughout the 1970s.

Management organization of industrial relations at house level varied widely between houses. However, there were three common features. First, a heavy reliance on management experience within the printing industry; of the 41 staff professionally concerned with industrial relations or personnel matters, only two had worked for five years or more outside printing, and 32 had never worked outside printing at all (Royal Commission on the Press 1976, p. 267). Secondly, a close integration of industrial relations into line management responsibilities: because of the strength of the production unions line managers had to devote considerable attention to industrial relations. This integration had many advantages, and accorded with trends which were to become more widespread in the 1980s, but also had disadvantages. Line managers lacked the time to stand back and reflect upon their practices, and were thus likely to preserve existing custom and practice. They were likely to dismiss new approaches suggested by industrial relations professionals or outside consultants as impractical. The limited range of experience and narrow focus meant that when senior management suggested new approaches production managers were ill equipped to evaluate them with the necessary sympathetic scepticism. Finally, the close involvement of senior managers in industrial relations led middle managers to complain that they lacked sufficient authority when dealing with chapel officials, and frequently found their decisions undercut.

Historically, four major production unions dominated the organisation of production in the industry: the National Graphical Association – NGA (renamed NGA '82 in 1982 following the merger with SLADE and renamed again as the Graphical Paper and Media Union in September 1991, following a further merger); Society of Lithographic Artists, Designers and Engravers – SLADE (merged with NGA to form NGA '82 in 1982); National Society of Operative

Printers, Graphic and Media Personnel – NATSOPA (merged with SOGAT in 1982); and Society of Graphical and Allied Trades – SOGAT (merged with NATSOPA to form SOGAT '82 in 1982). In addition, the AEU and EETPU organized their respective areas. Editorial staff have been organized throughout primarily by the National Union of Journalists, although a small minority belonged to the non-TUC Institute of Journalists. Relations between unions have varied in cordiality, depending upon organizational interests and economic circumstances; relations between the NGA, NATSOPA and SOGAT were traditionally difficult because of tensions between skilled and unskilled workers over differentials and promotion opportunities, with particular sensitivity between machine managers and machine assistants in the press room. Within each union relations between national and local officials also varied in cordiality, local officials and shop stewards (FOCs) in the NGA having a particularly strong local orientation, especially in London. Union branches and chapels in Fleet Street had substantial financial resources outside the control of national officials.

It is against this background that the attempt to introduce computerized photocomposition was made in the mid-1970s. The techniques of computerized photocomposition were developed initially in the United States in the 1960s. The central feature of computerized photocomposition is that it allows direct input of text by journalists and by telesales staff, thus eliminating the need for large amounts of keyboarding by compositors. It removes the need for the use of molten lead. Anthony Smith enthusiastically summarized the basic principles of computerized photocomposition in his book *Goodbye Gutenberg* (1980):

> the images of the pages are created photographically instead of physically in metal type. At the centre of the plant's operation sits the main computer ... which stores the copy digitally in its memory ... The computer performs the work of [hyphenating and justifying] the text, which used to be done line by line by the fingers of typesetters. The computer also makes the text corrections according to the orders of the editors and proof readers ... The business of 'text manipulation' (editing and transposing, correcting and rearranging the lines of print) grows easier and more flexible with every generation of equipment in the market ... In the case of the ad-taker, the vdu permits the operator, while taking down a telephone dictated classified advertisement, to check the credit worthiness of the customer, plan for future insertions of the advertisement in subsequent editions, inform the customer of the exact cost, and bill him automatically for his order.   (Smith 1980, pp. 86–7)

The product market and production system were equally suitable for the use of computerized photocomposition in Britain as in the United States. In both countries unit labour costs were relatively high and reductions in labour costs were a major objective. According to the Royal Commission on the Press, labour costs represented 44% of the variable costs of producing a national tabloid newspaper, and 40% of the variable costs of producing a broadsheet newspaper. 'First copy' or origination costs were especially high for broadsheet newspapers, which therefore had the greatest incentive to reduce origination costs, i.e. to reduce the number of compositors. In both countries the product market (both for advertising and sales) was becoming increasingly segmented, and the potential benefits of 'editionalizing' increasing: editionalizing involved the incorporation of specially targeted local advertising and editorial matter alongside standard copy. In both countries competitive pressures, both amongst newspapers and between newspapers and other mass media, were leading to the need for more frequent edition changes (if only to include the latest football results). In both countries there were obvious advantages in single keystroking, for example with telesales staff directly inputting material from material supplied over the telephone. In both countries the central production of editorial matter and its electronic transmission for regional printing offered the most economical and efficient combination of centralization and decentralization. Finally, in both countries compositors were the most strategically placed occupational group and therefore perceived as the greatest limitation on management control, although they were not invariably the highest paid (Rogers & Friedman 1980).

Management objectives from technological innovation

Managements and unions shared a common interest in restructuring and modernizing the industry to meet the threats from substitute products (especially television as an advertising medium): both recognized the importance of joint action to achieve this objective. But managements had four further objectives which involved conflict with union interests: reductions in labour costs; control over work allocation; monitoring and evaluating performance; and securing control over the recruitment process.

The specific means to achieve the reduction in labour costs differed between newspapers. But the common strategies included reducing the size of the composing room (by the introduction of 'single keystroking' by editorial and telesales staff where possible) and by buying out the piece-price system (the London Scale of Prices) for the remaining compositors. Management would thus increase control over labour costs by a comprehensive reduction in manning levels and by the establishment of a more predictable and more easily controlled payments system than the London Scale of Prices.

Secondly, management wished to regain control over the allocation of labour. Traditionally, operations in the composing room had been based on *de facto* labour only sub-contracting, the union chapels being the contractors, with a high degree of internal self-regulation: management control over the allocation of labour tasks was limited, the central role being played by the Father of the Chapel (FOC) rather than the nominal shop supervisor. Computerized photocomposition enabled, and required, a more coordinated and unified system, with the work flow more tightly coupled to a more technologically sophisticated and integrated production system. Management's objective was to create a more integrated composing room labour force, which would have fewer job categories and more flexibility than the traditional composing room. Under traditional forms of organization the composing room consisted of five major occupations: linotype operator (NGA); piece case hand (NGA); permanent time hand or stab hand (NGA); linotype assistant (NATSOPA); and proof puller (SOGAT). The linotype operator set copy in hot metal on linotype machines; the piece case hand set large type characters for headlines; the stab hand made up the pages by taking type from compositors and inserting it into a forme together with picture blocks on a table or stone; the linotype assistant acted as an assistant, for example in supplying machines with hot metal; and proof pullers took proofs of each story to the editorial or feature department (Royal Commission on the Press 1976, p. 15). The five occupations were capable of being reduced to one.

Thirdly, managements wished to improve monitoring and evaluation of performance. Traditionally performance in the composing room had been governed by the operation of the piece-price system: high levels of throughput were secured in exchange for high earnings (with wide differentials between subgroups). Introduction of computerized systems provided the potential for monitoring individual performance, which was especially necessary in the new circumstances with the ending of the motivation previously provided by the piece-price system.

Finally, computerized photocomposition enabled management to regain control of the recruitment process. Traditionally Fleet Street production workers were recruited via the trade unions: a pre-entry closed shop operated. In practice unions allocated jobs in the composing room in accordance with an internal seniority system. There were two justifications for this procedure: custom and practice and union guarantees of skill and quality. The result was an inevitable weakening of managerial prerogatives, and employee dependence upon union favour: control over recruitment inevitably strengthened union discipline. In discussions on the development plans for computerized photocomposition at the *Financial Times*, management's wish to assess the competence of employees wishing to work on the new systems rather than relying on union certification was an issue of controversy. Computerized photocomposition reduced the size of the composing room, and thus inevitably the number of employees covered by such a system. Equally importantly, it enhanced the role of clerical workers, whose recruitment was open and controlled by management rather than the unions (a post-entry closed shop then operated): the keyboarding skills required under the new system were widely available, unlike the keyboarding skills of traditional linotype operators.

Managements at different newspapers had different economic interests, reflecting different ownership structures, market situations and cost structures. The different ownership structures and market situations were reflected in the levels of investment individual newspapers were able to make in new technology, the price they were prepared to pay for new technology and the equanimity with which they could face potential conflict. Newspapers which were part of major international groups, like *The Times* which was then part of the Thomson Organization, were better able to contemplate major investments (and if necessary conflicts) than newspapers with more fragile financial structures like *The Guardian*. The different cost structures were reflected in the importance attached to new systems. Newspapers with large origination costs (notably broadsheets) had a major incentive to introduce comprehensive computerized systems, since origination costs represented a high proportion of production costs, whilst newspapers with higher press room and distribution costs (notably tabloid newspapers) had less incentive. These differences of viewpoint were reflected in discussions within the Newspaper Publishers Association. Such factors affected relations with individual unions. Employers wishing to institute radical changes were especially anxious to develop good relations with NATSOPA, since the clerical workers which it recruited would be critically important in implementing new systems. Despite differences in economic situation, all owners shared a common interest in securing agreement to the introduction of new technology with all unions at the same time.

Union policies

Owing to labour strength it was believed in 1975 that computerized photo-composition could only be introduced into the industry with union support. The unions in the industry adopted different approaches towards the issue: some in favour, some hesitant. The NGA, as the union representing compositors, was the most hesitant, rejecting the principle of introducing new systems which involved the elimination of dual keystroking. NGA national policy was clearly stated in 1978 to be: continued NGA jurisdiction over typographic input (whether computerized or not); no compulsory redundancy; reductions in the working week, extended holidays and improved sick pay and pensions arrangements. The effect of such a policy was, of course, to reduce the attractiveness of investment in new technology to employers. SLADE adopted a similar stance. NATSOPA, which organized primarily clerical workers and press room machine assistants, was more positive, seeing new technology as a means of breaking down long-standing

barriers between skilled and unskilled workers. According to the General
Secretary of NATSOPA, Owen O'Brien, 'speaking realistically, the lines of
demarcation (between NATSOPA and NGA members) will not be blurred, they
will be obliterated. The sooner practical people in all unions come to terms with
reality the better for all concerned ... instead of trying to maintain an outmoded
guild mentality that should be interred in peace.' SOGAT had less enthusiasm for
new technology than NATSOPA, since fewer of its members stood to gain directly,
but was generally supportive.

The NUJ was hesitant about new technology, different groups within the union
having different interests and views. However, large parts of the union leadership
were sceptical about the operational benefits of new systems and doubted
whether their introduction would improve editorial quality. As a clear expression
of a 'professional' viewpoint the NUJ view is worth quoting at length: '...
engineers designed these systems with a view to cutting costs by using less labour,
simplifying routines and work patterns, ensuring clearer and neater copy.
Editorial requirements are for spending money to get more or later information,
staffing up to guarantee capacity at peak times, bringing copy flow as close as
possible to deadlines, and providing for the greatest possible number of editorial
checks and balances' (NUJ 1980). Reflecting this view, NUJ chapels were
informed by the national leadership that they would have 'the right to refuse to
enter into negotiations on new working methods and will have the support of the
union in such a refusal ... any agreement reached must include a clause
guaranteeing existing chapels' members and freelances the right, if they so wish,
to continue working on traditional lines, without prejudice to pay and
promotion'. The union published detailed guidelines for negotiators on the issues
they were to be concerned with, and the objectives to be achieved: editorial
quality was to be safeguarded; working conditions protected; health and safety to
be maintained (there was particular concern with the effects of possible emissions
from computers on pregnant women); information was to be provided on the
scope of technological changes; and manning levels, recruitment and training
were to be protected. (For further discussion of NUJ policy on new technology in
the 1980s, see Noon 1991).

Negotiating strategies

Survey evidence indicates that managements normally consult employees directly
affected by technological innovation, either directly or via union representatives,
but are normally reluctant to take part in negotiations, since negotiations imply
employee rights in determining the organization of the production process,

usually seen by UK managements as an element in the managerial prerogative (Daniel 1987; Martin 1994). However, in the Fleet Street environment in the 1970s managements had little alternative to negotiation.

There were two alternative approaches possible to negotiations. The first was an industry level approach, in which all employers jointly attempted to reach agreement with all unions jointly: the agreement could be more or less precise, but greater rather than lesser precision was desirable if employers were to secure the benefits of collaboration. The industry wide approach represented a safety-first strategy. The potential consequences of new technology as a competitive weapon between owners were controlled, whilst no owner faced the risks of being the pace-setter. Similarly for the unions involved; industry level negotiations provided a means of controlling the pace of change, since unions recognized that maintaining unity amongst employers necessarily involved proceeding at the pace of the slowest, in practice the least financially secure owner. Union participants in industry level negotiations were likely to be national officials, and success would strengthen national union discipline. Both owners and unions recognized that an industry wide approach was likely to be fragile, because of the basically competitive structure of the industry, and that once unity was broken on either side the process of disintegration was likely to be rapid.

The second approach simply involved each employer developing an individual strategy, and conducting negotiations at company level. Such strategies were based on the interests of individual owners, and were developed to meet the particular production requirements of individual newspapers. Hence newspapers which had few small advertisements (like the *Daily Mirror*) were not concerned about integrating telesales with origination and could therefore focus on a different range of issues. However, any agreement reached in an individual house was likely to have implications for other houses, in view of the close links between different production chapels in the industry, and any 'first mover' was likely to face determined union opposition. Company level negotiations could be carried on with individual unions, according to local priorities, and with either national or chapel level officials.

It is obvious in retrospect that attempting to follow both strategies was always likely to result in the failure of both. Given the basically competitive structure of the industry, both employers and union negotiators were likely to be anxious about simultaneous discussions in other arenas, and suspicious of back door deals. Attempting a dual track strategy appeared to offer advantages, since one strategy could be aborted if the other succeeded. But the chances of success for industry level negotiations were substantially reduced by suspicions that individual employers were simultaneously attempting to secure better individual deals – or were willing to accept worse individual deals.

Managements and unions sought to secure industry wide agreement on the introduction of computerized photocomposition with the creation of a Joint Standing Committee of employers and unions and the negotiation of a framework agreement. The Newspaper Publishers Association and the unions in the industry (except the EETPU) signed a joint declaration, Programme for Action, in 1976. The agreement provided for improved pensions, voluntary redundancy, and an 'agreed strategy' for decasualization (i.e. to replace the practice of hiring different numbers of workers in the press room and distribution areas according to the number of copies to be produced by a more conventionally regulated system). It was anticipated that individual proprietors would negotiate supplementary agreements at company level to fill out the terms of the framework agreement and to link the agreement with their own requirements from new technology, but that the terms of the individual agreements would be consistent with the framework. The agreement was signed by the employers and by union General Secretaries. Employers saw the agreement as a strategy for reducing labour costs without widespread industrial disruption, if at a slower pace than some managements wished. The agreement was put to the members of all unions early in 1976, and rejected by a majority of the members involved in all the unions, except the NUJ, whose members accepted by a small majority. Union members' rejection of Programme for Action aborted the attempt to develop an industry wide approach to control the social consequences of new technology in the industry.

There were both short-term and long-term reasons for aborting Programme for Action. The initial support for the strategy amongst both employers and, especially, trade unions was based on a sense of crisis in the industry in 1974. In a national context of high inflation the industry's product markets (cover sales and advertising space) appeared to be in rapid decline; production costs were increasing; new technology was threatening to 'decimate' jobs; the industry was in the public eye, with the appointment of a Royal Commission on the Press concerned partly with labour relations issues. Managements believed that unions were on the defensive and that the opportunity existed to 'smash the unions', as one manager later declared. The sense of crisis gave meaning to the unions' traditional motto of 'unity is strength'. By 1976 the sense of crisis had begun to evaporate. No newspaper had closed; advertising revenues were beginning to pick up; the overall rate of inflation was falling and the rate of increase in production costs was declining; the initial phases of the introduction of computerized photocomposition in limited form in provincial newspapers had not had any disastrous effects on jobs and therefore the process could be left to normal procedures; managements appeared to lose their conviction of the need to smash the unions to ensure survival. Union leaders did not undertake energetic campaigns to persuade their members of the virtues of Programme for Action (except in NATSOPA), and the lengthy negotiations allowed opponents of the agreement to mobilize support: there was especially strong opposition in some Fleet Street NGA chapels, as was to have been expected. The decision to present the agreement only to the membership directly affected (rather than to all members of the relevant unions) reflected the special position of the Fleet Street membership in the unions concerned, but it also meant that there was no possibility of mobilizing non-Fleet Street members behind proposals which could be interpreted as protecting the long-term interests of the majority of union members at the expense of the immediate interests of Fleet Street members.

But the most important reason for the failure of Programme for Action was the dynamics of competition between employers and between unions in the industry: the failure to establish the complementarity of interests amongst employers and amongst unions on a firm basis was as important as the conflict of interests between management and unions. Without agreement amongst employers and amongst unions it was impossible to secure agreement between employers and unions. Competition between owners and a conviction amongst some owners that better deals could be secured with their own employees than via industry level negotiations undermined commitment to the NPA and to involvement in the Joint Standing Committee. Similarly, tensions between unions remained throughout the discussions on the Programme. Throughout the 1970s complex negotiations occurred on the creation of a 'single union for the printing industry', which would include NGA, NATSOPA, SLADE and SOGAT, but the negotiations were not to be concluded until the 1990s. The interests of skilled and semi- or unskilled workers were perceived to be different, both generally and specifically with regard to new technology. It is therefore unsurprising that the NGA General Secretary, Joe Wade, had the greatest difficulty in maintaining a clear standpoint throughout the discussions, or the greatest difficulty in securing significant support for the Programme within his union.

Discussions between individual newspaper managements and unions on new technology had preceded the industry level negotiations, most importantly at Mirror Group Newspapers and at the *Financial Times*. In principle – and to some degree in practice – these discussions were rolled up into the industry level negotiations. However, the managements concerned were anxious not to lose the benefits of the initiatives taken, and contacts were maintained between managements and national union leaders at company as well as at industry level throughout the Joint Standing Committee period. Moreover, discussions between middle level managers and FOCs continued at individual newspapers on the details of new technology and on possible means of handling its impact on pay, working conditions, the allocation of jobs, even redundancy. The emergence of the Joint Standing Committee inevitably impinged upon and confused house level discussions; the Committee's demise left the way clear for the house level discussions to be pushed ahead. Among newspaper groups which concluded company level agreements in the period were Mirror Group Newspapers and Times Newspapers Limited; the *Financial Times*, which had been an early mover, abandoned its plans partly for technical reasons.

Conclusion

Historically, labour organization has been strong in the UK newspaper industry: the operation of a pre-entry closed shop and strong work-place discipline ensured high levels of earnings and control over the pace at which new technology was introduced. Trade union influence was especially strong in Fleet Street, strengthened by the lack of unity amongst employers, the time sensitivity of the product and the fragmented but tightly coupled production system. Newspaper proprietors, like Lords Rothermere and Beaverbrook, accepted losses or low returns on capital to maintain continuity of production, regarding newspapers as a form of consumption rather than production: newspaper proprietorship provided status and political influence, as much as economic benefit. Reflecting these circumstances, union members were able to thwart the first attempts to introduce computerized photocomposition into the industry in the 1970s: a limited form was introduced by Mirror Group Newspapers in the 1970s, but neither the Thomson Organization nor the owners of the Financial Times were able to introduce the systems they envisaged. Computerized photocomposition was introduced into UK national newspapers in the 1980s, but in the context of Murdoch's closure of New Printing House Square and opening new production facilities in Wapping and the dispersal of newspaper production away from Fleet Street.

This chapter has provided a summary account of new technology and industrial relations in Fleet Street between 1975 and 1980, focused on management union relations. The objective has been both historical and analytical. The analytical objective has been to emphasize four themes. First, the contradictory pressures upon and interests within both managements and unions: it is rarely the case that a management totally united behind technological change is opposed by unions or employees totally opposed. The interest of some groups of managers may be affected positively by change, and others negatively. Second, the outcomes of attempts to introduce technological change may not be directly 'read off' from the objectives and interests of the parties: institutional structures and processes may have a major impact on outcomes. It was not inevitable that managements and unions should fail to agree on means of introducing new technology in Fleet

Street in the 1970s: the structure of management–union institutions and the dynamics of inter-institutional – and intra-institutional – bargaining affected the outcome. Third, 'encompassing' organizations are more effective at aggregating interests than 'particularistic' organizations and therefore more effective in negotiating flexibly. Finally, collective employee organization functions as a means of expressing 'voice' in relation to technological change.

Resistance to new technology may be prompted by fears of changes in work tasks, working conditions, pay, employment levels, union organizing boundaries or other factors. The resistance may be active, involving publicity, working to rule or other forms of industrial action. The resistance may have no influence, may modify the directions of change or may prevent change completely. In Fleet Street the issues raised by technological change were comprehensive: computerized photocomposition involved major changes in work tasks, in the division of labour and consequently in pay (and in particular differentials), employment levels, and union organization. The result was active, collective resistance, although the scope covered by the collectivity was limited. As the chapter has demonstrated, there were major inter-group differences in the response to technological change, reflecting its differential impact. The resistance was effective, in the short run. Employers proved incapable of introducing new production methods on their own terms, resulting in some cases in capital being invested without the predicted cost savings being made and in other cases in the complete redirection or suspension of technological change. Computerized photocomposition was subsequently introduced into UK national newspapers only in the context of major changes in ownership structure and substantial upheavals in industrial relations. It had proven impossible to introduce the technological changes required within traditional industrial relations structures, processes and conventions.

The experience of new technology in Fleet Street indicates the role of resistance as an expression of 'voice'. But the effectiveness of 'voice' depends upon the audience, whether the audience is the agency introducing new technology or the public. If the audience is unreceptive, either because the economic motivations for introducing change are so strong or because the public is indifferent or hostile to the groups affected, the voice is likely to remain unheard. In the case of UK national newspapers economic motivations were powerful and the general public were not sympathetic to workers whom they traditionally regarded as over-privileged; employees successfully resisted changes in the short run, to experience larger changes within five years.

References

COCKBURN, C. (1983).   *Brothers: male dominance and technological change.* London: Pluto Press.

DANIEL, W. W. (1987).   *Workplace industrial relations and technical change.* London: Francis Pinter.

DODGSON, M. & MARTIN, R. (1987).   Trade union policies on new technology. *New Technology, Work and Employment* 2(1), 9–18.

ECONOMIST INTELLIGENCE UNIT (1966).   *The national newspaper industry: a survey.* London: EIU.

FREEMAN, R. B. & MEDOFF, J. L. (1984).   *What do Unions Street do?* New York: Basic Books.

HIRSCHMANN, A. (1970).   *Exit, voice and loyalty.* Cambridge, Mass: Harvard University Press.

MARTIN, R. (1981).   *New technology and industrial relations in Fleet Street.* Oxford: Oxford University Press.

MARTIN, R. (1994). Innovation and industrial relations. In *Handbook on industrial innovation*, ed. M. Dodgson and R. Rothwell. Cheltenham: Edward Elgar.

NATIONAL UNION OF JOURNALISTS (1980).   *Journalists and new technology.* London: NUJ.

NOON, M. (1991).   Strategy and circumstance: the success of the NUJ in new technology policy. *British Journal of Industrial Relations* **29**, 2.

OLSEN, M. (1982).   *The rise and decline of nations.* Cambridge, Mass: Harvard University Press.

ROGERS, F. & FRIEDMAN, N. S. (1980).   *Printers face automation.* Lexington, Mass: D. C. Heath and Company.

ROYAL COMMISSION ON THE PRESS (1976).   *Industrial relations in the national newspaper industry.* London: HMSO.

SISSON, K. (1975).   *Industrial relations in Fleet Street.* Oxford: Blackwell.

SMITH, A. (1980).   *Goodbye Gutenberg: the newspaper revolution of the 1980s.* Oxford: Oxford University Press.

TRADES UNION CONGRESS (1979).   *Employment and technology.* London: TUC.

10

The impact of resistance to biotechnology in Switzerland: a sociological view of the recent referendum[1]

MARLIS BUCHMANN

Switzerland is an excellent case for studying public debates about new technology. The Swiss political institutions of direct democracy enable citizens to set substantive issues regarding new technologies on the public agenda and to subject them to a popular vote (Frey & Bohnet 1993). Given modern society's great functional differentiation and distinct relative autonomy of its various subsystems (see Touraine, Chapter 2), the institutions of the initiative and the referendum provide a means for securing citizens' direct access to the political arena and for voicing their concerns about social, political, cultural and technological developments.[2] The public discourse engendered by initiatives and referenda not only raises citizens' consciousness about the issues in question, but also assumes a monitoring function regarding the relatively autonomous activities pursued in society's various subsystems (see Bauer, Chapter 19). In this respect, initiatives and referenda concerning scientific and technological issues may be regarded as highly constructive forms of public resistance to scientific and technological developments because they stimulate debates in the public arena about the issues in question, enable citizens to state their preferences, and thus bind scientific–technological activity into democratic procedures. The concerns raised by public debates about scientific–technological developments may be interpreted as signals to the respective institutions to reconsider and re-evaluate their activities. These particular forms of public resistance to new technologies may result in significant adaptations of scientific–technological endeavours to address public concerns. Such effects are not only likely to increase the institution's legitimacy *vis-à-vis* the public, but may also engender scientific–technological innovations. Within this general framework, the purpose of this contribution is to examine the socio-political significance of the referendum on biotechnology and reproductive medicine accepted by the Swiss electorate on 17 May 1992.

In particular, I ask whether the public debates about these new technologies signal a recent change in public attitudes toward grand scientific and technological projects in so far as they raised the fundamental question about limits of scientific–technological growth. In examining the ways in which the regulations of biotechnology proposed by the referendum were evaluated, I then ask whether attitudes regarding the capacity of institutions to enforce such rules determine voters' behaviour at the poll. In order to answer these questions, I first describe the history of the initiative and referendum on biotechnology and reproductive medicine. I then assess the substantive meanings of the opinions expressed at the polls. Using data from a recent survey of the Swiss population about this referendum, I examine the individual motives that were advanced by supporters and opponents. Based on individual level data, I next investigate the voting behaviour of different social groups in Swiss society. The social distribution of supporters and opponents of the referendum is further examined by assessing the association between various contextual characteristics of communities and each community's voting behaviour. Based on community level voting data, I finally compare the plebiscite on biotechnology and reproductive medicine with those regarding various other political issues. Results of these investigations suggest that one of the basic socio-political problems that was at issue with respect to the constitutional amendment regulating the application of biotechnology and reproductive medicine concerns values of social modernization. In the concluding section of this chapter, I discuss the likely effects of this referendum on the future legislative process regarding biotechnology.

The history of the initiative and referendum on 'Biotechnology and Reproductive Medicine'

Over the last years Switzerland has seen much public controversy about biotechnology. The rapid development of this scientific discipline and the breathtaking demonstration of its possible application has engendered mixed public opinions: open admiration for specific applications as well as tormenting uneasiness about possible misuse. In view of the unparalleled possible consequences of these scientific endeavours for humankind, public doubts emerged regarding whether the professional rules of the medical association and the self-imposed restrictions promised by industry would suffice to prevent misuse in the application of genetic engineering and reproductive medicine. The call for state regulations, providing basic guidelines regarding both research in biotechnology and reproductive medicine and possible applications of respective results, therefore gained much public support. The public salience of this issue was heightened in 1987 by the successful launching of a popular initiative (i.e. a petition for a referendum) demanding an amendment to the Swiss constitution regarding the protection of human beings against misuse of biotechnology and reproductive medicine. This petition was launched by the bi-weekly magazine *Beobachter* in 1987. The *Beobachter*, established before the Second World War, is a magazine rich in tradition that has come to assume a special function within the Swiss media. Maintaining an extensive counselling service for its readers, the *Beobachter* represents the 'wailing wall' of Swiss society. Based mostly on readers' experiences, the magazine uncovers excesses of the Swiss social, political and business systems, reporting, for example, on injustices experienced by citizens as a result of inappropriate actions by politicians, bureaucrats and/or business people. In order to give public salience to particular problems, the *Beobachter* has several times used the political instrument of the initiative, launching petitions for referenda on various issues.

The *Beobachter* petition was signed by 126,686 Swiss citizens. It thus succeeded in collecting a minimum of 100,000 signatures necessary for submission to a popular vote. In response to this petition, the Federal Assembly (i.e. the Federal Government and the Parliament) presented an alternative proposition in 1991. While the popular initiative sought only to restrict biotechnological misuse of human beings, the governmental proposition asked for an amendment to the constitution extending protection to human beings, animals and plants. Otherwise, the Federal Assembly's proposal included all essential requests presented in the *Beobachter* petition for a referendum. In sum, the political mobilization organized by the *Beobachter* succeeded in setting biotechnology on the public agenda, demanding constitutional regulations of biotechnology and subjecting it to a popular vote.

The amendment to the constitution proposed by the Federal Assembly, Paragraph 10 of Article 24 (hereafter Paragraph 10), includes three paragraphs (Bundeskanzlei 1992). The first paragraph states that human beings and the environment are protected against misuse of biotechnology and reproductive medicine. In the second paragraph, the Federal Government declares the protection of human dignity, personality, and the family when regulating the use of human genetic material.[3] The final paragraph entitles the Federal Government to regulate the use of genetic material of animals, plants and other organisms with the obligation not only to acknowledge the dignity of the creature and the security of human beings, animals and the environment, but also to protect the genetic diversity of animals and plants. These three paragraphs clearly express the fundamental purpose of the constitutional amendment, that is, to set up some basic guidelines that would guarantee the future development of biotechnology and reproductive medicine while at the same time preventing possible misuse of these new scientific techniques. In essence, Paragraph 10 asks for a development of biotechnology and reproductive medicine informed by socio-political guidelines.

Given the substantive correspondence between the two proposals, the initiators responsible for the *Beobachter* initiative announced its withdrawal and declared support for the proposition presented by the Federal Assembly. As a result, the governmental proposition was the only one submitted to a public vote. The plebiscite was held on 17 May 1992. The Swiss electorate accepted the proposed amendment to the Constitution, Paragraph 10 of Article 24, by a majority of 73.8%. The ballot was cast by 38.6% of the Swiss electorate, representing an average voting participation.[4]

Assessing the specific meanings of voting preferences

On 17 May 1992 voters had to decide whether constitutional regulations should restrict research in biotechnology and reproductive medicine and the application of respective results. The amendment to the Constitution proposed by the Federal Assembly intended to prevent misuse in genetic engineering and reproductive medicine and hence to specify the conditions under which the future development of these scientific disciplines should be guaranteed.

Yes-votes may be motivated by two highly different sets of considerations. On one hand, they may basically express the opinion that these scientific endeavours should be continued on the condition that misuse is banned. Accordingly, these votes reflect conditionally favourable attitudes towards technological innovations. On the other hand, an affirmative vote may also reflect dissatisfaction with the status quo. Some voters may have supported the referendum because they preferred general statements in the constitution regarding the regulation of biotechnology to the then lawless state. The two groups of affirmative voters thus only agree on the necessity of regulating biotechnology. They may differ considerably in their opinions regarding how strictly biotechnology should be regulated and regarding their overall evaluation of biotechnology itself.

The assessment of opinions opposing the referendum is even less straightforward. Opposition to the referendum may arise for very different reasons. It can express opinions opposing any restrictions regarding the use of genetic engineering and reproductive medicine. Or it may reflect, by contrast, the attitude that the proposed amendment does not impose enough restrictions on the use of biotechnology and reproductive medicine. This latter attitude may again be motivated by very different political standpoints. From a conservative standpoint, more restrictions should be imposed on biotechnology and reproductive medicine because of alleged ethical (and religious) problems and imponderable consequences of these new technologies. Opposing votes based on these arguments may be characterized as innovation-averse attitudes. Radical-ecological and radical-feminist standpoints or those presented by the handicapped may favour a more restrictive use of genetic engineering and reproductive medicine in order to prevent instrumental and utilitarian approaches to handling human beings and animals. In fact, political opposition to the referendum arose among two opposite political groups demanding more restrictions on the use of biotechnology and reproductive medicine: the traditional conservative parties (i.e. Christian Democratic Party and Swiss Popular Party) on one hand, and radical-ecologist, radical-feminist groups as well as radical interest groups of the handicapped, on the other (VOX 1992).

In order to determine the validity of the meanings attributed to affirmative and opposing votes and to assess the relative weights of the various political standpoints opposing the referendum, we need to examine (1) the individual motives that were advanced by the supporters and the opponents of the referendum on Paragraph 10 and (2) the voting behaviour of different social groups in Swiss society. These analyses are based on two data sets. The first, the VOX survey, includes **individual level data** (VOX 1992). It is a representative sample of the Swiss electorate, surveyed some weeks after the plebiscite.[5] Interviewees were asked to answer questions about their motives for either supporting or opposing the referendum. This information is highly valuable because it provides substantive arguments for voting preferences, thus allowing us to assess the meanings voters attributed to the referendum. Information about interviewees' social characteristics enable us to investigate the social distribution of supporters and opponents of the referendum. However, the small sample ($N = 512$) sets severe limits on multiple breakdowns by economic, social, and political characteristics. For this reason, the individual level analysis is complemented by an **aggregate level analysis** based on contextual community data, including information on structural characteristics of communities and communal voting behaviour. The community data were provided by Rolf Nef of the Institute '*cultur prospectiv*' in Zürich. Nef's data set is based on the 2290 political communities in Switzerland. Communities are weighted by voters in order to ensure the proportional representation of communal voters.

The analysis of **individual motives** (Table 10.1) includes the major motives that informed the voting behaviour of the supporters of the referendum. The most prominent motive, supported by 42% of the yes-voters, refers to the necessity for some basic legal regulations regarding the development and application of

Table 10.1. *Motives of supporters* (N = 332) *and opponents* (N = 105) *of the Referendum on Paragraph 10 of Article 24*

Supporters' motives		Opponents' motives	
Motives	%^{a}	Motives	%^{a}
Legal regulations necessary	42	Fears of biotechnology and manipulation of nature	46
		Not enough restrictions imposed	7
Protection against misuse	32	Too many restrictions imposed	5
Research-related motives	33		
Protection against unrestricted experimental research	20		
Legal guarantees regarding future development of biotechnology	13		
Unrelated (i.e. false) motives	1	Unrelated (i.e. false) motives	10
No motives	1	No motives	6
		Unspecified general opposition	8

^{a} Multiple choices possible.
Source: VOX (1992).

biotechnology and reproductive medicine as opposed to the then lawless state. Thirty-two per cent were motivated to support the referendum by the desire to install a legal basis that would exclude manipulative misuse of genetic engineering and reproductive medicine. Concerns about biotechnological research motivated another 33% of affirmative voters. Twenty per cent of those concerned with research in biotechnology favoured some legal protection of human beings against unrestricted experimental research in biotechnology, and a further 13% considered legal guarantees regarding the further development of this key technology to be necessary. The motives advanced by supporters of the referendum confirm our assumption about the meanings attributed to yes-votes. The predominance of the motive expressing the necessity for some basic legal regulations suggests that a considerable proportion of affirmative votes reflect dissatisfaction with the status quo. Since research-related motives are also prominent among yes-voters, we may conclude that some supporters are open to these technological innovations under the condition that specified socio-political guidelines are respected.

Motives of the opponents, by contrast, are much less specific (Table 10.1). Forty-six per cent of the opponents were against Paragraph 10 because of diffuse fears of these new technologies. Fearing manipulations of human beings and nature, they are reluctant to accept the future development of these scientific endeavours. Seven per cent of the opponents stated that the proposed amendment to the constitution was not restrictive enough. It was only 5% of the opponents that regarded the referendum as imposing too many restrictions on the future advancement of biotechnology. It is worth noting that 24% of the opponents based their voting decision either upon unrelated (i.e. false) motives (10%), no motives (6%), or unspecified general opposition (8%). Among the supporters of the referendum, the corresponding motives make up only 2%. The motives predominant among opponents thus are either general scepticism about biotechnology and reproductive medicine or undefined opposition to the referendum. Previous Swiss research has shown that voting motives such as general scepticism and undefined opposition are likely to be found among citizens characterized by low social and political integration as well as low trust in political authorities (Sidjanski *et al.* 1975; Keller 1991; Sacchi 1991; Kriesi 1992). To determine whether no-votes are primarily grounded in low social and political integration or in value priorities, we need to examine the social distribution of supporters and opponents of the referendum.

Individual determinants of voting behaviour

It is a common sociological assumption that membership in social groups and political contexts shapes social and political preferences expressed by voters (Dalton, Flanagan & Beck 1984). Based on individual level data of the VOX survey, Figure 10.1 shows that education, occupational position and age greatly discriminate between supporters and opponents of the referendum, whereas sex and religious affiliation show modest influence on the voting behaviour.

Results in Figure 10.1 are presented as deviations from the sample mean of 73% yes-votes. Strongest opposition to the referendum arose among people who only completed compulsory schooling, farmers, the self-employed,[6] and the elderly (60 years of age and above) – traditionally associated with the less privileged segments of the Swiss population and those characterized by low social and political integration. It is well known that social groups located at the margins of society are rather reluctant to accept (technological) innovations, primarily out of fear that these new developments might increase their social marginality (Buchmann 1991, 1992). The strong opposition to Paragraph 10 based on general scepticism which motivated nearly half of the no-votes suggests that marginal segments of the Swiss population are not confident about the capability of the political authorities to regulate this new technology adequately. Likewise, opposing votes based on unspecified general opposition, false motives or

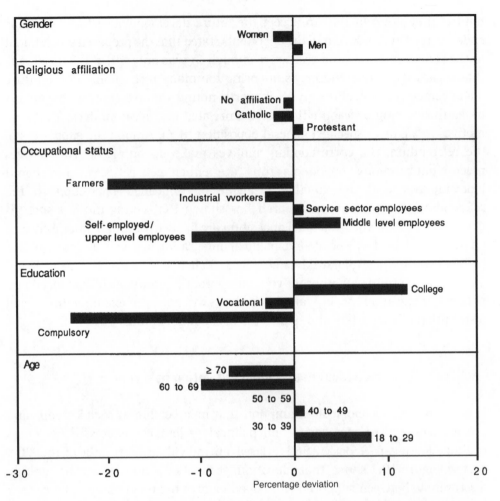

Figure 10.1. Voters' social characteristics and support for Paragraph 10 of Article 24.
Deviations from total percentage yes-votes (73%), individual level data.

no motives are typical of marginally integrated social groups (e.g. Kriesi 1992, p. 97). Often lacking the necessary information for making a sound decision, members of these groups tend to oppose any change subjected to a popular vote. The radical-ecological and radical-feminist opposition to the referendum thus does not appear to be particularly salient. Strongest support, by contrast, came from college-educated people, employees in middle level managerial positions, and the young (less than 30 years of age) – commonly identified as the more privileged and integrated groups in society. These results suggest that members of privileged groups show greater confidence in the capacity of institutions to enforce regulations on biotechnology. A considerable proportion of these voters also supports scientific and technological developments under specified conditions and opposes unconditional scientific and technological growth. Compared to the unquestioned belief in scientific endeavours predominant in the 1950s and 1960s (Buchmann 1993), the conditional support for scientific and technological innovations signals a recent change in attitudes towards grand scientific and technological projects. Setting the evaluation of scientific–technological activities on the public agenda was itself an important achievement of the biotechnology initiative. Overall, the social distribution of supporters and opponents of Paragraph 10 suggests that the degree of social and political integration greatly determined people's voting behaviour.

Political preferences and voting behaviour

Compared to individuals' social group memberships, party identification and left–right political orientation are directly related to individual value priorities. Given the elaborate multi-party system in Switzerland, the collective articulation of political ideas and ideological preferences is highly differentiated. Party identification may thus be regarded as an adequate indicator of individuals' value priorities. While party identification allows us to consider how nuances in political profiles affected the voting behaviour, individuals' self-assessment regarding left-wing/right-wing political orientations shows us how one of the most powerful political cleavages shaped the opinions expressed at the poll.

Figure 10.2 shows that political identification with the Independent Party, the Social Democratic Party, or the Green Party is associated with strong support for the referendum (expressed as deviations from the sample mean of 73 % yes-votes). Regarding their ideological profile, these parties have been strong advocates for the implementation of new political measures (e.g. environmental measures) and the corresponding social measures. Against this ideological background, support for the referendum, which opposed unlimited advancement of biotechnology, expresses their critical attitude towards mainstream social developments. However, given the strong support for the referendum by voters dissatisfied with the status quo regarding biotechnology, we assume that the proportion of yes-votes that are in fact based on these ideological considerations is rather small. Political identification with the Christian Democratic Party or the Swiss Popular Party, especially, is associated with much lower support of the referendum. The traditional constituency of the Swiss Popular Party, and to a lesser extent, of the Christian Democratic party, is the *petit bourgeoisie*. Supported by segments of the population that are – structurally speaking – the stagnant social groups in society, we may again assume that the low social and political integration of a considerable proportion of these voters prompted them to vote no. Doubting the

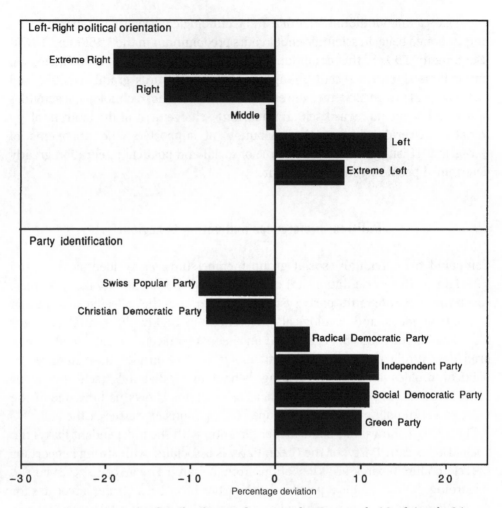

Figure 10.2. Voters' political values and support for Paragraph 10 of Article 24.
Deviations from total percentage yes-votes (73%), individual level data.

authorities' capability of successfully implementing regulations regarding biotechnology, these voters opposed the referendum. Identification with the Radical
Democratic Party, the party that traditionally stands for 'free-market politics' and
'minimal state intervention', is associated with an average support for the
referendum, thus accurately reflecting the party's position in the contemporary
Swiss political landscape. Today, the Radical Democratic Party represents the
powerful mainstream political force, after having played a progressive role in the
historical development of modern Swiss society. In accordance with these results
is the finding that right-wing political orientations are strongly associated with
opposition to the referendum and left-wing political preferences with support for
it.

Contextual determinants of voting behaviour

The previous individual level analysis has shown strong support for the referendum by members of social groups structurally located at the higher echelons of the social hierarchy and weak support by members of social groups with restricted access to highly valued material and immaterial goods (i.e. income, education, or decision-making power). Using aggregate level community data, similar findings are generated with regard to the community's structural centrality or marginality. Figure 10.3 shows the strong association between the community's structural position and aggregate support for the referendum on Paragraph 10, measured as percentage of yes-votes in a given community. Based on three structural characteristics indicating the community's educational distribution, occupational structure, and income tax revenues, the community's location on the centre–periphery dimension is assessed. Discussing each structural indicator in detail, I will show that the strongest opposition to the referendum arose among communities at the structural periphery, those characterized by traditional educational and occupational structures, whereas modern communities, those at the structural centre, most strongly supported the referendum.

Results in Figure 10.3 are shown as deviations from total percentage yes-votes of 74%. With respect to education, the proportion of college-educated community members indicates the community's access to modern values. When these values are highly accessible, the community's support for the referendum was high. Communities characterized by more than 20% of college-educated members strongly supported the referendum. Weak support was provided by communities in which less than 10% of the population has a college degree. The community's occupational structure probably is the single best indicator for characterizing its economy. Based on the three-dimensional distribution of percentage self-employed, percentage industrial workers, and percentage service sector employees, the community's predominant economic activities were identified. Communities in which self-employed labour force participants are predominant are labelled 'self-employed communities'. Similarly, communities in which industrial workers and service sector workers are the dominant economic forces are characterized according to the relative weights of these two groups either as 'industrial worker/service sector employee communities' or 'service sector employee/industrial worker communities'. Acknowledging that modern society's economic modernization involved subsequent shifts from agricultural work (self-employed farmers) to industrial work and then from industrial work to service sector work, we may identify communities according to their level of economic modernization. Communities dominated by self-employed labour force partici-

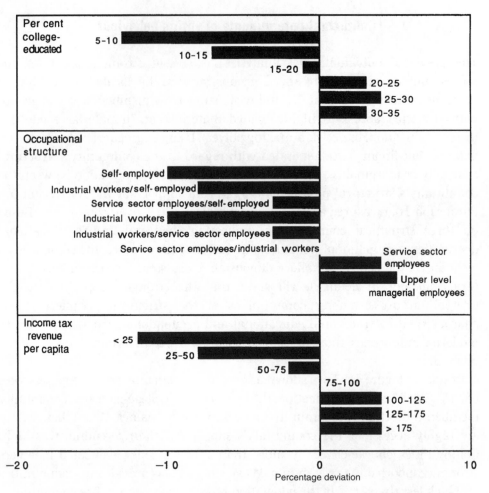

Figure 10.3. Community structural characteristics and support for Paragraph 10.
Deviations from total percentage yes-votes (74%), community level data.

pants are most traditional; those dominated by service sector employees or by
upper-level managerial employees are most modern. Figure 10.3 shows that
support for the referendum on biotechnology and reproductive medicine is
strongly associated with the community's level of economic modernity. Similar
distributions are found when looking at the community's economic capacities as
measured by its income tax revenue *per capita*. This is an indirect income
indicator, measured in percentage of the Swiss average income tax return *per
capita* and community.[7] Figure 10.3 shows that poor communities (less than 25%
of the Swiss average) supported the referendum considerably less than rich
communities (those above the Swiss average). Overall, the community level
distribution of yes-votes and no-votes shows that peripheral communities were
most reluctant to support the referendum. Our previous individual level findings

showing the effect of low social and political integration on voting preferences are thus also supported at the aggregate level.

The wider socio-political significance of Paragraph 10 of Article 24

Differences in social and political characteristics between supporters and opponents as well as differences in respective motives suggest that yes-votes primarily express the support for basic legal regulations regarding biotechnology. A considerable proportion of yes-voters also show interest in defining the legal context within which further advancement of this new scientific discipline should be possible. No-votes, by contrast, primarily express general scepticism towards this new technology combined with doubts regarding the authorities' ability to implement successfully the regulations regarding biotechnology. Based on this assessment of voting preferences, we may tentatively assume that a considerable proportion of yes-voters envisages a pattern of social development that integrates scientific innovations under the condition that specified socio-political conditions are respected.[8] These individuals are open to new developments, but, unlike the modernists in the 1950s and 1960s, they reject an unlimited and unrestricted modernization process. The validity of this interpretation can be indirectly tested by comparing the community voting behaviour regarding Paragraph 10 with other referenda. Because Switzerland's political system is based on direct democracy at the federal, cantonal and community levels, frequent public votes on major issues are common.

I will compare community voting outcomes on Paragraph 10 with those on twelve other national plebiscites held between 1975 and 1992. Each of these specific issues (e.g. relating to Switzerland's membership in the UN, the legalization of abortion, the introduction of a value added tax, etc.) can be assigned to a more general category of socio-political questions (e.g. relating to social inequality, modernization, etc.) (Nef & Rosenmund 1984). From a sociological point of view, these basic issues constitute the major socio-political cleavages or conflicts of interests in Swiss society. Based on this framework, votes on Paragraph 10 can be interpreted relative to these major socio-political cleavages and their social significance can therefore be determined (Nef 1980).[9] Using this approach, I examine the socio-political significance of the plebiscite regarding the use of biotechnology and reproductive medicine. The analysis is based on aggregate level community data of selected plebiscites indicating percentage of the communal electorate supporting the respective referenda.[10] Table 10.2 shows the Pearson correlation between the referendum on Paragraph 10 and selected plebiscites held in the 1970s and 1980s. These correlations do

Table 10.2. *Plebiscitary support for the Referendum on Paragraph 10 of Article 24 and selected other Federal referenda. Pearson correlation of percentage yes-votes based on weighted community level data, 2290 communities*

A. Animal protection issues	
1975: Protection of farm animals (constitutional amendment)	0.70
1978: Protection of farm animals (Laws)	0.63
1985: Regulations against vivisection	0.54
1992: Regulations against animal experiments	0.54
B. Environmental protection issues	
1984: Higher taxation of truck traffic	0.63
1987: Protection of a high moor (Rothenturm)	0.51
1988: Comprehensive traffic conception: rail transport goods	0.63
C. Modernization issues	
1978: Promotion of higher education	0.33
1985: Family and inheritance laws (gender equality)	0.37
1986: Membership of the United Nations	0.53
D. Individual rights issues	
1977: Legalization of first-trimester abortion	0.46
1985: Anti-abortion law (Recht auf Leben)	−0.60

The following referenda were accepted by the total electorate: Protection of farm animals 1975, 1978; Higher taxation of truck traffic; Protection of a high moor; Family and inheritance laws.

not necessarily imply that a majority of the supporters of Paragraph 10 voted in the predicted way with regard to the selected referenda (especially where yes-votes represented a small share of total votes).

The first four plebiscites, listed in Table 10.2A are labelled 'Animal protection issues'. Here, the first two referenda asked for legal regulations insuring the humane treatment of cattle, sheep or other useful animals raised on farms. The other two referenda requested far-reaching restrictions on the use of animals in scientific research, be it vivisection or animal experiments in general. Communities' support for any one of these four animal protection referenda is closely associated with approval of the referendum on Paragraph 10, the regulations regarding the use of genetic engineering and reproductive medicine. This striking relationship indicates that these referenda refer to a common socio-political conflict concerning how and to what extent (natural) resources should be used. Support for these referenda thus expresses preferences for a moderate use of resources and limits on human interventions into nature.

A similar socio-political conflict is involved with selected plebiscites labelled 'Environmental protection issues' (Table 10.2B). The first and third referenda address traffic problems. Both referenda were concerned with regulating (individual) traffic in order better to protect the environment. Increasing costs of goods transported by truck and promotion of rail transport of goods should help to limit traffic mobility. The second referendum listed in Table 10.2B addresses a classic environmental issue; it asks for protection of endangered natural landscapes, such as high moors. The strong positive correlation coefficients indicate that the community's voting behaviour with regard to ecological issues corresponds with the political opinions expressed at the polls regarding the referendum on biotechnology and reproductive medicine. The basic conflict of interest involved here is the cost of economic growth.

Similar communal voting patterns regarding genetic engineering and reproductive medicine on one side, animal protection and ecology on the other, suggest that positions on these referenda may help differentiate values regarding society's development. Support for these referenda expresses preferences for a balanced socio-economic development, one that is not at the expense of animals and the environment. No longer favouring unlimited economic growth in today's advanced industrialized countries, support for these referenda indicates critical assessment of the post-war modernization model based on unrelenting economic growth (Inglehart 1977, 1984, 1990).

Further support for this interpretation is provided by the rather strong positive associations between acceptance of Paragraph 10 and some referenda directly addressing socio-political development questions labelled 'Modernization Issues'. This is shown in Table 10.2C. Three socio-political issues were selected for this comparison: promotion of higher education, revised family and inheritance laws, and membership of the United Nations. Addressing problems of broader access to higher education, gender equality in family and inheritance laws, and increased international political engagement of Switzerland, these measures represent new elements in an advanced industrialized society's development. Support for these modern changes goes hand in hand with acceptance of Paragraph 10.

This interpretation is further corroborated by the strong associations between this referendum and those labelled 'Individual rights issues' (shown in Table 10.2D). The major political issue debated around the referendum on the legalization of first-trimester abortion is whether decision-making powers regarding this matter should be transferred to individuals or remain in the hands of corporate groups (e.g. church, medical profession). Support for liberalizing abortion indicates a preference for more individual rights. The respective positive and negative correlation coefficients show that support for Paragraph 10 is associated with preferences for more individual rights.

Conclusions

The purpose of this chapter was to examine the socio-political significance of the referendum on biotechnology and reproductive medicine accepted by the Swiss electorate on 17 May 1992. The Swiss biotechnology referendum is a rare example of direct democratic participation in the evaluation of a new technology. Setting the evaluation of these scientific-technological activities on the public agenda was an important achievement of Swiss citizens. The successful biotechnology initiative prompted the Federal Assembly to propose an amendment to the Constitution (Paragraph 10 of Article 24) demanding regulations regarding the protection of human beings, animals and plants against misuses of genetic engineering and reproductive medicine. Respecting the initiative's major goals, the intention of Paragraph 10 is to prevent misuse of these scientific and technological innovations and to specify the conditions under which advancement of these disciplines should be guaranteed. Examining the individual motives that were advanced by supporters and opponents of the referendum and investigating the voting behaviour of different social groups and communities, the analysis shows that yes-votes were motivated by an interest in constitutional regulation of biotechnology and by their desire to define the legal context within which further advancement of this new scientific discipline should be possible. No-votes, by contrast, primarily expressed general scepticism towards this new technology combined with doubts regarding the authorities' capability of successfully implementing regulations regarding biotechnology.

In conclusion, these findings underscore the great importance of Swiss direct democracy in enabling citizens to set substantive issues regarding new technologies on the public agenda and to state their respective preferences. The broad media coverage of biotechnology in the months preceding the vote documents the extensiveness of the public discussion (Schanne & Caspar 1992). This discussion and the voters' preferences expressed at the poll show the wide agreement among the Swiss public on the necessity of regulating biotechnology. This public statement about biotechnology may be interpreted as a strong signal to the respective institutions to consider the citizens' concerns in their future activities. There is no doubt that the voters' preferences expressed at the poll will have an impact on the future bills specifying the regulations included in the constitutional amendment.

Notes

1 Acknowledgement: I would like to thank Rolf Nef for providing the data on the structural characteristics of Swiss communities and the communal voting behaviour regarding various referenda. I have also benefited much from his helpful comments. I would also like to thank Maria Charles and Stefan Sacchi for providing both valuable suggestions and assistance. The author bears sole responsibility for the opinion expressed herein.

2 At the Federal level, the compulsory (*obligatorisch*) and the optional (*fakultativ*) popular referenda are institutionalized. Among other things, all changes in the constitution must be submitted to a compulsory public vote. The 'optional referendum' includes all legislative proposals, general federal decrees, or constitutional emergency decrees. They are submitted to a public vote if a certain number of citizens or cantons so requests within a given period of time (Katzenstein 1984). At the Federal level, initiatives require 100,000 signatures. The issues advanced by initiatives are then subjected to a popular vote.

3 The second paragraph includes seven subparagraphs specifying bans and restrictions related to human genes (e.g. banning the mixing of non-human with human genes).

4 In the 1980s, the electorate's participation in Federal referenda averaged around 40%.

5 The sample distribution of supporters and opponents of the referendum deviates by only 1% from the voters' distribution, which provides evidence for the representativeness of the sample.

6 Unfortunately, the VOX survey combined the self-employed and employees in upper level managerial positions into a single category. Using aggregate level community data, I will document below the great differences in voting behaviour between communities characterized by a high proportion of employees in upper-level managerial positions and those dominated by self-employed individuals.

7 Taxation rates vary by community. Since poor communities tend to have higher rates than rich communities, this indicator is a conservative estimate of the communities' economic capacities.

8 My caution in advancing this hypothesis stems from the observation that a considerable proportion of yes-voters were dissatisfied with the status quo regarding biotechnology.

9 This interpretation is possible because the communal participation differences remain constant, although communal electorate's participation rates vary across referenda as do citizens who vote at each election (Rolf Nef, personal communication).

10 The data were again provided by Rolf Nef. His data set includes the communal voting behaviour of all Federal referenda held since 1970. Communities are again weighted by voters in order to ensure the proportional representation of communal voters.

References

BUCHMANN, M. (1991).   Die sozialen Ursachen der Technikfeindlichkeit. *Neue Zürcher Zeitung*, Technologie und Gesellschaft, August 1991.

BUCHMANN, M. (1992).   Soziale und politische Konflikte im Sog des beschleunigten sozialen Wandels. *ETH-Bulletin*, February 240, 3–6.

BUCHMANN, M. (1993).   Die widersprüchliche Vielfalt von Werten in der gegenwärtigen Gesellschaft. Swiss Federal Institute of Technology, unpublished manuscript.

BUNDESKANZLEI (1992).   *Volksabstimmung vom 17. Mai 1992. Erläuterungen des Bundesrates.* Bern: Bundeskanzlei.

DALTON, R. J., FLANAGAN, S. C. & BECK, P. A. (ed.) (1984). *Electoral change in advanced industrial democracies: realignment or dealignment?* Princeton, NJ: Princeton University Press.

FREY, B. S. & BOHNET, I. (1993).   Democracy by competition: referenda and federalism in Switzerland. *Publius: The Journal of Federalism* **23**, 71–81.

INGLEHART, R. (1977).   *The silent revolution: changing values and political styles among Western publics.* Princeton, NJ: Princeton University Press.

INGLEHART, R. (1984).   The changing structure of political cleavages in Western society. In *Electoral change in advanced industrial democracies: realignment or dealignment?* ed. R. J. Dalton, S. C. Flanagan and P. A. Beck, pp. 25–69. Princeton, NJ: Princeton University Press.

INGLEHART, R. (1990).   *Culture shift in advanced industrial society.* Princeton, NJ: Princeton University Press.

KATZENSTEIN, P. J. (1984).   *Corporatism and change: Austria, Switzerland, and the politics of industry.* Ithaca, NY: Cornell University Press.

KELLER, F. (1991).   Autoritärer Populismus und soziale Lage. In *Das Ende der sozialen Schichtung,* ed. V. Bornschier, pp. 274–303. Zürich: Seismo Verlag.

KRIESI, H.-P. (1992).   Bürgerkompetenz und Direktdemokratie: Die Schweizer StimmbürgerInnen, Stimmbeteiligung und Wahlentscheidung nach neuen Vox-Umfragen. *Widerspruch* **12**(24), 92–100.

NEF, R. (1980).   Struktur, Kultur und Abstimmungsverhalten. *Schweizerische Zeitschrift für Soziologie* **6**, 155–90.

NEF, R. & ROSENMUND, M. (1984).   Das energiepolitische Plebiszit vom 23. September 1984 zwischen Entwicklungserwartung und Wachstumskritik. Ein Beispiel ereignisorientierter, raumbezogener Gesellschaftsanalyse. *Schweizerische Zeitschrift für Soziologie* **10**, 689–722.

SACCHI, S. (1991).   Politische Orientierungen und soziale Schichtung im verselbständigten Handlungssystem. In *Das Ende der sozialen Schichtung,* ed. V. Bornschier, pp. 235–73. Zürich: Seismo Verlag.

SCHANNE, M. & CASPAR, R. (1992).   Gentechnologie in schweizerischen Tageszeitungen. Ein interkultureller Vergleich am Beispiel der Berichterstattung zur Abstimmung über den Verfassungsartikel über Fortpflanzungs- und Gentechnologie (17. Mai 1992). *Beitrag zur Session 22 'Environmental Risk Literacy' der 2. Internationalen Tagung und Ausstellung über Umweltinformation und Umweltkommunikation.* Ecoinforma '92, Bayreuth.

SIDJANSKI, D., ROIG, C. et al. (1975).   *Les Suisses et la politique.* Bern: Lang.

VOX (1992).   *Analyse der eidgenössischen Abstimmung vom 17. Mai 1992.* Vox-Analyse Nr. 45, Schweizerische Gesellschaft für praktische Sozialforschung Zürich; Forschungsstelle für Politische Wissenschaft, Universität Zürich.

PART III
International comparisons

11

The politics of resistance to new technology: semiconductor diffusion in France and Japan until 1965

ANTONIO J J BOTELHO

Introduction

The invention of the transistor in 1947 ushered electronics into a new era.[1] Although access to transistor technology was relatively open, France and Japan responded differently to the opportunities offered by the new technology. This chapter compares their national responses to semiconductor technology, particularly transistors, up to the mid-1960s. Semiconductor diffusion occurred faster in Japan than in France and this had important industrial competitiveness consequences.[2] Partly as a result of historical patterns of state action (or lack thereof) and societal arrangements, by 1972 there was no French firm among the world's top ten semiconductor firms, and only one among the top twenty (Webbink 1977, p. 22). In contrast, there were two Japanese firms among the top ten and five among the top twenty. By 1960 Japanese transistor production was larger than French. Early technological choices shaped the future competitiveness of the national industry in semiconductors.

Conventional views portray Japanese society and state as forward looking and fascinated with material progress. Since the Meiji restoration Japanese society has been eager to adopt modernity and build upon it. In contrast, France, her society and state, are portrayed as conservative with regard to new technologies at least until World War II. Formalistic explanations which analyse the cultural form of resistance hide the historically contingent and political nature of that resistance. To identify cultural, market or societal differences may be a sufficient explanation of resistance from a functionalist viewpoint but it fails to uncover the process characteristics of the phenomenon, which is an important resource for collective action.

Besides conventional state technology policy, the political shaping of the market and institutional arrangements among firms, state, and academic actors

Table 11.1. *Semiconductor technology imitation lags, France and Japan*

	Average imitation lag (years)		
	United States	France	Japan
For all innovations	0.1	2.8	2.5
For all transistor innovations except surface barrier	0.1	2.8	2.1
For 1950s innovations	0.1	3.0	3.4
For 1960s innovations	0.0	2.6	1.2

Source: Adapted from Tilton 1971, p. 27.

played key roles in the diffusion process. From an industrial governance and strict technological diffusion perspectives either France or Japan could have become a major force in the new technology, France and Japan being considered the classic 'strong states' (Zysman 1983; Campbell *et al.* 1989). In both countries, post-war technocratic elites with a modernizing vision sought to determine economic and social choices to create progress (see Touraine, Chapter 2). At the end of the war both industries lagged behind that of the United States by at least a decade; France was initially closer to the semiconductor research frontier than Japan.

Japanese and French patterns of innovation in semiconductor technology were similar until the 1960s. As shown in Table 11.1, there was little difference in imitation lags between France and Japan, except for the 1960s, when the lag of France becomes much larger than the one of Japan. In the early 1950s, the Japanese electronics industry (JEI) was smaller than the French electronics industry (FEI) and had fewer and smaller specialized firms, the potential carriers of diffusion. In 1955, JEI's production was small and exports were negligible. By 1959, the JEI was the world's second largest producer of transistors and had the second largest global electronics industry, together with West Germany.

This research shows how in France in the 1950s the fragmentation of professional groups and institutions and conflictual business–government relations inhibited a coherent state policy for semiconductors, and electronics in general. More importantly, the efforts of French semiconductor firms to ride the production learning curve were stunted by the French electronics market, biased towards professional electronics. This bias favoured basic, as opposed to applied and engineering, semiconductor R & D. The French state, furthermore, did not react to the opportunities of this new technology until the late 1950s, when it adopted a basic research strategy. A semiconductor industrial policy did not emerge until the mid-1960s, and then only halfheartedly, when higher entry barriers and a changed international market further constrained a catch-up strategy. This contrast with Japan, where occupation policies promoting democratization and the stubborn permanence of old wartime networks among business, government and academia contributed to the development of a semiconductor industry in the 1950s. The network's industrial vision and production-driven semiconductor policy was consensual (Samuels 1987, p. 8). Even more important was a large domestic market for consumer electronics which drove the early growth of the semiconductor industry.

Existing accounts of the diffusion of semiconductor technology are often unpolitical and ahistorical. They hinge on economic assumptions about the firms' vested interests – economic and cognitive – in the previous electron tube technology that produces an expected pattern of resistance to the new technology. For example, Malerba (1985) erroneously assumes that after World War II, the European electronics industry did not have any disadvantage in relation to US industry, in contrast to the Japanese. The fact that the US electronics industry accounted for close to 90% of global electronics production casts serious doubts on his claim. Next, he asserts that after the invention of the transistor, in Europe and Japan 'only large, established, vertically integrated electron tube producers were present.' However, as entry barriers in the years following the transistor invention were quite low, small firms both in Europe and Japan began producing semiconductors. While in Europe the semiconductor market and national policy often precluded the growth of these firms, in Japan the first producers of transistors were small and large firms alike such as Sony, Kobe Industrial, and Hitachi (Tilton 1971; Malerba 1985, pp. 3–5).

Studies of resistance to new technology have looked at bureaucratic structures as a target of resistance. Resistance to new technology takes various forms, including state policy (see Bruland, Chapter 6 or Bauer, Chapter 5). However, too formalistic an approach is problematic as state policy is a process that cannot be reduced to either societal choice or a target of resistance. In fact, as this research will show, state policy also structures the environment, in this case the market, in which the resistance to new technology is played out. Bureaucratic structures can be the resistant actor. The study of the politics of new technology reveals how political power may shape the conditions and the very object of resistance (see Jasanoff, Chapter 15). Therefore, in order to analyse the impact and to identify the possibilities of resistance to a new technology one has to study its underlying networks of actors and institutional arrangements in historical perspective.

This comparative analysis of the diffusion of semiconductor technology in France and Japan puts on its head the assertion that the form of resistance may be an independent variable to analyse the dynamics of resistance to new technology (see Rucht, Chapter 13). Diffusion and resistance are two sides of the same coin. The political structuring of markets for new technology, which affects its pattern of diffusion, may also affect the patterns of resistance to new technology. National industrial policies shape the structure of an industrial sector and the pattern of its competitiveness, thus constraining future preferences and outcomes. Organizational and cognitive characteristics of research networks, business–government relations, and vision of professional groups are key factors for explaining differences in cross-national patterns of diffusion of new technologies.

In order to provide a baseline for latecomers the first section dwells briefly on the history of US pre-war electronics and the early evolution of the transistor. Its aims are to show that first, even before the transistor, electronics was a relevant industry and, second, that structure of the market shaped the pattern of technological change. The next two sections examine the multifaceted responses of France and Japan to semiconductor technology. The conclusions concern our understanding of the differential patterns of resistance to new technology, and explore the long-term effects of resistance.

The American electronics industry: from the electron tube to the transistor[3]

For half a century the electron tube dominated a yet ill-defined electronics industry. Radio, radar and TV became the main markets for the electronics industry. Military demand played a crucial role in the development of that industry. During the First World War new applications for military requirements led to improvements and to mass production techniques. Radio broadcasting was the most important market for the electron tubes and dominated electronics until 1940. During the 1930s the price of receiving tubes used in radio sets declined and the range of industrial applications expanded dramatically. During World War II, radar was the driving product of electronics. The war quickened the pace of innovation and productivity, and produced innovative academia–industry–government arrangements. Sales of receiving tubes skyrocketed. The industry's revenues jumped from about $250 million in 1941 to nearly $4 billion in 1951 (Fink 1952, p. 81). The post-war years saw television, which became a key market for electronic components.

The invention of the transistor was the watershed of accumulated knowledge of solid-state and quantum electronics physics, and the potential technological demands of telecommunications. However, electron tube based electronics was so established that, even in the US, it took a few years for the transistor to capture the public's hearts and minds (Editors of *Electronics*, pp. 74–5). Military support of engineering development was critical for the rapid diffusion of this new technology (E. R. Piore in Zahl *et al.* 1951, p. 86; Tilton 1971, pp. 75–7; Misa 1987). The open military market was crucial for small semiconductor firms, which grew on the strength of their engineering and production R & D capabilities. Texas Instruments was the first to produce silicon grown junction transistors and its production flexibility was critical in an environment of rapidly changing technology and markets.

In the meantime tube production continued to grow, aided by the 1950s TV boom. Poor market prospects, intense competition and transistor high price discouraged the transistorization of TV sets. Microwave, power and special purpose non-receiving tubes also experienced steady growth. In the mid-1950s, the computer had not yet embraced the transistor for technical and cost reasons (Beter *et al.* 1955). During this period, the computer and telecommunications industries experimented with a variety of promising solid-state devices (e.g. tunnel diode and parametron) that fitted their particular switching needs. Most were superseded by later products and never found a market.

Transistorized computers diffused only in the late 1950s, helped by the federal government boosting procurement. In fact, as late as 1957 consumer electronics (CE) accounted for the lion's share of the $12 billion electronics market. A trend, however, was the stagnation of consumer electronics – except auto and portable radios – until further development of low priced colour or fully transistorized battery-operated TV sets. CE was then the major market for germanium transistors, followed by commercial-industrial, and military; the latter was the main market for silicon transistors. In the late 1950s transistor production doubled almost every year thanks to computers and missiles applications. Yet, price remained the major barrier to further market expansion.

In the 1960s the military market overtook all others as the Pentagon's transistor demand projections led to cheaper production processes. The military emphasis on aircraft and guided missiles deepened the interest in miniaturization. Molecular electronics, or later microelectronics, would give us the integrated circuit (IC). Owing to the growing computer market the semiconductor technology gradually detached itself from military demands. The commercial announcement of the IC was met with scepticism, but a government report remarked how future design would call for a fundamental change in fabrication and in approach to circuits (Ceruzzi 1989).

By the early 1960s, with semiconductor sales at half a billion dollars, the semiconductor industry faced uncertainty and was torn by competition, oversupply, price cutting, narrowing margins, and mounting barriers to entry. Firms responded with automation, a strategy of high-volume producers for the military, and with new forms of cooperation between transistor suppliers and computer manufacturers. The importance of R & D declined relative to marketing, distribution, production and quality control. Production technology became critical and new innovative firms were created around the new concept such as Fairchild Semiconductor with its revolutionary planar technique, first to market the monolithic IC.

US imports of electronic products increased steadily through the late 1950s, led by radio sets and components. In 1959, with US electronics exports registering modest growth, imports doubled, with over half coming from Japan. In 1962, when semiconductor sales finally equalled electron tube sales, the industry faced further problems: imports and unstable markets for its new products. Failures and mergers were rampant. Fear of foreign competition at the product level led to calls to restrict the import of transistors, mainly from Japan. Imports of transistor radio sets might undercut US radio production and thus overall demand for transistors.

France

It has been suggested that the slow diffusion of semiconductors in France was due to manufacturers' reluctance to adopt electronic devices that lacked reliability, and consumers' cultural resistance to the new technology (Croset 1966). While Japanese firms were producing millions of transistorized portable televisions both for the domestic market and for export, a French industry executive stated as late as 1965:

> ... the interest of transistorisation does not appear as pressing for a television set as it does to radio sets ... There is of course the autonomous portable TV set with a small screen. For this type of appliance, it is obvious that one ought to economize on the energy required from the batteries and thus transistorisation is a must ... But this type of appliance is just considered, at least in Europe, as a luxurious solution, destined to amuse privileged customers.   (Fagot 1964)

The Centre National d'Etudes des Télécommunications (CNET), a governmental R & D institution, produced point-contact transistors and developed applications within a year of BTL's invention (Sueur 1949). French firms grasped the importance of semiconductor technology and during the 1950s kept up with the pace of technological change (Gee 1959, p. 456). Yet by the early 1960s, they were plagued by the mounting inability to catch up.

Historical legacy
France pioneered the development of electron tubes during World War I, but fell behind in the inter-war period. France made important research accomplishments in electronics but neither in basic nor in applied research were there

scientific breakthroughs or coordinated innovations. Its electronic research was fragmented and piecemeal and its innovations lacked a scientific basis. State institutions rarely conducted any applied research. Shinn's survey of 34 French companies between 1880 and 1940 estimated that less than one-fourth had created in-house R & D departments by the 1930s (Shinn 1980). A large firm such as the 'Compagnie générale d'électricité' (CGE) did not set up an R & D laboratory until the late 1930s. However, in quantitative terms, France did not fare badly compared with Japan. The main difference lay in the significant number and size of Japanese government institutions doing applied and industrial research. The French state never took seriously the question of industrial research in the period (Shinn 1980, p. 619). In France, there was no institution similar, except perhaps the much smaller Bellevue Laboratories, to the Institute of Physical and Chemical Research (Riken), a joint business–government institute created in 1917, or the Electrotechnical Laboratory. Most applied and engineering research was carried out by foreign companies' laboratories in France. In telecommunications and radio broadcasting the state quasi-monopoly and erratic investment did not incite private R & D (Libois 1983, pp. 248–54). On the eve of World War II, French telecommunications was underdeveloped and its equipment industry was dominated by foreign firms.[4]

The State and French electronics

The CNET team that manufactured the first transistor in France was made up of researchers from its Transmission Department and from the Société des Freins et Signaux Westinghouse (SFSW). SFSW and Compagnie Française Thomson–Houston (CFTH) produced semiconductor detectors for radar applications in France. Two SFSW researchers had semiconductor applied research experience in wartime Germany, H. Welker and H. F. Mataré. Welker (1912–81) was a German physicist who, together with the physicist K. Seiler, had played a critical role in the wartime German radar research at Telefunken and Siemens, respectively (Eckert & Schubert 1990, pp. 80, 116, 156, 192–3).

The planners of the French post-war economic miracle did not put priority on electronics until the Third Plan. It was not a priority in the first 'Plan Monnet', although it noted the importance of the electronic tube and components sectors and recommended their modernization (Giboin 1950). A report of the Third Plan advanced that continuous basic and applied research and stable and growing markets were prerequisites for the expansion of the electronics industry. Nevertheless, the expansion of the industry's two main markets, professional equipment (PE) and consumer electronics (CE), were dependent on the state: the former on 90% of its demand, and the latter on investments in national radio and TV broadcasting.

In 1952, the FEI was placed among the best, having just quadrupled its exports. The CE market was small and biased towards radio; the TV market lagged relative to its neighbours. There were one million sets in the UK and just 40,000 in France. State monopoly over television broadcasting and state planners' unwillingness to promote consumer goods blocked the growth of the market (Mallet 1954; Emery 1969, Ch. 14). The strength of the FEI lay in professional equipment (PE) for military purposes. For example, the CFTH, up to 1945 a successful producer of CE, strengthened its pre-war broadcasting equipment production and diversified into radars and microwave equipment. In order to insure competitiveness in these areas and goaded by the French policy of technological autonomy it became engaged in research and manufacturing of power electron tubes (Thomson–CSF, n.d.).

There was a major difference in the evolution of CE and PE before and after the war. While the former stagnated in the post-war period, the latter grew. In 1952, PE accounted for 37% of industry's revenues against 32.5% for CE. In 1938 these shares had been reversed, at 57% and 13%, respectively. The good performance of PE had a troubling implication: dependence on the protected military market with only a small percentage in the industrial market. Through the mid-1950s, the FEI remained heavily dependent on the PE market.[5] By 1955, CE's share of the French electronics market had fallen to just 25%. The radio market remained flat; the TV market had doubled, but remained smaller than elsewhere. TV set producers, to counteract the pattern of state inaction, set up a new organization to promote standards and plan the sector's long-term growth.

The FEI was highly concentrated, and bureaucratic management hampered its strategic flexibility. Half a dozen firms accounted for three-quarters of total revenues, large conglomerates with a high degree of vertical integration which produced the whole range of PE and CE, and controlled almost the total production of tubes, components and devices. Next, there was a group of specialized medium-sized firms producing a few types of CE goods. Finally, there was a number of small firms producing CE in an artisanal fashion, especially TV sets, which carried high margins.

By 1960, despite continued growth, the French CE industry lagged behind those of Germany and Japan. By that time the market structure was reversed again. In spite of the 1957–8 economic crisis, CE grew faster than PE, accounting for 40% of the market, pushing up sales of standardized tubes, and parts and components. Helped by the franc devaluation, electronics exports produced an all-time high sectoral trade surplus. A stagnation of PE resulted from the government's austerity policy. The FEI's investment capacity declined sharply and imports rose faster than exports, leading to a sharp reduction in margins, and accelerated the industry's restructuring. Contemporary observers, acknowledging the pressure of international competition, still defended FEI's good health with its positive trade balance and fast export growth. Moreover, they remarked that the FEI had 98 % of all American offshore orders for electronics, supplied over 50 countries, and had the largest number of foreign licensees, being the largest French exporter of 'matière grise' (Mériel 1960). However, French foreign (outside the franc zone) exports were mainly of PE, and CE exports went largely to captive markets of former or existing French colonies; both markets with limited potential.

The diffusion of transistors in CE took off in the late 1950s, when transistor radios accounted for half of the radio production. The Compagnie générale de Telégraphie sans fil (CSF) and La Radiotechnique, a Philips subsidiary, made heavy R & D and production investments to face the competition from imports; both firms opened new semiconductor plants and Radiotechnique semiconductor business grew 130% in 1957. In general, however, due to FEI's specialization in professional equipment, its electronic devices production and technical capabilities were focused on electron tubes. Two local specialists reporting on the state of French electronics then claimed: 'In the field of semiconductors French production is still relatively low' (Goudet & Grivet 1960, p. 81). A late 1950s study by a telecommunications professionals group on the causes of the lag of the FEI singled out: (1) fragmentation of state demand and budgetary constraints, (2) underdeveloped radio and television infrastructure, (3) insufficient standardization of parts and components, affecting quality and series size, and (4) excessive dispersion of production capacities and insufficient specialization. The study also mentioned the industry's fiscal and social charges and the high luxury tax on TV sets. In line with the French autonomy vision and with its professional tradition, the group recommended the expansion of PE and of military-funded R & D, to avoid dependence on foreign firms. It called for the production in France of semiconductors currently imported and the launching of a rapid state intervention through CNET commissioned research contracts targeted at PE.

After the Rome Treaty, which brought tariff reductions and the Common Market competition, American semiconductor firms established assembly operations in France. IBM France, which until then obtained most of its electronic components from French subcontractors, began producing its own devices in France, which in turn became the main supplier for IBM's global operations outside the US. By the mid-1960s, integrated circuits (IC) had barely been commercialized by French firms, and experts noted that IC research was difficult because of the large number of types and the high initial production cost (Walravel 1966).

The ENS network and semiconductor research

Complex manufacturing processes and productivity levels were a permanent source of frustration for local pioneers. The French rationalist training and the elite 'polytechnicien' generalist scientific culture were at odds with the interdisciplinary, messy and unpredictable nature of semiconductor technology and manufacture. Not surprisingly, the initial response was to emphasize the theoretical aspects of semiconductors at the expense of experiments and industrial applications. Semiconductor solid-state physicists were headhunted by private and public laboratories and given a free hand in their research (Griset 1990, p. 289). It was the golden age of the 'Ecole Normale Supérieure' (ENS) solid-state physics laboratory (LPS) which trained a whole generation of researchers.[6] P. Aigrain, who among many others was sent abroad for graduate training after the war, would become the 'patron' of French semiconductors. Aigrain's laboratory at the ENS trained a whole generation of semiconductor solid-state physicists, who went on to head laboratories in engineering schools, science faculties, and private and public research organizations. This network of professionals, however, did not engender an industrial vision for French semiconductors.

A typical student of Aigrain's laboratory, who extended the network's reach into the corporate world, was physicist R. Veilex. He recalls that around 1957, when he joined the laboratory at ENS, the Germans had begun making the first radio sets with industrial transistors, and French firms wished to match them. In line with the historical legacy, French firms' semiconductor R & D was orientated towards sophisticated high-performance devices. A few firms such as CSF realized this trap (CSF was highly dependent on radar and telephone equipment, and parts and components, the latter by itself a quarter of revenues) and attempted to modernize the French CE market. Thus, in 1958, Veilex was sent for an internship at CSF's semiconductor laboratory, directed by Aigrain's former ENS colleague C. Dugas, to carry out research related to the development of the 'solistor', a high-frequency GaAs transistor, central to CSF's first portable radio set. However, given the ambitious design characteristics of the new device, CSF repeatedly faced production problems and eventually discontinued its production.[7]

Among private firms, CSF was the leader in semiconductor research. CSF was France's top electronics firm, with 25% of the industry's 1960 revenues; it had R & D at the centre of its corporate strategy. Its orientation towards basic research was shaped by military demands for technological autonomy and by its researchers' scientific orientation. The 1951 semiconductor department at its central basic research laboratory was directed by C. Dugas, who had been a colleague of P. Aigrain, first at Carnegie–Mellon where they did graduate work in solid-state physics, and later at the ENS. The laboratory kept in close contact with Aigrain's group at the ENS. CSF ties with the ENS physics group went back at least one generation, as CSF's technical director and future president (1960), M. Ponte, had been a colleague of the school's physics 'grand patron', Y. Rocard, who had in the past worked for one of the group's companies. Ponte was a respected physicist and member of the French Academy of Sciences. CSF controlled the Compagnie Générale de Semi-Conducteurs (COSEM) which started mass production of transistors in 1955. From the start, however, there were engineering and manufacturing problems amplified by clashes between plant engineers and central laboratory researchers.

The CFTH, France's second largest electronics firm, faced internal resistance from its tube people, and initially did little semiconductor research. A serious effort did not start until 1956, with the help of General Electric Co. (USA), which would lead to a semiconductor joint venture in 1961, SESCO. Smaller electronics and telecommunications firms, particularly those working for the military, carried out semiconductor research on an *ad hoc* basis, and sometimes developed capabilities in niche markets such as power transistors.

During the 1950s the CNET did applied semiconductor research but a separate semiconductor laboratory was not established there until the late 1950s. The interdisciplinary and craftsmanlike technical work of applied semiconductor research was looked down upon by the elite telecommunications engineering 'corps'. Not surprisingly, as late as 1958, the engineer in charge of the PCM department (I. Franke) was a foreign contractual engineer.

However, by the late 1950s, the erosion of the competitiveness of the French electronics components industry and the growing demands created by technological state programmes (nuclear bomb, supersonic fighter aircraft, etc.) put an end to the blue sky research era. As Franke, head of a newly created semiconductor research department at CNET, put it:

> Only a strict limit on the researches to do can respond to the needs of an efficient program, which nevertheless will remain incomplete and of which success is difficult to assess. The great temptation that the researcher has to be always open to new ideas ought to disappear in the face of the need that presses upon him of solving scientific problems for pursuit of a particular technical goal.   (Franke 1958, p. 5)

Summarizing, semiconductor research in France was defined mainly as theoretical solid-state physics. This definition was made jointly by the 'elitist' ENS research vision and the French state's autonomist policy. When new demands emerged in the late 1950s, the network's dominant position and its vision of basic research blocked the emergence of an alternative professional mission linking semiconductor research to the industry's destiny. The rising prestige and political influence, symbolized by Aigrain's meteoric career in the state administration, further legitimized the network's cognitive, as opposed to industrial, definition of semiconductors.

CNET's semiconductor group in a subordinate position and CNET's inward looking culture failed to generate an alternative vision. Further, the rising costs of semiconductor manufacture led them to abandon a switching equipment technology based on in-house developed semiconductors. Later, the choice of leap-frogging an intermediate switching technology in favour of a fully digital switching system gave France a critical head start in telecommunications in the late 1970s (see Miles and Thomas, Chapter 12). However, to make use of fully digital technology meant to take access to the latest American semiconductor.[8] This option foiled CNET's development of microelectronics technological research and had an adverse impact on the sector's long-term capabilities. In 1986, telecommunications equipment accounted for 26% of the IC demand in France. The telecommunications establishment did not launch a sustained semiconductor research programme until the late 1970s (Poitevin 1986, p. 2). Similarly, political pressures to legitimize the 1966 'Plan Calcul' (PC) gave birth to a strategy for bringing French computers to the market. That led also to abandonment of the autonomist semiconductor project embodied in the 'Plan Composants', in favour of devices available in the international market. As hinted above, military funding was crucial to the development of solid-state physics research projects at university and firms' laboratories. However infighting, organizational resistance and lack of coordination prevented the French military from playing a coherent role to promote semiconductors in France. It was only in the 1960s that the Air Force sought to coordinate research on silicon transistors and IC micro-electronics.[9]

Japan

> Transistors are the brightest field in Japanese electronics; a half a dozen manu-facturers – the tube producers ... plus Sony and Sanyo – may now be producing more transistors of the entertainment type than any other nation in the world. Rapid development in circuit design in the last year sees transistors being used in Japan in communications equipment, computers and industrial controls as well as consumer goods. Indeed transistorization has somewhat the same faddist appeal in Japan that it has in the US.   (Leary 1960, p. 65)

As in France, Japanese firms and state institutions realized early the importance of the new technology. However, the Japanese responded differently. State policies for the expansion of the communications and broadcasting markets contributed to a boom in the domestic CE market, which grew at an average annual rate of 60% from 1955 to 1960. This growth generated a large demand for semi-conductors, particularly transistors, so that by the late 1950s firms had developed creative transistor applications. Japan's transistor production was driven by consumer electronics (CE). Over the next decade, the state bowed to the firms' aggressive semiconductor strategy and implemented a long-term policy, including negotiated protectionism, which helped to ride successfully the mid-1960s technological and market fluctuations. These actions, and those promoting and protecting an infant computer industry, were pivotal for the Japanese successful diffusion of the transistor.

Accounts of the diffusion of semiconductors in Japan that overlook the early push into CE, miss the shaping power of this critical market (Okimoto, Sugano & Weinstein 1984; Gregory 1986; Freeman 1987; Fransman 1990). During this phase firms accumulated critical experience in semiconductor mass production and incorporated technological capabilities and capital that allowed their later push into computers.[10] Spurred by a strong domestic radio set demand, semiconductor production grew 125 times in the second half of the 1950s. By 1959 Japan was already the world's second largest producer of transistors, behind the US; in 1962 it was close to matching US production.

Historical legacy

Already before World War I, encouraged by economic expansion several electrical firms had set up R & D laboratories, which were for the most part small and poorly equipped, as firms depended on foreign licensing for their technology. The industry grew by developing receiving tube radio and communications equipment for military and consumer uses.[11] Somewhat contrary to the conventional account, according to which the Japanese militarization overtook electronics, we find that as early as the 1930s the Japanese military sponsored developments in electronics, such as the Yagi antenna, widely used during World War II in both Japanese and Allied radar equipment.[12]

Military research, concentrated in specialized laboratories, grew as the Pacific War progressed. Government control of the economy promoted close cooperation with leading enterprises in sectors necessary for war mobilization. The expansion of research organizations occurred mainly in government-run laboratories related to the military. Hitachi, in close cooperation with the military, became the leader in electrical machinery by investing in quality control and setting up specialized research units. Specialized firms joined forces, as for example in the formation of Toshiba by Tokyo Electric and Shibaura Engineering. Toshiba's superior management techniques, borrowed from GE during the inter-war period, made it the largest supplier of electrical equipment to the armed forces (Leary 1960). It was the first firm to set up an independent electronics research laboratory; while other firms reorganized and expanded their R & D laboratories. By 1938 Hitachi's Central Research Laboratory employed over 350 workers and Toshiba's Mazda and Tsurumi Research Laboratories employed 160 and 170, respectively.[13] Most Japanese electronics companies, large and small, had by then considerable manufacturing and, to a lesser extent, R & D capability in pre-war electronics and telecommunications. Firms benefited from state funding and large industrial contracts.

Following the outbreak of the Pacific War, research centralizing measures were taken and new institutions were established (Agency of Science and Technology, Science and Technology Council, etc.), with lasting consequences. One institution was the Japan Scientific Research Council which promoted cooperative research between enterprises, government laboratories and universities. Scientific associations, such as the Japan Association for the Promotion of Scientific Research, with large public and private research funds, established committees that brought together academic and state scientists and engineers. Another measure consisted of associated basic research laboratories in the Imperial universities (30 new ones between 1941 and 1945) which did research for the Army (Hirosige 1963; Home & Watanabe 1990). These moves helped break the isolation of academic research.

Wartime demand for domestically developed technology gave a boost to 'applied' basic research. Laboratories under civilian ministries, including the Ministry of Commerce and Industry (forerunner of the Ministry of International Trade and Industry, MITI) and the Ministry of Communications (MoC, forerunner of the Ministry of Post and Telecommunications, MPT), benefited greatly from the wartime expenditures. The Electro-technical Laboratory (ETL) of the MoC more than doubled its budget between 1938 and 1944. Similarly, Riken's 1944 budget of 14.7 million yen was alone over half of the whole revenue of the research activities for military industry (see Cusumano 1989; Coleman 1990). The number of scientists and engineers per 10,000 of the Japanese population practically doubled between 1940 and 1947 (Home & Watanabe 1990, pp. 337–8).

Firms used high margins from military procurement to consolidate and establish specialized laboratories. The new integration and management of research is exemplified in Nippon Electric's (NEC) General Laboratory with a staff of over 1000. Another was the 1941 Hitachi's Central Research Laboratory, next to an existing research facility. A third type was the research foundation, developed by new firms and smaller zaibatsus. A group of firms in a small zaibatsu with limited resources would set up a research foundation, such as the Furukawa Physical and Chemical Research Laboratory. A fourth type was the product laboratories such as Toshiba's Mazda Laboratory, split into eight different research laboratories, including Electronics and Electric Apparatus. The expansion of research organizations peaked in 1942, when the First Naval Technical Arsenal employed over 33,000 people and the Central Aeronautical Research Laboratory over 1500. In electronics, the civilian ETL in 1945 employed almost 3000 people and had the largest budget among comparable institutes. In the private sector,

Hitachi's applied research-orientated Hitachi Research Laboratory employed over 1000 while its Central Research Laboratory had a staff of 360 during 1943–4. Both Toshiba and NEC had total research staffs of over 1000 at the end of the war (Kamatani 1963).

Throughout World War II research was steered by state institutions connected to the military. Its legacy was the dissolution of barriers among academia, business, and public laboratories, and new arrangements for cooperation among these actors. Fuji Tsushinki's (later Fujitsu) Technical Development manager after the war recalled that his interest in computers came out of his participation with military communication systems during the war. Fujitsu built the first Japanese computer and became one of the world's leading computer manufacturers. Sony's founder, A. Morita, worked in Navy radio projects during the war (Fransman, 1990, p. 15). The Navy Technology Research Institute, for example, having concluded that the atomic bomb would not be completed before the end of the war either in Japan or in the US, directed its efforts to radar research (Low 1990, p. 351). Japanese wartime R & D did not produce any major innovation, but it rather successfully copied Western products and techniques and achieved considerable manufacturing capabilities, developing expertise 'upon which a major new export industry could be based in the post-war years'. While this remark was made in relation to optical equipment, it could well be applied to electronics.[14]

The State and the Japanese electronics industry

Almost half of all Japanese electronic component firms in existence at the end of the 1960s were founded in the decade following the war (Takahashi 1993). Initially small operations catering to the amateur radio market which boomed after the war, the late 1940s austerity measures and economic recession forced small firms to become subcontractors to large firms. The JEI underwent major changes in the wake of the Korean War boom (1950–3), the beginning of radio and TV broadcasting, and the ensuing depression. It was during this later phase that large electrical equipment firms such as Hitachi and Toshiba began manufacturing electrical and electronic home appliances, including radio-sets. They were not totally foreign to the industry, as they previously manufactured receiving tubes and were creditors of the bankrupt small radio set makers and component firms. They also could make use of their extensive sales network to distribute the new products. As a result most small firms left the CE market and produced components.[15]

The structure of JEI in the 1950s was divided between large, electrical equipment firms, often part of a large diversified pre-war zaibatsu (Toshiba, Mitsubishi and Japan Radio, Hitachi), and smaller companies more specialized in electrical and related products, such as telecommunications equipment (NEC, Matsushita, Kobe Industrial, Oki, Sony, Sanyo, Yaou, Fuji Electric). Toshiba was in the mid-1950s the largest producer of exclusively electrical and electronic equipment. In contrast, Matsushita (National, JVC and Panasonic brands), the pre-war leader in electrical household appliances, was an aggressive newcomer to consumer electronics armed with Dutch know-how from a joint venture with Philips and a keen marketing strategy. Sony and Sanyo entered this market segment in the 1950s. These middle-sized companies, outside the purview of large risk-averse commercial banks, did the most advanced R & D and set new product trends (Leary 1960, pp. 60–2).

During the Occupation, authorities ordered the production of four million radios to diffuse the ideals of democracy. In 1948, the number of radios in households exceeded pre-war levels. They also stressed the rebuilding of Japan's communications infrastructure. Japanese technicians received training in advanced communications technology, which, patterned on American military standards, improved quality. In 1950, under the guidance of SCAP, the Radio Communications Industries Association (later EIAJ) set out to improve the quality of radio receivers and their components. A couple of years later a somewhat similar group was established by the NHK Research Laboratory to improve TV sets (Takahashi 1993). In contrast with France, Japan's successive telephone expansion programmes (1946–9, 1953–8 and 1959–63) assisted a group of preferred suppliers, NEC, Hitachi, Fujitsu and Oki, to become major producers of transistors by 1960.

Contrary to what happened in France, where state radio and television broadcasting policies as well as consumer credit policies conspired against the growth of the CE market, Japanese policies induced the development of domestic CE. Radio broadcasting was freed from central control in 1951, when private broadcasters were allowed to compete with the state's networks. By 1960 there were over 300 AM broadcasting stations, reaching 85% of the households, and FM, barely three years old, already had a dozen stations. Television broadcasting began in 1953, with the Japan Public Broadcasting Company (NHK) competing with private stations. Rapid expansion propelled TV broadcasting equipment production. By the mid-1960s TVs were in about 85% of the households and had 1239 stations, of which 456 were commercial. In 1960, Japan was one of the first countries to adopt the US NTSC colour TV standard. Colour TV broadcasting commenced soon after Toshiba's completion of the first colour TV set plant. In 1963, Japan was second only to the US in the number of TV sets owned (Kumata 1959, p. 490).

Employing consumer electronics products, first radios and then TVs, as a lever, Japanese production of transistors and electronic components grew quickly. In

1958, over 95% of transistor production went into radios. The industry profited from two 'Kandem Booms', home electrical machinery booms (Satofuka 1993). The first, from 1955 to 1963, spread the transistor radio. Almost all portable radios produced (61% of total radios) in Japan in 1958 were transistorized. Already in 1958, TV production (32% of electronics production) was higher than radio (19%). In 1958, TV sets and other electronic consumer goods accounted for 49% of total sales of electric household appliances; one year later for 60%. Japanese production of household electric appliances was second to the US, having overtaken that of European countries (Sato 1963). In 1960, the JEI broke the $1 billion mark, with consumer goods (radio+TV, about 40%) and communications equipment (25%) at its basis. At the same time, Sony and other firms began marketing transistorized black and white TV sets. A couple of years later, Sony introduced the first low-priced small transistorized battery-operated TV set, using high-frequency mesa and power transistors. The second boom, from 1966 to 1970, spread the colour TV receiver.

Mass production of semiconductors started in 1955, the same year that Sony introduced a portable radio with five junction transistors, that revolutionized the market. Sony's successful production of transistorized radios and the firms' difficulties in procuring high-gain transistors for radio sets in the US market mobilized large firms to enter transistor production.[16] In 1958, transistors accounted for the bulk of semiconductor production and the largest producers (under license from American firms), the firms that produced most radios, began exporting. Japan's formidable transistor production expansion was driven by the dynamic CE market.

Exports dovetailed this pattern, increasing 200% during the first boom and 150% during the second. Half of the exports went to the US; Europe and developing countries followed. By 1958, Japan had captured 30% of the portable radio and almost 40% of the transistorized radio markets in the US, becoming the fastest growing segment of Japanese exports. Exports of TV sets to the US began in the early 1960s, and soon included portable transistorized and low-cost large colour models. With just 10% labour costs, cheaper materials and components explain the lower prices; components suppliers made constant efforts to reduce production costs.

These exports further contributed to an end market for Japanese semi-conductors. Moreover, transistor exports to the US, mainly entertainment grade, exploded from 11,000 units in 1958 to 1.8 million units during the first nine months of 1959, despite a 15% tariff. This sudden jump in Japanese transistor exports to the US (25% of total US unit production) prompted the US Electronics Industry Association (EIA) to request a government investigation. Great attention was paid to quality. Tax breaks were accorded for the import of testing equipment and the testing of devices and equipment to US military specifications was encouraged. These efforts paid off as Americans began to perceive Japanese products as good as US products.

Firms, however, did not quietly bow to MITI's plans. MITI efforts to stop firms' investments in semiconductor plants to avoid overcapacity in the late 1950s were unsuccessful. In the end, MITI had to assist them in developing new markets to absorb the excess production. So, in 1958, having made large investments in transistor production, Hitachi and Toshiba received large transistor orders from US firms. Also at the end of the 1950s, MITI's plans for a mixed corporation to manufacture computers faltered, given the firms' reluctance to cooperate. In another episode, MITI and the Japanese Foreign Office, fearing US retaliation on large radio exports, unsuccessfully set up export floor prices on transistor radios. MITI played a key role in protecting segments of the Japanese market, where domestic firms were not competitive. In 1962, it postponed the liberalizing imports of digital computers and colour TVs. It used taxes and foreign exchange controls to stave off imports. For almost a decade, it stalled TI's request to establish semiconductor operations in Japan and when permission was given it was conditional upon TI's transfer of advanced IC technology to several Japanese firms. As part of the deal, MITI talked NEC into sharing its planar technology licensed from Fairchild in 1962 with other firms.

Semiconductor electronics R & D

As shown above the technological basis of the post-war JEI developed in the 1930s. During the Occupation, domestic technological innovation was limited by income constraints, limited diversification, and by the firms' narrow economic ability. By the early 1950s, despite occupation policies, most pre-war zaibatsu had re-emerged and large commercial banks regained their influence. The banks' attitude discouraged electronics firms from innovating research, which partly explains the large number of technical assistance and license agreements with foreign firms. As a result innovative research was carried out in a few public laboratories, half a dozen universities (often in collaboration with government laboratories), and small and medium-sized firms out of reach of the banks. Further, in the context of the delicate balance of Japanese post-war politics, these firms through their associative institutions – professional, regional and sectoral – were able to extract support from MITI, such as the above mentioned grants-in-aid for research (Friedman 1988). The open research atmosphere of small companies attracted engineering talent. Only later in the decade, smaller firms, thanks to the new royalty remittance policy, passed agreements with foreign firms, often small and medium-sized.

The two successive booms of the 1950s were aided by state policies for electronics that reduced the dependence on foreign technology and promoted the television and transistor radio markets. State institutions involved the Ministry of Post and Telecommunications (MPT) and organizations under it: the public enterprise Nippon Telegraph and Telephone (NTT) and its Electrical Communication Laboratory (ECL) and the NHK radio and TV broadcasting public company; and MITI's Electrotechnical Laboratory (ETL). Semiconductor research being out at first at ECL, research on point-contact diodes began in 1949, and soon shifted to point-contact and junction transistors. By the mid-1950s several firms did transistor research; mass production began in 1955. They collaborated with private laboratories in medium-term projects and divided up research in long-term projects. For example, ETL was in charge of developing transistor applications while ECL was in charge of parametron applications.

MITI officials, initially, did not target the electronics sector. MITI strategies in the early 1950s centred on heavy industry sectors similar to French policy, although grants-in-aid for research financed 40 projects in small electronic components firms until 1955. MITI in 1957 promulgated the Electronics Industry Promotion Special Measures Law which stressed the production of electronic components. Yet research funding was relatively small. In 1959 its R & D subsidy was only $0.6 million and its 1956 loans represented less than 15% of total investment in electronics.

ETL did research in electronics, followed technical developments abroad, and set quality standards. Its affiliated Japan Machinery and Metals Inspection Institute was established in late 1957 and began inspecting transistor radios and spare parts in April 1958. More importantly, ETL carried out independent research and engineering work in those areas where imports run high and advised MITI on technological growth areas. ETL together with the Japan Electronics Industry Development Association (JEIDA) supported industrial R & D in magnetic recorders, instruments and production controls, and computers. Similarly, ECL did semiconductor-related research for telecommunications. University–industry collaboration was common among medium-sized companies such as Murata, Kobe Kogyo, Yaou and Sony. The giants, Matsushita, Hitachi, Toshiba and NEC, leased much of their development efforts to smaller companies and abroad. From 1952 to 1968, several firms, including Matsushita, Sony, Hitachi, Toshiba and NEC, passed 251 licence agreements on radio and television technologies with Westerns firms. Table 11.2 gives an overview on foreign ties held by Japanese electronic companies before 1965.

In 1961, sixteen electronics firms joined with academic sponsors to form a semiconductor research institute led by two Tohoku University (Sendai) professors, with long-standing records in electronics and materials research and in industrial collaboration. Tohoku University, with one of Japan's foremost

Table 11.2. *Japanese electronics industry foreign ties*

Company	Foreign tie	Nature of tie
Fujitsu	Siemens	Communications equipment (pre-war)
Hitachi assistance (1952)	RCA	TV tubes licence and 301 computer licence (1961)
		IC licence (1965)
	Western Electric	Transistor and diode licence (1953)
		Electronic switchboard (1957)
	N. V. Philips	–
Kobe Kogyo	RCA	Technical assistance (1952)
	Western Electric	Transistor licence
Matsushita Electrical	Western Electric	Transistor licence
	Arcair	
Matsushita Electronics	N. V. Philips	30% ownership (1952)
Japan Victor	RCA	Research
Mitsubishi Electric	Westinghouse	4% of stock
		X-ray licence
	RCA	Radio and TV sets licence
	Western Electric	Transistor licence
	Collins Radio	Jet plane radio gear technical assistance
	Servomechanisms	
	N. V. Philips	
	Corning	
NEC	Standard Electric	21% ownership
New NEC	Sylvania	10% ownership
		Licence to produce electron tubes
Sanyo	RCA	Transistor radio marketing outside US under RCA
		Victor trademark and in US by Channel Master
Shimadzu	Westinghouse	X-ray licence
Sony	Western Electric	Transistor licence (August 1953)
Standard Radio of Tokyo	Emerson	Transistor radio marketing
Toshiba	General Electric	7.4% ownership
	RCA	
	EMI	
	Western Electric	
	Raytheon	
	Hoffman	
	N. V. Philips	
	Motorola	Transistor radio sales
	International GE	Transistor radio marketing outside the US

Source: Author's elaboration from trade journals and companies' histories.

electrical engineering departments, was a spin-off from Tokyo University, and had close pre-war links to Matsushita. It was only in the early 1960s that large firms stepped up their R & D investments. An *Electronics* special report on Japan in 1962 confirmed the excellence of Japanese semiconductor R & D.[17] Their R & D aimed at cheapening and expanding production, at the development of high-frequency silicon transistors and rectifiers, and switching semiconductor devices. At Toshiba's Kawasaki transistor plant, for example, 'one of the most dynamic engineering and management groups works constantly and enthusiastically to perfect transistor manufacture ... engineers have been set free to develop new techniques for performing the drudgery of transistor assembly' (Leary 1960, p. 61). Transistor assembly operations used the latest techniques, often developed by small firms with the technical and financial assistance of the larger firms.

Conclusions

By the mid-1960s, the ratio of demand of semiconductors for electronic components in France was on average one-third smaller than in Japan, an indication of the slow diffusion of transistors in France. French and Japanese ratios converged in the late 1960s, as foreign firms established operations in France. From 1956 to 1968, on average, the value of Japanese semiconductor production, in percentage of both active component production and receiving tube plus semiconductor production, was significantly higher than that of France (Tilton 1971, p. 32). French semiconductor production and technological expertise remained behind the Japanese. The average imitation lag in France for such innovations in the 1960s was 2.2 times larger than Japan's as shown in Table 11.1.

Several policy choices configured the resistance to new semiconductor technology in France: the market skewed towards PE due to the voluntary underdevelopment of consumer electronics and communications markets; low barriers to entry for foreign semiconductor firms in the domestic market; national firms' high levels of bureaucracy; the basic research orientation of its semi-conductor R & D; and the state's passivity delayed the emergence of an active electronics (not to mention semiconductor) policy well into the 1960s, when it was already too late.

France's resistance to the adoption of semiconductor technology had a negative effect on its future competitiveness, as semiconductors increasingly became strategic inputs for other electronics sectors such as computers and telecom-munications.[18] The French mid-1960s nationalistic computer policy was largely undermined by the country's lag in semiconductors; it was abandoned in 1974.

On the demand side, the late blooming of the French CE market weakened the competitive position of the FEI; French firms were unable to fend off foreign competition in this sector. Furthermore, the policy to expand telecommunications, launched in the mid-1970s, did not provide the critical demand for the national French semiconductor industry; needs were met by imported semiconductors or those produced by foreign subsidiaries.[19] The national French semiconductor industry was never capable of closing the lag. A professional group of researchers with a clear vision of problems and markets did not emerge in the political arena until well into the late 1970s, when semiconductors finally became the core of Paris's electronics strategy. Curiously, around the same time the French telecommunications authority launched the videotext service Minitel. The state financial backing to the project, which distributed the equipment free of charge, eventually proved to be crucial to its success in comparison with similar experiences in other European countries (see Miles and Graham, Chapter 12). State officials initially saw Minitel as an end market for French semiconductors and a new market for flattering French electronic firms. In the end, once again, firms' need for quick profits and officials' need for political legitimacy led to the abandonment of the nationalistic semiconductor strategy.

Malerba (1985) correctly argues that the Japanese success in electronics can be mainly attributed to the protection of its large home market from foreign imports and direct investment. While this is undoubtedly a crucial element, his argument does not take into account that markets are also political creations; there is an innovative role for small firms. Their initial success and their collective action prompted larger firms to enter the market and state officials to support the industry (Tilton 1971, pp. 141–3).

The functional interpretation of resistance to new technology as a process of social selection (see Bauer, Chapter 5) demonstrates that there is no determined trajectory of technological change. This analysis suggests that, at least in the sphere of international competitiveness, national lags in the diffusion of new technology have serious economic and social effects, as they structure future technological choices. This chapter suggests, based on comparative case studies, that a better understanding of national patterns of resistance to new technologies requires: first, a multidimensional analysis to capture the contingent linkages between the type of market expansion and the dynamics of sectoral industrial growth; and secondly, a non-deterministic historical analysis of the complex relationship of social actors and state institutions in shaping the market, the sector's R & D, the industry, and the industry's organizational behaviour.

Notes

1 Semiconductors fall into two broad categories. Active components, which define electronics equipment in contrast to electrical equipment, include discrete semiconductors such as transistors, diodes, rectifiers and special devices. Other devices in this category are the previous generation of electron tubes – receiving, cathode ray, and special purpose. Integrated circuits, monolithic and hybrid, are both active and passive as they incorporate elements in their structure from both categories. The materials most commonly used for semiconductor manufacturing are germanium and silicon.

2 Accounts of semiconductor technology diffusion tend to focus on the American invention of the transistor and to narrowly emphasize scientific factors (Wiener 1973; Hoddeson 1981; Braun & MacDonald 1982; Hoddeson, Braun & Teichman 1992), relative to engineering, manufacturing and institutional factors (Morris 1990).

3 This section is, unless otherwise noted, based on Aitken 1985; Kraus 1968; Maclaurin 1949; Morris 1990, Ch. 2; and unsigned articles in the trade publication *Electronics*.

4 France's low telephone density fell outside the range common in nations of equal development (Nouvion 1984, p. 77). An argument on the historical 'state malthusianism' is in Griset 1983, pp. 103–4.

5 Mallet 1954; and Robert A. Mallet, interview, Paris, 12 April 1989.

6 Julian Bok, interview, Paris, 14 March 1989. On the development of solid-state physics in France, see Guinnier 1988.

7 Interview, Robert Veillex, Paris, 12 December 1990.

8 Libois 1983, pp. 172–3, and interview, Paris, 21 June 1989.

9 Marc Chappey, interview, Paris, 12 May 1989.

10 A circular efficiency imperative assumption shared by most accounts, in different guises, consistently dictates an homogeneous behaviour for all electronics firms. It is alternatively labelled cost reduction, technological innovation, environmental pressure, or sectoral industry 'mind-set' (Shearman & Burrell 1987).

11 *Electronics Industry Association of Japan, Thirty Years History of Electronics Industry, Daiyamondo-sha*, 1979, II-1 (in Japanese), cited in Satofuka & Rahman 1987, p. 69; see also Chokki 1987; Yamamura 1986.

12 H. Yagi had founded in 1924 the Saito Laboratory of Electrical Communication at Tohoku University, which developed close ties with Matsushita (Home & Watanabe 1990, pp. 332, 337). Hirosige (1963) calls this period a turning point in the country's history of scientific and technological institutions.

13 See the US Naval Technical Mission to Japan report series, such as E-19, *Japanese Electronic Equipment Construction Materials* (1945).

14 US Naval Technical Mission to Japan report E-19, 'Japanese Electronic Equipment Construction Materials' (1945), cited in Home & Watanabe (1990, p. 339).

15 The exceptions were Pioneer Electronics and Trio (presently Kenwood), which became middle-sized producers of stereo equipment (Ibid).

16 MacDonald 1954; Wada 1960. The account of the start of transistor production in Japan is a sore point of debate among opposing camps of students of Japanese political economy development: free marketeers and neo-statists. For a simplistic presentation see Zinsmeister 1993; and the aggressive replies and counter-reply in Letters, 84–95. A stimulating survey of the analytical issues involved is Noble 1989.

17 'Japan, Revisited', *Electronics*, 19 October 1962, p. 3.

18 Neo-Schumpeterian economists of technological change have suggested that semiconductors is a 'cumulative' technology, that is, each generation of technological development appears to lay the groundwork for the next (Malerba 1985, p. viii).

19 A former director of CNET commenting on the French digital electronic switching system
    choice of components said: 'We wanted to build a French system without concern for the
    origin of the components. It did not shock us to use foreign components. It was similar to the
    Minitel, which at the beginning had 65–70% of foreign components' (J. Dondoux, interview,
    Paris, 22 June 1989).

References

AITKEN, H. G. J. (1985).   *The continuous wave: technology and the American radio,
1900–1932.* Princeton, NJ: Princeton University Press.

BETER, R. H., BRADLEY, W. E., BROWN, R. B. & RUBINOFF, M. (1955).   Directly Coupled
Transistor Circuits. *Electronics*, June, 32–134.

BOTELHO, A. J. (1994).   The industrial policy that never was: French semiconductor
policy, 1945–1966. *History and Technology* 11, 165–80.

BRAUN, E. & MACDONALD, S. (1982).   *Revolution in miniature: The history and impact of
semiconductor electronics,* 2nd edn. Cambridge: Cambridge University Press.

CAMPBELL, J. C., with BASKIN, M. A., BAUMGARTNER, F. R. & HALPERN, N. P. (1989).
Afterword on policy communities: a framework for comparative research. *Governance* 2,
86–94.

CERUZZI, P. (1989).   Electronics technology and computer science, 1940–1975: a
coevolution. *Annals of the History of Computing* 10, 257–75.

CHOKKI, T. (1987).   Modernization of technology and labor in pre-war Japanese electrical
machinery enterprises. *Japanese Yearbook on Business History*, 1987, 26–49.

COLEMAN, S. K. (1990).   Riken from 1945 to 1948: the reorganization of Japan's
Physical and Chemical Research Institute under the American occupation. *Technology
and Culture* 31, 228–50.

CROSET, M. (1966).   La silice – agent de passivation des elements semiconducteurs au
silicium, presented to the 'Demi-journée d'Etudes: Optoélectronique et Semiconducteurs',
26 January 1966. Reprinted in *L'Onde Electrique* 46, 590–608.

CUSUMANO, M. A. (1989).   Scientific industry: strategy, technology, and management in
the Riken industrial group, 1917 to 1945. In *Managing industrial enterprise: cases from
Japan's prewar experience*, ed. W. Wray. Cambridge, Mass: Harvard Council on East Asian
Studies/Harvard University Press.

ECKERT, M. & SCHUBERT, H. (1990).   *Crystals, electrons, transistors: from scholar's study to
industrial research.* New York: American Institute of Physics.

EDITORS OF *ELECTRONICS* (1981).   *An age of innovation. The world of electronics
1930–2000.* New York: McGraw-Hill.

EMERY, W. B. (1969).   *National and international systems of broadcasting – their history,
operation and control.* East Lansing: Michigan State University Press.

FAGOT, J. (1964).   Les deux demi-journées SFER sur la transistorisation des téléviseurs.
L'Onde Electrique 44, 637–9.

FINK, D. G. (1952).   Cross talk. *Electronics*, February.

FRANKE, I. (1958).   La recherche sur les corps semi-conducteurs et sur leur utilisation
dans la technique des télécommunications. *L'Echo des Recherches* 31, 4–5.

FRANSMAN, M. (1990).   *The market and beyond – cooperation and competition in information
technology development in the Japanese System.* Cambridge: Cambridge University Press.

FREEMAN, C. (1987).   *Technology policy and economic performance: lessons from Japan.* London: Pinter Publishers Limited.

FRIEDMAN, D. (1988).   *The Misunderstood Miracle.* Ithaca, NY: Cornell University Press.

GASCHI, J., GROSVALET, J. & PEYRACHE, G. (1971).   Les composants électroniques. *L'Onde Electrique* **51**, 631–9.

GEE, C. C. (1959).   World trends in semiconductor development and production. *British Communications and Electronics,* June, 450–61.

GIBOIN, M. (1950).   Les travaux de la Commission de Modernization des Télécommunications du Plan Monnet. *L'Onde Electrique* **30**, 341–4.

GOUDET, G. & GRIVET, P. (1960).   France (Electronics research and development around the world). *Electronics,* 12 February, 79–81.

GREGORY, G. (1986).   *Japanese electronics technology: enterprise and innovation.* New York: John Wiley & Sons.

GRISET, P. (1983).   La Société Radio-France dans l'entre-deux-guerres. *Histoire, Economie, Société* **1**, 83–109.

GRISET, P. (1990).   Les relations du CNET avec l'industrie. In *Le Centre National d'Études des Télécommunications, 1944–1974. Genèse et croissance d'un centre public de recherche,* eds. J.-P. Bloch and P. Bata, pp. 283–313. Paris: CRCT.

GUINNIER, A. (1988).   The development of solid state physics in France after 1945. In *The origins of solid state physics in Italy: 1945–1960,* Vol. 13, ed. G. Giuliani. Bologna: Italian Physical Society.

HIROSIGE, T. (1963).   Social conditions for prewar Japanese research in nuclear physics. *Japanese Studies in the History of Science* **2**, 80–93.

HODDESON, L. (1981).   The discovery of the point-contact transistor. *Historical Studies in the Physical Sciences* **12**, 41–76.

HODDESON, L., BRAUN, E. & TEICHMAN, J. (ed.) (1992).   *Out of the Crystal maze. Chapters in the history of solid state physics.* Oxford: Oxford University Press.

HOME, R. W. & WATANABE, M. (1990).   Forming new physics communities: Australia and Japan, 1914–1950. *Annals of Science* **47**, 317–45.

KAMATANI, C. (1963).   The history of research organization in Japan. *Japanese Studies in the History of Science* **2**, 1–79.

KATZENSTEIN, P. J. (1978).   *Between Power and Plenty: Foreign Economic Policies of Advanced Industrial Nations.* Madison: University of Wisconsin Press.

KRAUS, J. (1968).   The British electron-tube and semiconductor industry, 1935–62. *Technology and Culture* **9**, 544–61.

KUMATA, H. (1959).   Japan the search for expression. In *Nationals and International Systems of Broadcasting – Their History, Operation and Control,* ed. W. B. Emery, pp. 480–91. East Lansing: Michigan State University Press.

LEARY, F. (1960).   Electronics in Japan. *Electronics,* 27 May, 53–100.

LIBOIS, L.-J. (1983).   *Genèse et Croissance des Télécommunications.* Paris: Masson.

LOW, M. (1990).   Japan's secret war? 'Instant' scientific manpower and Japan's World War II atomic bomb project. *Annals of Science* **47**, 347–60.

MACDONALD, W. W. (1954).   Cross talk. *Electronics,* December, 129.

MACLAURIN, W. R. (1949). *Invention and innovation in the radio industry.* New York: Macmillan.

MALERBA, F. (1985). *The semiconductor business – the economics of rapid growth and decline.* London: Frances Pinter.

MALLET, R. A. (1954). *Apperçus de l'Electronique Française.* Crédit de l'Ouest.

MÉRIEL, Y. (1960). L'industrie électronique (II). *L'Economie,* 24 June, 4–5.

MISA, T. J. (1987). Military needs, commercial realities, and the development of the transistor, 1948–1958. In *Military enterprise and technological change – perspectives on the American experience,* ed. M. Roe Smith, pp. 253–87. Cambridge, Mass: MIT Press.

MORRIS, P. R. (1990). *A History of the World Semiconductor Industry.* IEE History of Technology Series, 12. London: Peter Peregrinus Ltd, on behalf of the Institution of Electrical Engineers.

NOBLE, G. W. (1989). The Japanese industrial policy debate. In *Pacific Dynamics,* ed. S. Haggard and C. Moon, pp. 53–95. Boulder, Colo: Westview.

NOUVION, M. (1984). L'automatisation du réseau téléphonique français – des origines à 1945. *Télécommunications* **50,** 76–85.

OKIMOTO, D., SUGANO, T. & WEINSTEIN, F. B. (1984). *Competitive edge – the semiconductor industry in the U.S. and Japan.* Stanford, Calif: Stanford University Press.

PICARD, J.-F. (1991). L'organisation de la science en France depuis 1870: un tour des recherches actuelles. *French Historical Studies* **17,** 249–68.

POITEVIN, J. P. (1986). Composants: le rôle stratégique du CNET. *France Télécom* **60,** 1–7.

SAMUELS, R. J. (1987). *The business of the Japanese state. Energy markets in comparative and historical perspective.* Ithaca, NY: Cornell University Press.

SATO, K. (1963). Electric machine industry today. *Contemporary Japan* **27,** 438–60.

SATOFUKA, F. (1993). The case study of the Japanese Electronics Industry. Paper presented to the conference European Responses to the Post-war Electronics Challenge: Research, Business and Politics, Paris, Centre de recherche en histoire des sciences et des techniques, 27–28 May.

SATOFUKA, F. & RAHMAN, A. (1987). Post war science and technology policy of Japan: a view from a developing country. *Historia Scientiarum* **33,** 61–76.

SHEARMAN, C. & BURRELL, G. (1987). The structures of industrial development. *Journal of Management Studies* **24,** 325–45.

SHINN, T. (1980). The genesis of French industrial research 1880–1940. *Social Science Information* **19,** 607–40.

SUEUR, R. (1949). Le transistron triode type P.T.T. 601. *Onde Electrique* **29,** 389–97.

TAKAHASHI, Y. (1993). Progress in the electronic components industry in Japan after World War II. In *Technological Competitiveness: Contemporary and Historical Perspectives on the Electrical, Electronics, and Computer Industries,* ed. W. Aspray. New York: IEEE Press.

THOMSON-CSF (n.d.). *Historique THOMSON de 1893 à 1977.* Paris.

TILTON, J. E. (1971). *International diffusion of technology: the case of semiconductors.* Washington, DC: The Brookings Institution.

WADA, W. (1960). Japan. *Electronics,* 12 February, 93–5.

WALRAVEL, P. (1966). Quelques technologies nouvelles en traitement digital de l'information. *Les ordinateurs et l'informatique*. Institut Industriel du Nord, Bulletin de l'association des ingénieurs de l'IDN.

WEBBINK, D. W. (1977).   *The semiconductor industry: a survey of structure, conduct, and performance*. Washington, DC: Staff Report to the Federal Trade Commission (Economic Report).

WIENER, C. (1973).   How the transistor emerged. *IEEE Spectrum*, January, 24–33.

YAMAMURA, K. (1986).   Japan's *Deus ex Machina*: Western technology in the 1920s. *Journal of Japanese Studies* **12**, 65–94.

ZINSMEISTER, K. (1993).   MITI Mouse – Japan's industrial policy doesn't work. *Policy Review* **64**, 28–35.

ZAHL, H. A., PIORE, E. R. & MARCHETTI, J. W. (1951).   Defense Department plans for basic research. *Electronics*, August, 82–7.

ZYSMAN, J. (1983).   *Governments, Markets and Growth: Financial Systems and the Politics of Industrial Change*. Ithaca, NY: Cornell University Press.

12

User resistance to new interactive media: participants, processes and paradigms

IAN MILES AND GRAHAM THOMAS

Introduction: resistance and reception

New consumer IT (Information Technology) sometimes appears to be an unmitigated success in the marketplace. Even products which have been written off may break through in the longer term (e.g. the laser videodisc, now reappearing both as a vehicle for movies and in the new form of interactive Compact Disc); some products which are apparent losers have bounced back in a new form (e.g. videogames consoles, for a while seen as being displaced by home computers as a medium for games); yet others' deaths have been much exaggerated (e.g. a drop in home computer sales, in the wake of a boom, was heralded to be the passing of a fad).

Videotex is the exemplary failure to realize expected consumer markets for new IT – and even here, Britain's Prestel does not tell the whole story, as witnessed by the large markets established by France's Minitel. A less familiar failure in consumer telecommunications involved the collapse of the first CT2 (telepoint) portable phone systems: but this resulted from a combination of greedy pricing, incompatible standards and confusing signals from competing suppliers, and industrial belief in the consumer potential for such services in the UK was still being displayed by Hutchinson's eventually unsuccessful efforts to secure a foothold for their Rabbit system – after all, similar technology had taken off in Hong Kong. At the same time, cellular phone operators are pitching 'low cost' services at consumer markets, and the new generation of PCN (Personal Communication Network) products is soon to be launched.

Probably a great many failures, stemming from smaller suppliers, are less visible. For example, many of the products that appear in mail order catalogues are highly ephemeral. (Who, if anyone, has actually acquired an Omnibot home robot, or a device to disguise one's voice on the phone?) More significantly, there are many instances of one particular design configuration losing out in the battle between competing innovations. For example, cellular and PCN systems may well render CT2 redundant.

Market success or failure is, however, not co-terminous with public acceptance and resistance to new technologies. Products may be acquired reluctantly or enthusiastically; once acquired they may be used intensively or infrequently; only certain functions may be employed. There may not be active user resistance so much as resistance from 'passive users', by analogy with passive smokers. The purchaser is not necessarily the user – parents often buy home computers for their children, for example, and may well envisage these products being put to different uses than the child intends and implements. The home computer example is a good one: even when these products are acquired, there may be resistance to their use from other family members. These members may refuse to have anything to do with computers themselves, they may not allow them to be connected to the domestic TV or to be used in the living room, they may limit children's periods of use and the types of software that are bought. And children may reject educational software in favour of computer games (see Haddon 1988; Skinner 1992).

Potential acceptors/resisters to new consumer technologies may thus span the intended user, who may or may not be the purchaser; other household members (who are liable to be affected by at-home use of the product); members of the user's immediate social network; and broader social groups who may be concerned about use of the product on moral, political or other grounds (as in the scares about 'video nasties' and 'computer addiction'). All of these are important, but in this paper we mainly draw on evidence concerning users/purchasers, and political lobbyists, and will be looking especially at the acceptance/rejection of **interactive IT products**.

New IT, with its dramatic decreases in the costs of information processing, has made it possible to embed computer power within consumer products. As well as those which are clearly computer-like, for example pocket electronic diaries and organizers, this technology underpins audiovisual innovations like the Compact Disc (CD), communications products like mobile phones, and even household appliances like the microwave oven and toys such as talking dolls. All sorts of devices have been equipped with such facilities as memories, clocks and timers, remote controls, programmability, and so on (cf. Miles 1988). However, it is widely felt that many products are 'overfeatured', and that most users only bother to find out about and use the most basic features: this complaint is particularly prevalent in the cases of video cassette recorders (VCRs) and new telephone systems (especially in the office, but increasingly also in the home).

This highlights the uncomfortable fit between acquisition and acceptance of innovations. The challenge, here and elsewhere, is one of finding acceptable – in the sense of marketable – applications for the power of new IT. The use of IT in consumer products has generally been seen as a selling point: advertisements, and even the names assigned to products, have stressed such features and deliberately given the product a 'high-tech' image. But some manufacturers, fearing consumer resistance to complexity and to demands for user skills, have sought to distance their products from such imagery, especially from that associating their products with computers.

Looking from the point of view of purchasers, the issue is not really one of consumer acceptance of, or resistance to, new consumer IT in general; nor even one of responses to isolated products. Rather, the typical case – we shall be seeing some exceptions below – involves consumer choice between **alternative design configurations**. At any one time there are likely to be several competing innovations which offer overlapping functionalities and design features. For example (simply to line up similar innovations alongside each other), teletext provides some of the features of videotex in terms of presenting page-based information in a large print format. In addition to this instance, and the options for mobile phones mentioned above, we might cite VCR versus laser discs (and several standards **within** each), home computers versus games consoles, and so on.

Hesitation over the purchase may not reflect resistance, so much as bewilderment over the range of choice. Which functionalities or features are expected to be the key ones (both immediately and with sustained use)? How far do alternative products offer these? In which combinations? How can reliable information to inform one's choice be obtained in an easy way? The fact that functions and features can overlap means that a decision to buy a new product might also involve a decision to scrap, or at least downgrade, an item already in use – which could make it that much harder to justify a purchase. Other factors which can lead to hesitancy over purchase include uncertainty about which format and/or standard will prevail, and about whether to expect near-term price reductions and product improvements.

Configurations of interactivity

Our focus in this chapter is on consumer audiovisual and communications goods and services – such as hi-fi systems, TV and radio, and the telephone. Several interrelated technological trajectories are apparent in these areas, as in Figure 12.1.

Interactivity is a term increasingly applied to new consumer products in these

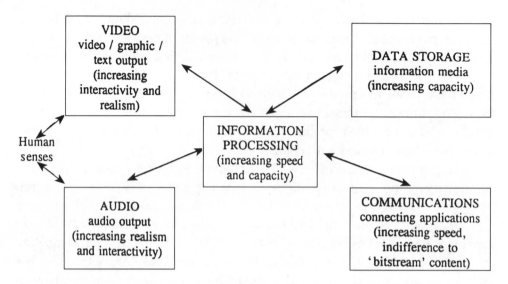

Figure 12.1. Interrelated trajectories in new interactive media.

areas. It is used with respect to telematic services (for example, a Prestel manager remarked in the late 1980s that the objective in the consumer market was to sell interactivity – in contrast to earlier promotion of the videotex service as 'a world of information at your fingertips'). The term is also applied to stand-alone devices (such as interactive Compact Disc – CD-I – players), and to software (some of the more sophisticated adventure games for PCs are described as interactive fiction; one of the main retailers of games software is Virgin Interactive Entertainments).

With interactive products, the flow of information is to some extent under the control of the user, rather than following a single sequence pre-determined by an author or information supplier. New messages are, in effect, generated in response to user inputs, meaning that even a published product has more of the attributes of a **conversation** than of the one-way flow of information typical of mass broadcast and recorded media – and to some extent of communications with long time-lags between messages (e.g. conventional mail).[1]

Alternative media permit different aspects of the flow of information to be controlled, to different degrees, by users. This may be critically important to particular users or applications. Some readers consider texts printed on paper – against which the new media are commonly contrasted – actually to be ideally interactive. Given a book, for example, they can rapidly flick through it, examine pages in sequences of their own choosing, take time off to make notes, and so on. (Such acquired skills in using a traditional medium may lead us to regard new media as lacking familiarity and friendliness.) The stored information is under the manual control of the user, and the printed page allows for (fairly) rapid search

– probably giving the reader more governance of the information flow than is available, say, to viewers of a conventional TV broadcast or even a videotape. An exchange of conventional letter post permits profound dialogue between the parties concerned: even if individual letters present information in an unchanging linear sequence, exchanges of letters can be quite conversational. But this interchange is liable to be slow and drawn out. Interactivity is not an all-or-nothing phenomenon, then. We can identify gradations of interactivity, as follows.

Minimal interactivity

At one extreme in the control of the flow of information are public address systems. These represent vanishingly low levels of interactivity – we are forced to be 'users', and can at best ignore them (or drown them out with personal stereos – which may be someone else's imposed, non-interactive, medium). Traditional broadcasting media display low levels of interactivity; you can at best change channels, even if this is more easy with a remote controller. As the number of channels grows, the choice of material grows, but the absence of storage means that the information is not interruptible, nor can it be reviewed, or accessed in any sequence other than the linear sequence originally supplied.

Moderate interactivity

Teletext is a new type of broadcast medium, displayed alongside TV broadcasts. Managers of a teletext system whom we interviewed portrayed this service as an interactive one. The teletext user may have no direct influence on the information provider, but the system provides a vast span of relatively rapidly accessible information, with informative menus setting out what is available. The 'feel' for users as they summon up specific 'pages' on their TV screens is quite similar to that of extracting information from an interactive videotex service – even though the teletext systems are continuously broadcasting numerous pages of information, while the videotex system sends out pages on demand.

The sense of interactivity gained from teletext (and books) stems from the opportunity to view and review particular pages, in an order of one's choosing. Books and, especially, teletext are poor where it comes to users affecting the information displayed on any particular page; but a sense of interactivity can be augmented by innovations (or skills) that give greater, or more convenient, control over a medium. The remote control 'zapper' for TV sets, for example, making it more convenient to change channels, has reputedly led to much greater levels of hopping across TV channels. Selection of teletext pages is an analogous operation, and would undoubtedly have been more tedious and uncomfortable if requiring manipulation of controls located on the TV set. Ability to control output without changing one's accustomed seating for TV viewing, was a vital component of teletext's success as a consumer innovation.

Videotape can be used in non-linear ways, too, but problems with constructing menus, and with rapidly searching for and accessing material from videotape, mean that it is hard to recompose the video 'text' in novel ways. Thus the main uses are for watching pre-recorded (linear) movies, and for time-shifting – watching programmes at one's convenience, rather than when they are broadcast (making it possible to view programmes and chunks of programmes in different orders, and to skip through and interrupt them). Still, it is common to find people making use of freeze-frame facilities, or reviewing specific portions of a programme.[2]

High interactivity

New IT can supply information tailored to user's requirements. **Online databases**, for example, provide bibliographic or other material in response to users' statements of their information requirements. Videotex was originally seen as a consumer version of such databases, providing data on train times, leisure facilities and the like. The information provider supplies the raw material for the database, and the interfaces, search languages and capabilities that are to be employed; the technological system proceeds to select and organize material on the basis of the user's interaction with it. Users thus have considerable control over information flow.

The classic online database would be used in short bursts by users, as they sought to locate particular key items of information (e.g. some business users would seek repetitive regular updating of key facts and figures like exchange rates, share prices, weather conditions). With the wide diffusion of PCs, disc-based **electronic publications** which can be explored in much the same manner have been established. Many early electronic publications – computer club newsletters for example – were little more than conventional text documents presented on a floppy disc, perhaps with a simple menu at the front end. But there is little advantage in reading a linear text from a screen (especially the poor quality screens that characterized earlier PCs). Growing attention has been paid to supplying text and other material in new forms that can take advantage of the opportunities provided by computer mediation of the material, such as **hypertext** and **hypermedia** 'documents'. The user can interact with the material over hours, or even days, as in the case of games and educational packages delivered via CD-I or CD-ROM. CD-ROM has already achieved rapid uptake in libraries, where it often complements online databases, and is now beginning to proliferate in offices and to find home computer applications. It is likely that we are just at the beginning of a wave of new electronic publications based on optical storage media, although conventions about appropriate pricing still have to be established.

The main information processing tasks may be carried out by mainframe computer 'hosts' or by users' PCs, but in either case the expectation is that an electronic publication will be accessed in diverse ways, so that software is provided to facilitate precisely this.[3] The 'author' is required to provide fine texture to the information flow, so that the user can change the course of events at several different layers of the informational hierarchy – select from main menus or submenus, browse forward or backward through texts, follow up particular leads or search for definitions of obscure terms, and so on.

Higher levels of interactivity

Advanced IT systems can be used to generate completely new material in response to inputs. For example, some decision support systems are beginning to become available, which are based on expert systems that 'learn' in response to substantial information inputs from users. Such systems can modify the rules they develop, with the result that similar information requests may generate quite different outputs at different times. But most interactive systems currently in use or on the market fall far short of this level of interactivity. They simply select among items of material supplied by information providers, according to user requests for information. There may be novel combinations of material, but modification of the underlying material as a result of interaction with individual requirements is very restricted. Typically, modifications are relatively trivial ones, such as presenting data in alternative forms (e.g. financial information services may present data as text, spreadsheet graphics, etc. – these may **look** very different to the user), addressing the user by name, providing records of previous inputs by the user (e.g. earlier shopping lists on a teleshopping system), and the like.

These new media, by offering much more active information processing in the IT system, provide 'publications' offering forms of interactivity more like those provided by interpersonal contact. Highly interactive products are **intimate** ones, with users expected to shape what happens next by being engaged with and responding to a flow of events. There may be 'default' modes, in which a linear flow of information is presented if the user takes no action, but the design of the products typically anticipates interactivity. This quality of intimacy may be of importance in the design of products which are intended to be highly interactive. In a recent talk, the multimedia marketing manager of a rival firm criticized Philips' CD-I as being based on TV visual output, and thus fundamentally associated with a product which is normally kept at a distance from the consumer, implying a barrier to intimacy. She argued that a PC environment, in addition to higher quality screen resolution than that provided by standard TV, provides a certain closeness to the application which gives the right interactive 'feel'. We should make allowances for bias in this case, and keep an open mind about whether the designers of CD-I (and similar product designs) will succeed in convincing potential users of the value of their product. But the general point about the interdependence of form and function in design remains valid. These characteristics of intimacy and control which are associated with highly interactive applications cannot, of course, be achieved without high levels of user acceptance. Acceptance depends at least in part on whether the design of the product is appropriate to the intended level of interactivity.

As well as electronic publications, other new media provide communications facilities which permit changes in the way of interpersonal contact itself – that is, the direct (potentially two-way) interaction of people, rather than human-machine interaction. Thus, mobile phones remove many of the 'black holes' – or are they precious havens? – in which people could not initiate or receive contacts. Electronic text and voice messaging systems allow for asynchronous com-munication and rapid delivery of text, freeing users from some of the limitations of traditional media. There may be scope for one-to-many and many-to-many communication in addition to the traditional one-to-one format. And some electronic media allow for anonymous contact to take place in 'cyberspace', by virtue of the fact that individuals happen to be accessing the same service, rather than because one has deliberately called the other. On CB radio, bulletin boards and chatlines, individuals may adopt pseudonyms and even role-play new identities; and in some multi-user role-playing games users both explore an electronic landscape and interrelate with each other, thus combining human and IT interactivity. We shall consider interpersonal communications in the next section, and then return to the topic of electronic publications.

Interactive communications media

Videotex

National differences are striking in the development and diffusion of many communications media. Perhaps these are more a result of the monopolistic role of national public telecommunications operators (PTOs) than because of cultural differences. **Videotex** forms a particularly interesting example of the variety of national paths in consumer IT use, as well as indicating the difficulties associated with developing interactive products for consumers.

In the UK, Prestel is now often written off as a great IT disaster, but its experience is highly instructive. Like the PC, it was seen as providing the general public with access to the computer power previously available exclusively to large organizations. In this case, public access to mainframe computers and their databases would be mediated through two familiar domestic technologies, television and the telephone. The presentation was intended to appeal to non-professionals, with relatively large text characters and simple but colourful graphics, laid out in attractive pages on the TV screen; it would be easy to learn, involving navigation via simple menus. Consumers would use a small and very basic keypad to input uncomplicated requests for action in terms of these menus. This design assumed that data transfer from the videotex system would be rapid, and that from the consumer slow: Prestel was to be a medium for electronic publication, then, rather than for communication.

The small early Prestel keypads made it difficult for users to input more than a few symbols at a time; but with the emergence of home computers, by the early 1980s, many potential consumers were using far more sophisticated devices. The Prestel design underestimated the scope for cheap processing power in the home provided by progress in microelectronics, and while few people had forecast the creation and popularity of home computers, this also reflected the public service provider mentality then prevalent in the PTO. That is, processing power was assumed to be concentrated in the service provider's mainframes, not distributed around the system. Home computer pioneers wanted to provide mass information-processing capabilities through powerful microprocessors and software, while network operators thought more of opening up access to centralized information resources. It was assumed that consumers would commandeer domestic TV sets,[4] tie up their phone lines, and spend time with cumbersome menus to locate data. But the data turned out to be of very uneven quality. The PTO was more concerned with establishing the technically advanced system quickly by extending its services to reach as many potential users as possible, than with seriously assessing consumer demand and the adequacy of information providers. Information providers were inexperienced with producing material in user-friendly and attractive ways, and often seem to have been half-hearted in their commitment to the new service. Finally, 'gateways' to real-time trans-actional services such as teleshopping were slow to materialize.

Prestel was an effort to establish a framework for consumer interactivity with a new computer medium, in the absence of much prior experience – the design was evolved before the 1980s introduced not only home computers, but also videorecorders, CD players, infra-red remote controllers and other IT-based products into large numbers of households. Also, the building-block household technologies of telephone and TV were ones where elaborate infrastructures delivered information products to their final users. In the 1970s these were thought of as public utilities, with a mission to provide collective services – and the telecommunication and broadcasting authorities were supposed to know what was best for consumers.

But such authority did not guarantee an authoritative view of consumers' response. Forecasts of millions of family users rapidly evaporated as, despite advertising campaigns and public displays, considerable uncertainty and little enthusiasm was generated about the new service's functionality. After more than a decade of searching for elusive consumer markets, Prestel withdrew its major remaining consumer services in late 1991. (These were mainly oriented to computer hobbyists, who would interface their home computers to Prestel via a modem.) Prestel remains as a business service, allowing relatively untrained staff to access information on flight availability or insurance rates, and to book seats or order automobile components. Many businesses have actually abandoned Prestel itself, using proprietary videotex systems set up by major companies in their sectors.

With Prestel's failure as a consumer medium major elements of its design fell. The role of the TV as part of the package has all but disappeared, to be replaced by the computer screen or dedicated terminal; more sophisticated keyboards have been introduced for user inputs; and faster communications in **both** directions are now standard. But certain videotex design features did prove successful in the case of **teletext**, where 'pages' are accessed via a TV set using a small remote control keypad. Like Prestel, teletext use required consumers to acquire new TV sets – but the teletext price premium was low, and there was no need to fit new telephone sockets or to pay for telephone network use, let alone paying subscription charges (other than the regular TV licence) or access charges for particularly interesting pages. Teletext information might be basic, but it was free: it contained TV listings; it could be accessed via the TV set without lengthy interruption of regular viewing. Its limited interactivity meant that a sophisticated remote controller (also becoming popular at about the same time) could be used rather than a complex keyboard. From a design point of view, teletext showed that, for certain classes of information at least, the limited resolution of the videotex/teletext screen format was not an insuperable barrier to user acceptance, but it also showed that the ensemble of design features must fit the application in question: the teletext screen format was acceptable because other design features did not get in the way, whereas with Prestel they did.

The great success in consumer videotex services, whose experience is often contrasted with that of Prestel, is of course the French **Minitel** system.[5] In establishing its design, the French PTO paid substantial attention to experiment and interaction with users, whereas Prestel's innovators assumed that they could anticipate the design criteria for interactive services.[6] While sharing many features of videotex design, Minitel differed in significant ways from Prestel.

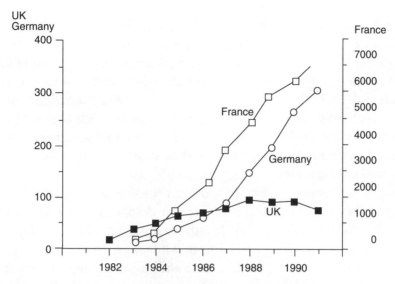

Figure 12.2. Videotex subscribers/terminals in the UK, France and Germany, 1982–91. Source: Thomas, Vedel & Schneider (1992).

Online search was facilitated by keyword search rather than by progression through menus. There was no subscription charge, and a simple, time-based cost structure was used. Gateways to service operatives allowed for teleshopping and related applications from the outset – indeed, the French PTO encouraged service providers to hook up to its network, whereas the British PTO had sought to control the facilities by running them on its own host computers. (This was supposed to ensure quality control, but these intentions were not notably fulfilled.) The French PTO distributed, free of charge, a standard simple computer-type terminal with an alphanumeric keyboard and built-in VDU. Users were motivated to gain experience: rather than assume that people would be motivated to invest money – and time – in searching for large volumes of online information, an important information resource (the telephone directory) was shifted to the new medium. Figures showing the growth (or decline) of the major national videotex systems in the UK, France and Germany (which is closer in conception to Prestel than to Minitel, although there are some important differences) are presented in Figure 12.2.

As is well known, messaging, rather than database access[7] proved to be the first major consumer application of Minitel. The messaging facility was designed in, after it emerged as surprisingly important in pilot studies. A large number of 'messagerie' services were established, with the notable successes being those which allowed anonymous communication on erotic topics; messaging proved popular on Micronet, too, the section of Prestel used by home computer hobbyists. So what emerged was not a large consumer market for interactivity with online databases, but one for new forms of interpersonal communication. The messaging facilities were rather primitive, being based around exchange of simple text: but individuals could adopt new personae and pseudonyms, even changing their apparent gender if they wished; they could rapidly make contact with other individuals with little of the awkwardness, sense of social distinctions and risk of embarrassment familiar in traditional meeting places. With a critical mass of terminal users established by the free distribution of terminals, other users could be located at any time of day or night, and even infrequently used information services could often generate enough revenue to remain viable.

This use of messageries in France attracted considerable attention and some censure, but the service has remained popular despite controversy. In the UK, however, the PTO had gone to some lengths to restrict similar uses of messaging facilities. Chatline systems on Prestel were withdrawn, to the annoyance (and with the subsequent loss) of many users; and the possibility of such services proving embarrassingly dominant was a factor cited when BT finally ruled out the notion of following the French in subsidized distribution of terminals, in 1989. However, the concern here seems to have stemmed more from the experience of voice chatlines on conventional telephones, rather than from particular problems generated in computer media. Let us briefly examine some of the new voice-based services involved here. We refer to these as **audiotex** services (here we are discussing what are also known as Premium Rate Services (PRS), although some definitions of audiotex include freephone services and some other kinds of computer-enhanced telephony).

Audiotex
Audiotex services have proved extremely successful in the UK in terms of market growth, though they have not been popular in all quarters. This success can be directly counterposed to the reasons for the failure of Prestel (and to some extent the success of Minitel). Firstly, with audiotex there was no chicken-and-egg problem to be solved relating to the supply of terminals and the number and quality of new services: nearly all households already possessed the necessary terminal for the reception of audiotex services, in the form of the normal telephone handset.[8]

In contrast to Prestel, audiotex was introduced without any huge publicity fanfare. The overlay network used for audiotex PRS was originally developed to supply freephone services, and from the beginning independent service providers were allowed to use their own equipment and to develop their own service formats.[9] In terms of user costs, audiotex uses a billing system similar to that which proved so popular in France with Minitel users: no subscription is required, and the cost of the services is based simply on the length of the call and the time of day (peak or off-peak) that the call is made.[10] Audiotex services are billed as part of normal telephone bills, with the revenue being shared by the network operator and the service provider.[11]

A problem which has helped create both acceptance (use) of, and resistance to, audiotex concerns whether the user is actually the subscriber paying the bill. A major factor in the initial popularity of audiotex was the possibility of trying out the new services at someone else's expense – whether that be children using their parents' telephone to call chatlines or employees using their work-place phones to listen to sports results, horoscopes or sex lines. When the first bills eventually arrived, their size contributed to a general outcry about the pernicious effects of audiotex. This aspect of the system design has led to substantial restrictions being placed upon audiotex services, and has posed difficulties to regulators in many countries.

The privacy and physical closeness of the telephone handset are a guarantee of one kind of intimacy – indeed, the audiotex interface provides a level of intimacy which, while perhaps inappropriate for the limited interactivity of most services, is not unsuited to the personalized – and sometimes personal – nature of the information being conveyed. The success of audiotex, which grew in the UK from nothing in 1986 to a £130 million sector in 1988 (total revenue for PRS), and then more slowly to £200m by 1991, shows that the consumer market is prepared to accept innovations in new interactive media if the services are useful or entertaining and the interface is appropriate. The word 'interactive' should be used with caution here: many of the services are merely tape recordings which are simply listened to (although others offer live conversation, televoting or financial information chosen by tone-dialling after the call has been set up). In 1991, around 350 UK service providers offered over 15,000 separate services, broken down roughly into information and sporting (45%), entertainment (26%), 'adult' (18%) and live conversation (11%) (ICSTIS *Activity Report*, 1989).

Interactive electronic publications

The challenge that suppliers face is one of determining what sorts of information product consumers might want to have in an interactive form. The failure of Prestel as a mass medium for text information, and the success of Minitel for consumer messaging, form compass points for industry thinking. So, in the home computer field, does the unexpected failure of practical housework applications (e.g. home accounts) and the success of computer games. Computers as interactive products have become a point of reference, though interactive publications may themselves challenge existing models of computer use. For instance, there is renewed attention to the idea of 'electronic books'. One significant development takes the CD-ROM player from being an adjunct to a PC to being a stand-alone and even portable device (incorporating a computer, in effect). Thus small CD-ROM player/viewer devices, such as Sony's personal stereo-like **Dataman** (envisaged as a lightweight option for people on the move who need to consult extensive reference material) may be a route to mass markets for 'electronic books'. This is not to say that we will shortly be reading novels from CD-ROMs, but rather that certain types of book may be emulated (or even improved upon, in terms of search facilities) by the new media.

There has also been substantial commercial effort to develop and market new multimedia products which use optical discs to deliver video as well as text, graphic, audio and animated material.[12] Suppliers have sought to develop new consumer electronic products which will play through TV sets like videorecorders, but which offer the sorts of interactivity that consumers have become used to with computer games and professional users with CD-ROM systems. The user is able to interrupt the action to search for more detail on a particular point, or to replay an episode, and to determine the sequence of events by means of menu choices or 'zapping' objects on the screen.

Delays occurred with the Philips/Sony CD-I system (while waiting for chips that could deliver full-screen, full-motion video), so that Commodore was able to hit the market first with its CDTV system in 1991 (this also lacked video, though it provided fairly high-quality animations). These multimedia products are being marketed as educational and entertainment media.[13] Business users, who need to acquire skills or task-relevant information, may be readily motivated to learn to use such products, but for consumers, the design may be more critical in determining market success. The challenge to suppliers is to design products (especially their interfaces) where the main selling point, interactivity, is not a familiar feature of consumer electronics. A major debate among multimedia innovators has been whether to design the interface around a home computer keyboard or a TV remote control. The implications for how consumers view the product are substantial (cf. our earlier discussion on 'intimacy').

CD-I consumer products released to date are designed to resemble video-recorders and CD players, with interfaces like TV remote controls. The innovative element of the design is to incorporate small pointing devices on the remote controls, so that menu choices can be made conveniently, and the user can interact with depictions of objects on the screen. The designers have drawn upon the almost iconic significance of remote controls in the 1980s, both as popular artefacts and as supposed sources of a more interactive 'zap culture' based on viewing small chunks of many diverse programmes. But not too many people can operate a single remote controller at once: if this renders interactivity focused on one user, this may be more like solitary home computer use than TV use. The suppliers hope that users will be able to work through educational and other practical material together, but it remains to be seen how satisfactory this will be in typical home settings.

It remains to be seen whether, in seeking to avoid the connotations of home computers as male- and youth-orientated, technically complex yet frivolous devices, the right choice has been made. It is likely that we will see much evolution in the interface over the coming years. Quite possibly a market will develop in alternative forms of interface, which might range from the large joy-stick, through penpoint systems, to the virtual reality helmet. It is interesting to note that CDTV already seems to have migrated back to being marketed as a peripheral to the home computer ('Commodore Amiga CDTV'), rather than as a new part of the TV system.

We anticipate increasing efforts to market 'electronic books' in coming years. Already a significant toehold has been gained by portable electronic dictionaries, and bibles and other such products are available. Unlike Dataman, these products are very restricted in the sorts of information they carry, and some are dedicated to one source of data built into the device. However, the push to portable computers (often disguised as organizers and diaries) has the possibility of opening the way to new publications. The selling points will be interactivity and mass data storage, since the 'feel' of electronic text is still generally regarded as inferior to that of a traditional publication. If users do resist screen-based media as modes of accessing information, their success will depend upon these media conferring significant benefits which traditional media lack.

These additional features may include access to video and audio data (as in CD-I encyclopaedias), and hypertext features.[14] Another possibility, the 'electronic newspaper', is being explored by several suppliers of information services. The aim is to provide users with news on those topics that are of special interest to them. Several commercial initiatives have been taken in terms of supplying customized newspapers (of a specialized kind) to business users by fax and even by a combination of FM radio broadcast and PC reception and selection of material.[15] The appeal of such newspapers is likely to be strongest for those who require specialized services, in connection with work or strong lifestyle interests. After all, the well-established 'interfaces' (layout, headlines, etc.) of conventional news-papers mean that they are well understood by most readers, and – together with 'topping up' from radio and TV news broadcasts, and now teletext – probably provide most readers with all the urgent news they require. Unless product designs are appropriate to the intended level of interactivity (or that actually preferred by users), consumer resistance will force either further modification of designs or else lead to the demise of the product.

Acceptance and resistance revisited

Concern **for** – rather than **from** – users of innovations is encountered in several of the product areas discussed here. Media scares feature the violent (and sometimes sexual) content of video and computer games; the lack of creativity as compared to traditional pursuits (despite the contrast between interactive games and passive TV); loss of previous, healthy activities, and other health issues such as possible epileptic fits induced by screen flicker; fears of addiction to the games; complaints about the pricing policy of the suppliers. Scares often concern young people – could this be why some advertisements appear to be targeted at more mature users? These concerns have a familiar ring – they have been raised about TV and numerous other contemporary pursuits – and are usually based on anecdotal and partial evidence; but this does not mean that they should be dismissed out of hand. Similar, but probably more pronounced, concerns are voiced about new telecommunications media.[16] Let us return to the audiotex example.

Unlike videotex, the development of audiotex led to the creation of a separate regulatory body in the UK, following on from controversies about video pornography, and the establishment of regulatory authorities for cable TV, which emerged after the launch of Prestel. Concern was expressed in Parliament, in the press, and via consumer organizations and trading standards bodies, over both the structure and the content of audiotex services. This concern over audiotex was in fact a response to the *lack* of initial resistance to the innovation, in terms of market success. If audiotex had not been so successful there would not have been such a fuss.

The main concern with the **structure** of audiotex was the billing problem. That the user of services is not necessarily their purchaser is a feature of telephony generally – as shown by parents' oft-voiced complaints about teenagers hogging the telephone, or tales of burglars/spurned lovers calling a far-distant Speaking Clock and then leaving the phone off the hook. But the problems with audiotex were felt to be more severe – most teenagers' calls are at local rates, and other continents' Speaking Clock numbers are not at our fingertips. Concerns over audiotex billing eventually led to a directive ordering the network operators to provide free call barring to PRS numbers where this was technically convenient (i.e. on digital exchanges), and to the creation of compensation funds to reimburse bill payers afflicted by the unauthorized use of their phones to call 'live conversation' audiotex services.[17] In addition, worries were expressed about the promotion of audiotex services, and restrictions have been placed upon advertisements.[18]

On the question of **content**, the main objections concerned one of the main reasons for the initial popularity of audiotex, the prevalence of so-called 'telephone sex' services. Although some of these involve live conversation on a one-to-one basis, they are mostly recordings of people 'talking dirty'.[19] In this respect, the debate is similar to that in France over the effects of the 'messageries roses' accessible via Minitel. With chatlines, there were worries about content: especially concern about what children and teenagers might talk about in group conversations (although others expressed the view that school playgrounds were just as likely to be sites of verbal corruption). There was also a fear that paedophiles might use children's chatlines as a way to initiate contact. The introduction of employees to supervise, and if necessary censor, chatlines did not completely remove such worries (and indeed led to other debates about civil liberties and censorship), and chatlines proved to be a persistent concern for audiotex regulators. Worried about its image, BT dropped its own highly successful chatlines in 1987 (and as noted earlier, subsequently discontinued the textual equivalent on Prestel). In 1992 audiotex chatlines were banned completely by Oftel.

Audiotex more generally has found it hard to shake off a rather dubious image, to the chagrin of network operators and respectable service providers. This image stems partly from the sex lines, and also because of various frauds in which service providers illicitly relieved either end users or operators of their money (e.g. cases of audiotex services advertising bogus offers of employment).[20]

Such problems of audiotex structure and content have been faced, with variations, in all countries where PRS have been introduced. A variety of regulatory models has been suggested or created to deal with them. These range from the placing of responsibility for audiotex structure and content on the network operators (e.g. in the USA, a measure which has led to one major operator refusing to handle PRS altogether), through governmental regulatory commissions (some European countries have considered this option, although it has generally been feared that a statutory body would prove to be too inflexible in dealing with a fast-moving target), to the intermediate UK model, where an independent regulatory body, ICSTIS (Independent Committee for the Supervision of Standards of Telephone Information Services) is funded by topslicing audiotex revenues. It both reacts to consumer complaints and initiates its own investigations. Its powers are enforced through the licences issued to the PTOs – effectively, if ICSTIS cannot come to an informal agreement with a service provider it deems to be infringing its Code of Practice, it can order BT or Mercury to withdraw one or more PRS lines used by that service provider.

For regulation to be flexible, and governed by an independent body, may help in allowing innovatory opportunities to be developed in the light of experience and new technical possibilities, while deterring features which would disadvantage or antagonize significant or vulnerable sections of the user community. Such adaptive regulation can minimize the likelihood that resistance to new technologies will arise from socially unacceptable side-effects of their introduction: it can help focus issues of acceptance or rejection on the design features of (competing) products and on the structures of the industries and markets which supply them.

This is not to argue that the ICSTIS model is perfect. It may be that the flexibility of the 'independent committee' model can be used for arbitrary censorship, or involve inequalities in the ability of service providers to challenge decisions (with only the biggest having the ability to mobilize countervailing forces). Conversely, one Member of Parliament (who has tabled a Private Member's Bill on the topic) feels that, without statutory powers, such an organization is not strong enough to regulate the industry. It remains to be seen whether debate on audiotex in future will indeed be more focused on design issues than on the content of the services provided.

In conclusion, we would argue that there is a need to unpack the concepts of acceptance and resistance in relation to new technologies. First, we need to distinguish classes of people who might be involved in acceptance or resistance processes, and we need to take account of the social situations (and in particular, household structures and processes) into which new technologies are inserted. Purchasers are not necessarily the only, or even the prime, users. Both active and passive users, as well as broader social groups and organizations, may have legitimate concerns and intervene in the processes of the introduction and use of new technologies in consumer markets.

Secondly, resistance (and by implication its converse) needs to be classified and 'unbundled'. Resistance ('moving against' something, seeking to prevent it) is not necessarily the same thing as rejection ('moving away from' it, avoiding it). Yet the inverse of 'acceptance' would seem to be 'rejection'. What antinomy does 'resistance' call up: 'collaboration'? 'endorsement'? 'acquiescence'? 'welcoming'? In the case of interactive products, there is significant scope for different levels of engagement (which may mean different degrees of resistance, but not inevitably). We can think of various 'levels' at which resistance/rejection might be expressed.

1 Using a particular **feature** of a product (e.g. reluctance to learn about sophisticated ways in which a telephone handset or videorecorder can be used, while willingly employing its basic functions).
2 Using a particular **product design** (e.g. refusing to learn to use a traditional computer interface after having experienced the 'user-friendliness' of graphical user interfaces).

3 Applying a particular class of technology to a particular **type of application** (e.g. not using home computers to record personal data, on the grounds that technology should be kept out of such matters).

4 **Any use by oneself** of a particular class of technology (e.g. personal avoidance of inoculations, or of animal products).

5 **Any use at all** of a particular class of technology or type of application (e.g. principled resistance to all nuclear power or genetic engineering).

In the IT case we rarely encounter level 5 resistance – though it is always possible that future developments in artificial intelligence, robotics or virtual reality will be anathema to some people. The most common processes of resistance and acceptance fall in the intermediate levels of the above classification. They are not concerned with large classes of technology *per se*, but with specific implementations and combinations of technology used for particular applications.

Design paradigms, which guide notions of appropriate models for the applications, may break down as new interactive products blur received distinctions between consumer technologies. It is not always obvious which are the appropriate models on which to base a given application. One factor which needs to be taken into account in the design of such products is the level of interactivity that will be required by an active user – and this is associated with the distance or intimacy between that user and the application which the new medium is being used to realize.

Notes

1 For more on the conversational nature of interactivity, see the discussion between Brand and Lippman in Brand (1987). The levels of interactivity identified below are based on those introduced by Thomas & Miles (1989) in their analysis of telematic services.

2 Opportunities to recompose audio 'texts' have been facilitated by the use of discs, allowing for more rapid search-and-retrieval than tapes: thus audio CD systems (effectively random-access devices) can be 'programmed' to select tracks according to user preferences, rather than following their conventional linear sequence.

3 In any case, the distinctions between large remote hosts and small local computers is being eroded by new kinds of networking architectures – both client-server and peer-to-peer – and trends towards 'downsizing' and distributed computing. Innovations such as the World Wide Web are providing sets of hypertext documents which may be located anywhere in the global Internet (see Krol 1992, Ch. 13).

4 Early Prestel advertising showed a nuclear family clustered round the TV set, with the mother and children looking on adoringly as the father pushed buttons on the Prestel keypad. Studies have shown that this is not exactly typical family behaviour! (See, for instance, Silverstone 1991.)

5 Strictly Teletel is the communications service and Minitel the terminal, but we shall follow most writers in eliding this distinction.

6 Recall that in the 1970s the British PTO could define even the colour and style of telephones, and had little experience of consumers who could opt for alternatives. A dramatically more

competitive environment surrounds communications services as we reach the end of the twentieth century.

7   With the exception of consulting the telephone directory, the most heavily used Minitel service.

8   Some audiotex services require the presence of DTMF 'touch-tone' signals to allow the user to make choices once connected. But these are a minority, and an increasing proportion of telephones incorporate tone-dialling.

9   In fact, a variety of arrangements is available to would-be information or other service providers, who can opt to do everything for themselves or to enter a managed service arrangement with either the network operator or an independent service bureau: for instance, many of the audiotex services offered by household names such as national newspapers (e.g. shares and sports information) are provided via service bureaux.

10 While this makes life simple for users, it does have an effect on the shape of the audiotex service market. Some services are not really worth the fixed extra premium, whereas other services would be profitable only at higher rates. It is expected that increased digitalization of the network will allow a proliferation of tariff levels. Multiple tariff levels already exist in some countries.

11 This billing system has some weaknesses – for instance, itemized billing for telephone services is not universally available and even in areas where itemized billing is offered BT excludes calls costing under approximately 50p.

12 cf. the set of articles on multimedia in the December 1991 issue of *Byte*. The Manchester *Guardian*'s regular Online section on Thursdays now frequently carries a CD-ROM review column.

13 The leading videogames companies have been quick to develop their own CD-based systems, which are now challenging other multimedia suppliers.

14 An interesting set of hypertext developments is the creation of annotated and interactive versions of familiar books – e.g. *Alice in Wonderland* – for the portable PowerBook range of Apple computers.

15 Examples of both are encountered in both the US and UK, with the radio mode being trialled in the UK as a way of delivering newspapers for blind people (thus eventually they are presented in Braille or speech synthesis).

16 Broadcast and recorded video, and other highly realistic products, are also sources of concern. International responses to video nasties have diverged; we are seeing smaller-scale panics about 'computer porn' – graphics and animations on floppy discs, CD-ROMs and computer networks – and we expect serious controversy about virtual reality products in the future.

17 The latter measure proved unsatisfactory. Imminent exhaustion of the compensation fund provided the direct reason for the later banning of chatline services in May 1992, although it has been alleged that this was more of an excuse than the real cause.

18 Other complaints about the structure of audiotex, for instance whether the network operator should also be allowed to be a major service provider in competition with the independents, were mainly voiced from within the industry and were of only peripheral concern to end users (who often did not bother to distinguish between the network operator and the service provider anyway).

19 Interestingly, some of the complaints about the promotion of audiotex services have involved the sex lines – often they are about the potential effect of the advertisements on children and other susceptible members of the population, but occasionally complaints allege that the content of the recordings does not live up to the lurid promises of the advertisements.

20 In terms of worry about audiotex content, there is also the possible impact of audiotex on the quality of viewers' and listeners' quizzes broadcast by TV and radio; whereas some programmes used to try to test the knowledge of their public, the introduction of audiotex as a public response medium has supposedly led to an emphasis on easy questions so as to maximise

revenue from telephone responses. This has not, however, been viewed as a matter for the regulators.

Glossary

CD-ROM	Read-Only Memory Compact Disc
CD	Compact Disc
CD-I	Interactive Compact Disc
CD-ROM	Read-Only Memory Compact Disc
CDTV	Commodore Dynamic Total Vision
CT2	Cordless Telephone 2 (telepoint)
ICSTIS	Independent Committee for the Supervision of Standards of Telephone Information Services
IT	Information Technology
PC	Personal Computer
PCN	Personal Communication Network
PRS	Premium Rate Services
PTO	Public Telecommunications Operator

References

BRAND, S. (1987). *The Media Lab*. New York: Viking.

HADDON, L. (1988). The Roots and Early History of the UK Home Computer Market. PhD thesis, Imperial College, London.

INDEPENDENT COMMITTEE FOR THE SUPERVISION OF STANDARDS OF TELEPHONE INFORMATION SERVICES (1989). *Activity Report 1989*. London: ICSTIS.

KROL, E. (1992). *The Whole Internet*. Sebastapol, Calif: O'Reilly & Associates.

MILES, I. (1988). *Home Informatics*. London: Frances Pinter.

SILVERSTONE, R. (1991). *Beneath the Bottom Line*. PICT Charles Read Memorial Lecture. Swindon: Economic and Social Research Council. (PICT Policy Paper No. 17.)

SKINNER, D. (1992). *Technology, Consumption and the Future*. PhD thesis, Brunel University.

THOMAS, G. & MILES, I. (1989). *Telematics in Transition*. Harlow: Longmans.

THOMAS, G., VEDEL, T. & SCHNEIDER, V. (1992). The United Kingdom, France and Germany: setting the stage. In *Relaunching Videotex*, ed. H. Bouwman and M. Christoffersen, pp. 15–30. Dordrecht: Kluwer.

13

The impact of anti-nuclear power movements in international comparison

DIETER RUCHT

Introduction

Nuclear energy's initial associations with weapons, war and death were balanced and eventually overcome by politicians and scientists promising 'Atoms for Peace'. That was the slogan to promote the civilian use of nuclear power in 1953. Now associated with a positive and bright future, nuclear power promised virtually infinite energy; according to Lewis Strauss, president of the US Atomic Energy Commission in 1954, nuclear power would be 'too cheap to meter' (Ford 1982, p. 50). Today, as we know, this bright vision has dimmed. Nuclear power is now associated with rising costs, problems of waste disposal, accidents, and fierce citizen opposition. Probably for the first time, a major online technology has been successfully challenged through a long and intense struggle including millions of citizens. The impacts of these struggles have been multi-faceted. The immediate substantial policy outcomes vary from country to country. In some places, despite promoters' efforts, nuclear power programmes were never realized. In others, programmes were realized only in part. In still others, nuclear power programmes were implemented in the face of significant opposition. Overall, however, the tide in the last two decades has turned and nuclear power has been discredited. Even in such countries as France and Belgium where nuclear power produces the lion's share of electricity supply, it is perceived as a necessary evil to be tolerated, at least for a few decades, because of huge capital investments.

In this chapter I will survey the scope and forms of resistance to nuclear power, the various effects of resistance on nuclear power as well as other dimensions, and I will offer some tentative explanations that could account for differences in the opposition outcomes.

The scope and forms of resistance to nuclear power: an overview

Early resistance

The civilian use of nuclear power was developed out of military applications, and as various surveys unveiled in the early 1950s, the attempts to promote civilian uses still suffered from the nuclear bomb's negative image. At best, the use of nuclear power was perceived as a Faustian bargain whereby benefits involve risks. By the time the first civilian reactors went into operation in the 1950s, increasingly positive views of advancing technology and technology's con- tribution to economic growth and welfare led many to perceive nuclear power to be a major tool of progress (Gamson & Modigliani 1989). Until the late 1960s, however, nuclear power advanced only slowly. This was due not to public distrust but to technological problems, and more importantly, to the reservations of electricity managers who were reluctant to adopt nuclear power for economic reasons. By the time these reservations began to fade, however, reservation and even outspoken resistance among scientists and ordinary citizens had begun to flourish. Although the anti-nuclear power movements in most advanced Western countries took off in the late 1960s and early 1970s, criticism of nuclear power had begun earlier. Let us take a brief look at earlier signs of opposition in three countries.

In the **United States** local opposition rose in the late 1950s, including demonstrations against the Shippingport reactor near Pittsburgh in 1957 and the test reactor near York, Pennsylvania, in 1959. A more sustained effort was made by the **Committee for Nuclear Information** formed in St Louis in 1958. Barry Commoner, one of the most conspicuous critics of nuclear power, was an active member. After a four-year struggle, plans for the Bodega Head reactor were cancelled in 1964. The site, close to a geological fault, was thought dangerous due to the risk of earthquakes, and in this conflict even the traditional **Sierra Club** joined the opposition – though they did not take a clear-cut anti-nuclear position. In many ways similar to the quarrels at Bodega Head, an application for a proposed nuclear plant at Corral Canyon, near Malibu, was in 1967 rejected by the **Atomic Energy Commission** (AEC), and eventually, in 1970, the electricity company withdrew its revised proposal. Furthermore, there was an intense debate over the Ravenswood plant proposed for New York's East River. Remarkably, this project stimulated opposition not only from the city council, two citizen groups, and the **Scientist Committee for Radiation Information**, but also from David Lilienthal, the former AEC president. He strongly opposed the siting so near millions of people. Moreover, Lilienthal after having become the spokesperson of the coal producers, attacked nuclear programmes in general for being expensive and dangerous. Owing to this widespread criticism and doubts raised within the AEC staff the

Ravenswood project was abandoned early in 1965. Criticisms of planned reactors in Oregon, Vermont and Florida, while not preventing the construction, at least resulted in cooling system improvements. Worthy of mention is the late 1950s resistance to nuclear waste disposal in the Gulf of Mexico and similar plans for Cape Cod in Massachusetts. In the latter area, local groups also successfully opposed the implementation of a major nuclear 'park' which would have included several reactors and a reprocessing plant.

In **France**, early local opposition to nuclear power was much less frequent and more dispersed. When the French government decided to build a nuclear bomb in the atomic centre of Marcoule, two dozen people from a nearby spiritual community, headed by the charismatic leadership of Lanzo del Vasta, engaged in an act of civil disobedience in 1958. At the same time, the teacher Jean Pignero became worried about the danger of radiation from the extensive use of nuclear material in power plants and medicine, but it was not until 1962 that he founded the first anti-nuclear association and produced a newsletter.

In **West Germany**, local opposition was reported at several places in the 1950s and 1960s. People were worried when nuclear plants or other nuclear facilities were built in Jülich, Karlsruhe, Gundremmingen and Nürnberg. Several attempts were made to block the construction of nuclear reactors in the neighbourhood or even city limits of Stuttgart, Berlin and Ludwigshafen. Early doubts were also raised against the nuclear waste deposit Asse near Braunschweig and uranium mining in the Black Forest. The first German anti-nuclear group was probably the *Kampfbund gegen Atomschäden* formed in 1956. Four years later, the German section of the rightist *Weltbund zum Schutze des Lebens* was established, a group still active today. Among the most prominent critics of nuclear power was Karl Bechert, a nuclear chemist and social democratic Member of Parliament, and the medical doctor Bodo Manstein.

Although this list of incidents of early opposition to nuclear power is not exhaustive, and examples from other countries could be added, we should not forget that these were local protests attracting relatively few people; they were moderate in tone – on a national level they were barely perceived. Most people welcomed, or were at least indifferent to, nuclear power.

The long wave of opposition

By the end of the 1960s doubts about nuclear power grew, especially among scientists in the US. The **Union of Concerned Scientists**, formed in 1968, became an important actor in public hearings and debates over nuclear risks studies. Also, individual scientists, for example John Gofman and Arthur Tamplin, specialists in low-level radioactivity, became heavily engaged in the nuclear debate. Popular books such as Novick's the **Careless Atom** (1969) fuelled the argumentative repertoire of citizen initiatives that, in many places, began to oppose nuclear power in public hearings and litigation procedures. A wave of nuclear criticism began to rise and soon crossed the Atlantic. Anti-nuclear arguments developed in the US spread among the growing opposition in France, West Germany, Switzerland and, with some delay, most other West European countries.

Emerging opposition patterns were very similar. At first, people did not reject nuclear power *per se*, but rather criticized such aspects as site selection, dangers of low-level radiation and insufficient containment. Local groups were not yet closely bound together. However, the growth of opposition at most places where nuclear reactors were announced for construction meant that citizen protest could no longer be ignored. The 1973 so-called oil crisis helped to put the energy issue, and thus nuclear power, on the general public agenda. Moreover, local citizen groups began networking on regional, national, and even international levels. By 1974–5 anti-nuclear groups in several countries had formed a bona fide anti-nuclear movement and were attracting more and more people. With the help of Ralf Nader, in 1974 the first national meeting of anti-nuclear groups met for the **Critical Mass** conference in the US. Scientists in various countries raised doubts about nuclear power and collected signatures to express their dissent. This, of course, provoked governments and the pro-nuclear lobby to engage in a counter-campaign in which, for the most part, the challengers' arguments were not taken seriously.

Since nuclear power plants were not likely to be stopped by mere argumentation, anti-nuclear groups began to think about other means of opposition. In the US, groups initiated referenda in Western states which all failed. In Wyhl in southwestern Germany, local groups engaged in a mass act of civil disobedience by occupying the construction site for several months. This extraordinary event inspired other groups in both Germany and abroad. In 1976 and 1977 conflicts intensified. In such places as Grohnde and Brokdorf in Germany and Creys-Malville in France, anti-nuclear groups and police clashed violently. In such other places as Seabrook (New Hampshire) and Diablo Canyon (California), civil disobedience was widespread (Epstein 1991). One should not forget, however, that beyond these spectacular events hundreds and thousands of citizen groups engaged in rather conventional forms of opposition by organizing information meetings, distributing flyers and pamphlets, collecting signatures, living together in anti-nuclear camps, demonstrating and rallying, etc.

Partly due to this unexpected wave of dissent the once firm pro-nuclear front became weakened in the second half of the 1970s in many countries. Whether this fuelled the anti-nuclear critique or took the wind out of its sails was dependent on specific circumstances that cannot be discussed here. The Three Mile Island accident in 1979 certainly supported the anti-nuclear critique, but I

Table 13.1. *Protest events and participants in protests against nuclear power in four countries, 1975–89*

	France	West Germany	Netherlands	Switzerland
Number of protest events	273	301	67	87
Number of participants per million inhabitants	11.000	28.000	15.000	24.000

Source: Koopmans 1992, pp. 63f.

doubt that it was decisive. Economically, nuclear power had already proved to be a dead end in the US some years before (Campbell 1988). Nevertheless, Three Mile Island provoked a last wave of mobilization in the US and fuelled anti-nuclear campaigns in several European states.

Opposition to nuclear power varied considerably among European countries. In West Germany, opposition was sustained for many years. In Switzerland, the opposition failed to achieve majority support for a complete phasing out of nuclear power, but in a referendum held in 1990 a majority voted against additional reactors for a period of ten years. In Sweden, a 1980 national referendum brought a decision to phase out nuclear power in the long term, and in 1988 the Swedish parliament decided to stop all existing reactors by the year 2010. In Austria, the only nuclear reactor, while completed, was not permitted to operate due to a slight majority (50.4%) registering their opposition in a 1978 national referendum. In Italy, nuclear power was not introduced due to a 1987 referendum. In the Netherlands, the nuclear programme was drastically reduced and in Denmark and Norway all nuclear power plans were completely abandoned. In other countries nuclear opposition was less pronounced. Great Britain, once a forerunner in nuclear matters, experienced a moderate opposition, but this reached neither the size nor the radicalness of some of its European counterparts. In Belgium opposition was marginal and the nuclear power programme was even expanded. France, despite considerable and fierce opposition, implemented its programme without significant delay or reduction in scope.

By the mid-1980s, the struggle over nuclear power was largely settled in most Western countries. Although the Chernobyl accident in the spring of 1986 awakened many, in most countries it was not politically decisive. Those countries which had already decided to stay out of the nuclear business, to phase out nuclear power, or at least to reduce it drastically, felt their misgivings confirmed by the accident. In countries already strongly committed to nuclear power (France, Belgium, Japan) or not facing strong anti-nuclear resistance (Great

Britain), responses were mixed, but not sufficiently negative to change nuclear policy fundamentally.

Whatever the specific national situation, by the end of the 1980s the anti-nuclear wave ran out, though for different reasons. In some cases, for example Austria, Italy and Denmark, the anti-nuclear critique had proved to be successful and there was no longer reason to protest. In other cases, for example in Germany, Great Britain and Switzerland, a nuclear stalemate had been achieved and only some local nuclear power conflicts continued. In still other countries, such as France, Belgium and Japan, protest has been in vain and anti-nuclear groups were simply too disappointed and too tired to continue. Even in those countries, however, opposition against the siting of nuclear waste deposits is still important.

An attempt to assess resistance to nuclear power

Is there a way to measure opposition to nuclear power in order to rank countries according to their degree of resistance? Personally, I think it is possible, although data collection is extremely time-consuming. Probably the best way to do it is to collect protest event data drawn from daily newspapers and/or archives. Two projects are under way that allow for a quantitative analysis of the scope and intensity of protest against nuclear power and other issues as well. The first project, undertaken by Hanspeter Kriesi and his collaborators Jan W. Duyvendak, Marco Guigni and Ruud Koopmans, provides data on four Western European countries (West Germany, France, the Netherlands, Switzerland) from 1975 to 1989. My own research will cover, among other things, protest events in West Germany, France and the USA from 1970 to 1992.

Drawing on data the Kriesi team has published so far I will give some basic information on the amount of protest (see Table 13.1). If one considers the absolute number of protest events it is clear that resistance was highest in West Germany. In relative terms, however, looking at the number of people mobilized per million inhabitants, Switzerland comes close to Germany, and both are followed by The Netherlands and France.

It may also prove interesting to look at the forms of action. Here I rely only on my own data on France and West Germany. As Table 13.2 shows, protests in France were significantly more aggressive. The share of violent protest was about 2.5 times as much as in West Germany.

Unfortunately we do not have equivalent data on protest in most other countries. Based on my reading of the relevant literature on nuclear power conflicts in various countries (see, for example, Nelkin & Pollak 1981; Falk 1982;

Table 13.2. *Forms of action of the anti-nuclear movement in France and West Germany 1970–92*

Forms of action	France (%)	West Germany (%)
Signatures	10.8	3.9
Demonstration	44.3	55.7
Protest in hearings	0.3	6.5
Litigation	1.6	3.0
Civil disobedience	14.2	18.5
Violence	24.3	9.7
Other forms	4.5	2.7
Total	100	100
	(N = 379)	(N = 994)

Source: Rucht 1994.

Table 13.3. *Resistance to nuclear power in cross-national comparison*

Country	Mass mobilization	Disruptiveness	Score
West Germany	+ + + + +	+ + +	8
France	+ +	+ + + +	6
Switzerland	+ + + +	+	5
USA	+ + +	+	4
Austria	+ + +	+	4
The Netherlands	+ + +	+	4
Denmark	+ + + +	0	4
Japan	+ +	+ +	4
Spain	+	+	2
Italy	+	0	1
Sweden	+	0	1
Norway	+	0	1
Finland	+	0	1
Great Britain	+	0	1
Belgium	0	0	0
Canada	0	0	0

Source: D. Rucht, own estimates.

Kitschelt 1986; Jasper 1990; Rucht 1990; Rüdig 1990; Joppke 1993; Flam 1994), as well as my own newspaper archive, I will try to give a crude measure of resistance to nuclear power in countries on which I have some information. I score these countries according to the degree of mass mobilization against nuclear power and the degree of disruptiveness, that is, the significance of acts of civil disobedience and damage of property. Though I would not claim to establish a highly reliable ranking of countries, I think that their grouping into three categories of high, medium and low resistance comes fairly close to reality. Table 13.3 shows that there is a wide gap between West Germany at the top of the resistance scale and Belgium and Canada at the end.

The effects of resistance to nuclear power

It is extremely difficult to assess the effects of resistance to nuclear power. One basic problem is to clarify the subject, that is 'Effect on what?' Of course, what comes immediately to mind is the effect on nuclear policy, for example, measured as the share of nuclear power in overall electricity (see Figure 13.1) or energy production.

But on a closer inspection, things are more complicated. We could easily add other meaningful measures, including the extensiveness of the nuclear chain (uranium mining, enrichment, fabrication of fuel, breeder reactors, reprocessing plants, waste disposal). Moreover, we could refer to existing commitments to future nuclear power development. In addition, one could go beyond substantial nuclear policy outcomes and include both indirect effects on other energy sources (e.g. solar energy) and procedural outcomes such as extension of formal citizen participation rights, access to information, inclusion of critical expertise, tightening of security standards, etc. Finally, one could take into account the effects of resistance to nuclear power that influence broader attitudes toward modern technology, and industry, and the state, their role as citizens, their behaviour as energy consumers, etc.

While we may obtain good, and multi-dimensional measures of various dependent variables, the more difficult problem is that of causal attribution. As a rule, many factors come into play, and they are hard to disentangle and weigh separately. Fierce anti-nuclear opposition may have a decisive impact on, say, substantial policy outcomes in one case, but in others different factors may be more important, for example, doubts expressed by experts, skyrocketing costs, improvements in a country's energy situation due to newly discovered oil or gas resources, etc. Moreover, some of these causal factors may interfere. For example, citizens litigating against the construction of nuclear power plants may cause

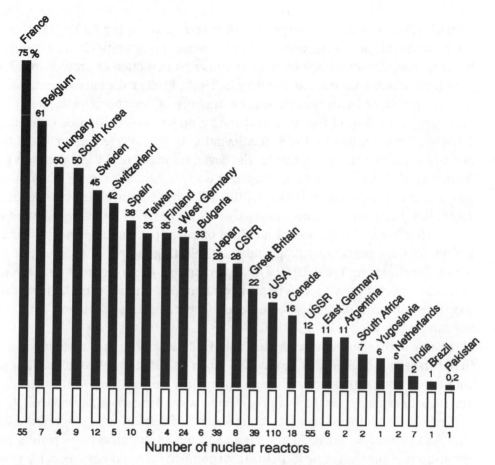

Figure 13.1. Nuclear power in comparative perspective; share of nuclear power of overall electricity supply and number of nuclear reactors in 25 countries. Source: IAEA; data from December 1989.

considerable delays which, in turn, raise the costs to levels where nuclear power is no longer attractive.

Obviously such combinations of effects and causes cannot be dealt with in a short essay. Therefore, I will limit myself to a few aspects only. Let me start with substantial nuclear policy outcomes in terms of the relation of what has been intended to what has been achieved.

The realization of nuclear programmes

Above, I have scored selected countries according to a simple index of anti-nuclear resistance combining the degree of mass mobilization and disruptiveness. We can hypothesize that the greater the resistance against nuclear power in a given country, the lower the degree of implementation of planned nuclear capacity.

In his *World survey of opposition to nuclear energy*, Wolfgang Rüdig (1990) has presented a table on the implementation of nuclear programmes in 39 countries. In comparing the total nuclear capacity under construction or ordered in 1974 with the realized total nuclear capacity in 1988, Rüdig calculated deviations as percentages. In order to understand the measure of deviation, I will give some examples. If nothing of the nuclear capacity under construction or ordered in 1974 was in operation in 1988, the deviation is -100% (e.g. Austria). If the planned capacity was fully achieved, the deviation was 0 (e.g. West Germany). If more than the originally intended capacity was realized, this results in a positive percentage (e.g. Japan, which had 14.989 MW under construction or ordered in 1974, but then had an actual capacity of 28.253 MW in 1988, an addition of 89%). This is a crude but useful measure of intentions and implementations. I will present an extract of Rüdig's table by selecting figures of those countries which I included in Table 13.3. The idea is to test the hypothesis that strong resistance correlates with low implementation. We should expect, then, that countries with high resistance scores tend to have a negative deviation which, in the extreme case, goes up to -100%.

Looking at the figures presented in Table 13.4, we find an inconsistent pattern. In the first group of countries, France and Japan do not fit the hypothesis. Though confronted with significant resistance to nuclear power, these countries were even able to expand their nuclear programme beyond the scope planned in 1974. In West Germany, the pro-nuclear forces succeeded despite having to confront the strongest resistance. Also, in the second group of countries, the results are inconsistent. Only the countries in the third group fit the hypothesis as, confronted with no significant opposition, they managed to go beyond (Canada) or even far beyond (Belgium) their original plans.

What conclusions can be drawn from this preliminary examination? First, and above all, strong resistance does not necessarily hinder nuclear power development. Obviously, intervening variables coming into play either allowed nuclear programme expansion despite significant resistance (France, Japan) or helped reduce or even cause the abandonment of nuclear power in the absence of powerful anti-nuclear movements (e.g. Italy). In the final section I will discuss these factors in a more general way.

Another conclusion could be that the measurement of deviation was too premature. Imagine that it takes a long time from the point when a nuclear plant is ordered until it is set into operation. Imagine, too, that resistance to nuclear power may not necessarily prevent the completion of reactors that are already under construction, but may affect decisions about the future of nuclear power in the next decades. In this regard, the consequences of opposition to nuclear power up to the mid-1980s could have only partly materialized by 1988. Nevertheless,

Table 13.4. *Resistance to and implementation of nuclear programmes in a comparative perspective*

Country	Resistance score	Deviation from projected capacity 1974–88 (%)
West Germany	8	0
France	6	+150
Switzerland	5	−48
USA	4	−57
Austria	4	−100
The Netherlands	4	−86
Denmark	4	−100
Japan	4	+89
Spain	2	0
Italy	1	−100
Sweden	1	+32
Norway	1	−100
Finland	1	+50
Great Britain	1	0
Belgium	0	+324
Canada	0	+66

Source for deviation: Rüdig (1990, p. 350).

resistance may have affected relevant decisions made before or after 1988 that concern the long-term future of nuclear power. As a matter of fact, the West German anti-nuclear power movement did have effects on the nuclear power programme in so far as, contrary to plans made in 1977 and 1981, nuclear power will not be expanded in the foreseeable future. This is not reflected in Rüdig's figures of 1988. In this light the hypothesis presented above works better than the first test showed. If we produced another table correlating resistance to nuclear power with the likely future of nuclear power in, say, the year 2010, most countries would fit the hypothesis nicely. Notwithstanding such conceptual improvement, France and Japan would remain as exceptional cases.

Side effects of resistance to nuclear power
As indicated above, resistance to nuclear power may have side effects beyond substantial policy outcomes. As far as procedural effects are concerned, the impact of nuclear power opposition seems to be modest in most countries. Only

slight extensions of participation rights have been achieved in some countries, and nothing has changed in the others. Generally, the most significant result has been the improvement of security standards (e.g. emergency cooling systems, reactor containment, control systems, etc.). This, in turn, contributed to rising costs and thus helped make nuclear power unattractive, in some countries at least. The tightening of security standards due to harsh criticism in country A could well have an impact in country B where no significant opposition was existent, but where it was sufficient to point at the standard set in country A in order to achieve an improvement. This, by the way, also indicates the need to look at diffusion effects between countries instead of evaluating separately the effects of resistance country by country.

Probably the most far-reaching side effect of resistance to nuclear power is subtle and submerged. Again, differences in countries cannot be neglected. As a rule, however, I would speculate that the long struggle over nuclear power, the intense public discourse, and the accidents that have occurred so far have affected the broader perception of modern technology. The first lesson learned by many, including some politicians, is that complex technologies may have disastrous consequences – consequences systematically underestimated by those who, by their very role as scientists, technicians or managers, have an inherent interest in putting forward technologies. The second lesson is that, in spite of good arguments, it is extremely difficult to stop a techno-industrial complex once it has gained momentum. The third lesson is that, in regard to certain technologies, learning by doing – which means learning by errors – is a strategy which humankind cannot afford. Consequently, then, it may be wiser to exert control, to assess potential risks in advance, to listen to dissenters, and probably to renounce making steps before trying to be at technology's forefront. For example, stopping the fast breeder reactors and commercial nuclear reprocessing in the USA in 1977 now clearly appears by no means to be a loss either in technological or in economic terms.

Explaining different policy outcomes

Having indicated the problems of causal analysis in a complex field such as nuclear power policy, it seems to be an act of some temerity to assess the impact of nuclear resistance beyond a narrow case study. I will not be able to present such a comparative analysis here, but I can briefly report on the results of a collective study coordinated by Helena Flam (1994) on the anti-nuclear conflict in eight Western European countries. In one of the evaluative final chapters, together with my Norwegian colleague, I have tried to describe and explain nuclear policy outcomes (Midttun & Rucht 1994). Let me sketch our approach and our main findings.

In describing nuclear policy outcomes we found it necessary not only to refer to one final outcome, but to distinguish between at least two time-specific outcomes. Outcome 1 was the situation in 1974–5, shortly after the so-called oil crisis. At that time the nuclear leaders were Great Britain and Sweden, followed by France, West Germany, Italy, the Netherlands, and eventually Austria and Norway. Outcome 2, which referred to the late 1980s, was different. Now, France had become the nuclear champion, Sweden was next, followed by West Germany, with Great Britain in fourth position.

According to our analysis, these time-specific outcomes were produced by very different circumstances. In the first phase, political economy factors (such as the resource base, techno-industrial structure, state engagement in electricity companies and nuclear industry, and the military complex) could explain a great deal of the variation between the countries. This explanation, however, scarcely worked in the second phase when nuclear opposition had emerged and when nuclear matters were disputed in the political arena. Then, we found a greater impact of such political structures as the access of challengers to the decision-making system, the coherence of the decision-making system and the coherence of the implementation system. Political structures were extremely favourable for nuclear power in France and, to a lesser extent, in Great Britain. But there also remained puzzling cases, in particular Sweden. Therefore, we turned to a third set of explanatory factors which we subsumed under the label political process factors. Here, we included attitudinal and behavioural factors such as consensus among elites, conflicts in crucial arenas (such as referenda and parliamentary activities) and contingent events (such as the Wyhl occupation in 1975, the disastrous demonstration in Creys-Malville in 1977, and Chernobyl in 1986). Only by drawing on these three sets of variables could we find satisfactory explanations for all cases. We learned that we needed highly sophisticated explanatory models in order to grasp adequately nuclear policy outcomes across time and across countries.

Conclusion

Assessing the impact of anti-nuclear power movements in comparative perspective is a difficult business. First, we have to take into consideration various dimensions of impact. Second, there are difficulties of measurement in most dimensions. Third, there is the serious problem of disentangling and weighing various factors that potentially come into play. Finally, factors that are relevant in a certain period or a certain country may be unimportant in another case and vice versa. Only a complex approach taking into account time-specific effects, covering a broad variety of factors of the political economy, political structure and political process can provide satisfactory explanations. Such approaches, as well as the simple correlations of degrees of resistance and degrees of nuclear power implementation presented here, demonstrate that to conceive anti-nuclear power movements' strengths as the crucial factor for the actual weight of nuclear power may be misleading. In one case, strong resistance may be relevant but in others it can be irrelevant. Overall, however, it is important to stress the fact that the efforts of the anti-nuclear power movement were not in vain. Not only has the movement won spectacular victories in some countries, but even where it failed to stop nuclear power it has had such important side effects as improving security standards and making people more sceptical of modern technology dangers.

References

CAMPBELL, J. L. (1988). *Collapse of an industry: nuclear power and the contradiction of U.S. policy*. Ithaca, NY: Cornell University Press.

EPSTEIN, B. (1991). *Political protest and cultural revolution: nonviolent direct action in the 1970s and 1980s*. Berkeley: University of California Press.

FALK, J. (1982). *Global fission: the battle over nuclear power*. London: Oxford University Press.

FLAM, H. (ed.) (1994). *States and anti-nuclear movements*. Edinburgh: Edinburgh University Press.

FORD, D. (1982). *The cult of the atom: the secret papers of the atomic energy commission*. New York: Simon & Schuster.

GAMSON, W. A., & MODIGLIANI, A. (1989). Media discourse and public opinion on nuclear power: a constructionist approach. *American Journal of Sociology* **95**, 1–37.

JASPER, J. M. (1990). *Nuclear politics: energy and the state in the United States, Sweden and France*. Princeton, NJ: Princeton University Press.

JOPPKE, C. (1993). *Mobilizing against nuclear energy: a comparison of Germany and the United States*. Berkeley: University of California Press.

KITSCHELT, H. (1986). Political opportunity structures and political protest: anti-nuclear movements in four democracies. *British Journal of Political Science* **16**, 57–85.

KOOPMANS, R. (1992). *Democracy from below: new social movements and the political system in West Germany*. PhD dissertation, University of Amsterdam.

MEZ, L. (ed.) (1979). *Der Atomkonflikt. Atomindustrie, Atompolitik und Anti-Atombewegung im internationalen Vergleich*. Berlin: Olle & Wolter.

MIDTTUN, A. & RUCHT, D. (1994). Comparing policy outcomes of conflicts over nuclear power: description and explanation. In *States and anti-nuclear oppositional movements*, ed. H. Flam, pp. 371–404. Edinburgh: Edinburgh University Press.

NELKIN, D. & POLLAK, M. (1981). *The atom besieged: antinuclear movements in France and Germany*. Cambridge, Mass: MIT Press.

NOVICK, S. (1969). *The careless atom*. Boston: Houghton Mifflin.

RUCHT, D. (1990).   Campaigns, skirmishes and battles: antinuclear movements in the USA, France and West Germany. *Industrial Crisis* **4**, 193–222.

RUCHT, D. (1994).   *Modernisierung und neue soziale Bewegungen. Bundesrepublik Deutschland, Frankreich und USA im Vergleich.* Frankfurt: Campus.

RÜDIG, W. (1990).   *Anti-nuclear movements: a world survey of opposition to nuclear energy.* Harlow: Longmans.

In the engine of industry: regulators of biotechnology, 1970–86

ROBERT BUD

Regulators' attempts to balance conflicting interests and to convert the public's and politicians' feelings about technology into coherent action have fascinated students of industrial policy (Rothwell & Zegveld 1981; Grant 1989). By contrast historians, though in principle sharing such interests, have tended to treat regulators either as external brakes or as lubricants in processes controlled more directly by market forces and technological logic. Thus historians' interest in the chemical industry which gave rise to studies of companies and processes has not engendered a corresponding corpus of works on such regulatory bodies as the FDA or Britain's Alkali Inspectorate (see the still unmatched study of DuPont: Hounshell & Smith 1988). Now, however, with a new interest in technological systems, we are acquiring a few historical case-studies from modern industrial society in which regulators have themselves been considered central parts of the process, acting, so to speak, as components of the engine itself.[1]

In particular, historians are now investigating policies which, since the 1930s, and latterly during the Cold War, encouraged the development of such state sponsored technologies as synthetic rubber, computing and semiconductors used by the space and military industries. The support of such academic categories as scientific instruments and cancer research is also being elucidated.[2] Rarely, nonetheless, is regulation yet seen as a whole: promoters of new technology have still been conceptually divorced from those responsible for its constraint and from those responsible for the protection of the public and of workers.

This separation is clearly inappropriate when studying such a recent decade as the 1970s, when regulators challenged by international competition and progressively shocked by a new environmental awareness and a tenfold rise in energy prices strove to make strategic technological choices (Williams 1973; Roobeek 1990). Technology and industry policy were integrated across the industrial world. Governments sought to make industry aware of long-term social needs through financial incentives, through such institutional innovations as fostering pre-competitive collaboration on R & D, and by moulding the climate of

opinion. It was in this period of urgent and energetic activity that modern
biotechnology emerged as a beneficiary of the regulators' newly-won zeal. There
have been several comparative studies showing the diverse ways countries
supported and controlled the subject (Jasanoff 1985). This chapter is intended to
suggest that those approaches were much more than merely different policy
approaches to the same given subject. They also reflected, implemented and
reinforced rather different visions of the potential nature of biotechnology itself.

Biotechnology

Although within a decade it would be seen as peculiarly threatening, during the
1970s biotechnology was promoted and demarcated by regulators responding to
concerns over the fate of other industries.[3] It seemed to provide an industrial and
growth-friendly way both of limiting pollution and oil use and of meeting
international competition in traditional industries. Following quickly from the
interaction of diverse concerns expressed in countries around the world, the
process by which biotechnology was being defined was as modern as its scientific
basis (Jasanoff 1985). Key regulatory regimes were put in place, and substantial
amounts of both money and work were put into promoting the concept of
biotechnology. During the next decade, as energy costs and environmental
concerns receded, scepticism about the role of regulation would grow. Certainly,
such key discoveries and innovations as insulin and monoclonal antibodies for
diagnostic purposes could not be put down to the activities of regulators. It was
their efforts, however, that gave apparent meaning to a wealth of activities that
had no philosophically necessary connection or implications. It was they who
created the concept of the biotechnological revolution and indeed also supported
the products that would express it. Their support world-wide led to the favouring
of 'gasohol' after the oil crisis. The development of single-cell protein foods was
supported; the development of what would become the Quorn™ mycoprotein,
now widely available in Europe, was subsidized by the British government.
Equally, regulators have limited the range of products that may be made or sold:
the administration of the milk stimulating hormone bST (produced by an
engineered organism) is still widely forbidden. In Japan, the most statist of
industrialized countries, the influence of MITI served to highlight the potential of
biotechnology to local industry. Meanwhile, MITI's perhaps legendary power also
led, world-wide, to a concern about this new core industry. The United States
established a regulatory regime for biotechnology generally more liberal than the
European, justified by a vision of its potentially wealth-generating and science-
validating role as well as by its benign qualities. In the 1970s, industrial-scale

production of genetically modified organisms had been permitted even though Congress had threatened to ban it. Within a few years those arguments intended to counter the regulatory threat would win enormous resources from local financiers and on Wall Street.

However distinct the concepts of biotechnology thereby promoted, however differing the underlying concerns, and however many other players, those regulators were explicitly constructing biotechnology as an answer to public resistance to hitherto developed industries. They had to engage with the public's worries about radical transformation of their environment as well as with strategists' long-term concerns about industrial health. In the period covered by this study, 1970–86, the regulators' concerns were predominantly environmental and economic, as they dealt with biotechnology in terms already framed for the chemical industry. Distinctive cultural issues associated with the autonomy of the natural, cyborgs, life, death, eugenics and the essence of the human being seem to have been marginalized. At an OECD meeting, representatives of both France and Canada expressed hopes that biotechnology could be 'banalized' (OECD 1988). Nonetheless these and other such wishes were frustrated. An undercurrent of distrust survived every legally binding reassurance.

By 1986, the imminence of widespread commercial recombinant DNA products had engendered an enormous, almost intolerable, tension between the basically optimistic vision of the new industry which would yield personal and corporate fortunes, and of the Fenris Wolf that needed to be bound for fear it would consume the world. Congresswoman Schneider complained: 'So I am feeling terribly impotent in terms of solving this problem. And I think it needs to be solved – we cannot continue to have the trade association which says everything is fine, and you [Jeremy Rifkin] say everything is horrendous.'[4] Such tension would highlight the need for specific regulations that would colour and indeed justify the industry.

The regulators

Biotechnology policy has distinctively incorporated both incentives and controls. Whilst discussions of regulation of release of organisms and of the practice of recombinant DNA have often been separated from those evaluating national support measures, many of the same officials and elected representatives were involved in both activities.[5] In the USA this linkage was formalized during the 1980s by the Biotechnology Science Coordination Committee and across the Atlantic by the EEC's Concertation Unit for Biotechnology in Europe, known as 'CUBE'. Though such organizations may have been wrecked by ultimately

destructive tensions and their power was limited, in general regulation in the
cause of safety has been modulated by concerns over regulation in the cause of
industrial change. A judicial framework granted 'biotechnology' an unam-
biguous existence. Laws and consequent committees of Congress have anointed
themselves with the title, such as the Biotechnology Science Coordination Act.
Regulatory agencies such as BSCC and CUBE defined their remit by the
name. They have been responsible both for the economic health and for the
environmental and safety aspects of the industry. Internationally there was
competition and emulation in biotechnology policy. Brock points out how the
Japanese tried consciously to emulate the British, the Americans and the Germans
around 1980 (Brock 1989). Later other countries were spurred on by the model
of Japanese initiatives. Much work went into formally defining the subject even
though every definition appeared curiously thin and remote from the prevailing
sense of the subject. A 1981 OECD study appended almost twenty different
definitions (Bull, Holt & Lilly 1982). Most used some variant of the idea of the
manufacture of useful things using natural organisms, illustrated by the European
Federation of Biotechnology definition of the subject as 'the integrated use of
biochemistry, microbiology and chemical engineering in order to achieve the
technological (industrial) application of the capacities of microbes and cultured
tissues and parts thereof' (EFB and DECHEMA 1982, p. 5). In the United States the
idea was refined in the OTA report on 'Commercial Biotechnology' identifying the
special category of 'new' biotechnology which incorporated just recombinant
DNA, monoclonal antibody and new fermentation technologies.

It is perhaps surprising that it has proved even harder to define or even
recognize biotechnology from an industrial point of view. Throughout the
century, exciting biological discoveries have been made with potentially
important industrial consequences. Around 1970 new nitrogen-fixing plants,
single-cell protein food to meet the market of an expanding world population, and
novel drugs seemed around the corner. Yet they did not themselves define a
separate industrial category. In general they were being developed as elements of
new product strategies within the chemical industry. Nor, even when the
invention of recombinant DNA techniques a few years later sustained a wave of
new enthusiasm, would genetic engineering alone do as a surrogate technical
definition (Krimsky 1991; Wright 1993). From the technology's perspective,
academe can be treated as a source of ideas to be raided just as the stock market
is a source of finance. Even if the word 'biotechnology' begs us to think of a
distinctive scientific base, even if the qualifier 'new' is attached, and even if it is
agreed that the core techniques are associated with the manipulation of DNA, the
technology would not be reducible merely to the science.

Production, marketing and finance raising so essential to the technology would be lost by an exclusive focus upon science. The distinguished organic chemist Carl Djerassi in testimony to Congress has repeatedly compared recombinant DNA techniques to welding, and no more than welding defines the car industry can a molecular biology technique in itself define an industry.[6] The technique has been used in mining, pharmaceuticals, agrochemicals and single-cell protein by an enormous diversity of companies – large firms clearly in the chemical industry, prospective pharmaceutical companies and research boutiques. With such close integration within a variety of diverse industries, technical qualities alone would not suffice to define a unique discrete biotechnology.

Europe and Japan

To this technology at least, therefore, the regulators have not been external, but instead have been intrinsic to its very formulation and conceptualization. This recognition should also inform our vision of the subject's history, rather than allowing it to be dominated only by such intellectual breakthroughs in the discipline of molecular biology as the discovery of DNA's structure 40 years ago or of the technique of recombining DNA 20 years ago. It is worthwhile also taking as a benchmark the quite separate history of industrial policy around 1970.

Although biotechnology had been named and discussed among intellectuals on several occasions earlier in the twentieth century, the late 1960s were key to its transformation into a modern super-technology (Bud 1993). This was not just because of the scientific and technological ferment which would enable its prosperity to be sustained, but because a variety of interest groups within national administrations particularly identified a need for it. Across the world, policy makers were trying to discern the new technologies that would prove acceptable in a culture rendered suspicious by the Vietnam War, and environmentally sensitized in the 'Age of Aquarius'. Even before the advent of recombinant DNA, there was therefore widespread public concern over the role of science and industry in general. This was reflected in calls for a clear cultural and indeed economic call for a new technological order. Reich's book *The Greening of America* (1970) was a best seller. Biotechnology was constructed to provide one answer. Essentially low energy and hopefully less polluting using mild conditions and waste biological materials its potential contrasted with that of the chemical and nuclear industries. During the 1970s a series of ecological catastrophes seemed to point to the failing of traditional technologies. In the USA the Love Canal was found to have been poisonously polluted. The nuclear plant at Three Mile Island suffered a partial meltdown. In Europe the chemical plant at Seveso exploded. What, more precisely, biotechnology would entail, and how it would be demarcated, depended upon the precise nature of the new order that regulators across the world needed to implement.

This argument from 'need-pull' becomes particularly evident if one looks first at those pacemakers of industry at the time: Japan and Germany. The modern use of the word Biotechnology can be seen to be derived from the title of an enthusiastic 1974 report by the German chemical manufacturers' organization DECHEMA produced for the research ministry BMFT.[7] In Japan, Life Sciences were promoted by the Science and Technology Agency at about the same time. While to the participants in more slowly growing economies these two countries were outstandingly successful, to many of their citizens, particularly the young and well educated, the costs of post-war prosperity seemed too great.

During the early 1960s an overwhelming 87% of prize-winning Japanese technologies had been devoted to the pursuit of efficiency. The pursuit of environmental integrity had attracted only 3%. By the early 1970s environmental issues had risen to 13% of awards, conservation won 7% up from 4%, and efficiency was down to 69%. In response the 1977 Science and Technology White Paper expressed the implications as bringing 'people to expect a science and technology to give assurances of "safety, conservation and environmental integrity", instead of "speed, low cost, quantity and convenience"' (Japan, Science and Technology Agency 1977, pp. 176–8). The emerging consensus suggested that there would be need for the development of new non-polluting technologies and a knowledge-intensive work-force. In 1980 biotechnology was identified, along with information technology and material science, as a core technology of the future by the influential industry agency MITI. Indicatively, as Brock has pointed out, the Bioindustry office was located in the agency's Basic Industries Division concerned with the structural renewal of the chemical industry (Brock 1989).

In Germany biotechnology also emerged during the late 1960s in response to concerns about the environment, at the same time as the Japanese came to be concerned about their society. Underlying the concerns of a variety of new Green pressure groups were eight emphases: Decentralisation–Participation–Reduction of Power–Sparing use of natural resources–Ecological behaviour–Healthy technology–Freedom from coercion–Multiplicity. Hence the emerging Green tendency favoured technology which would depend on renewable resources, was associated with low energy processes that would produce biodegradable products and waste, and was concerned with the health and nutrition of the world (Bossel 1978).

Thus in both Japan and Germany policy makers were driven by a sense of the inadequacy of the last generation of technologies. In other words biotechnology was called into existence as the motor of a third industrial age before its content was understood. What was known was that the previous generation would be an inadequate basis for future growth. It was apparently maturing and the succession was far from clear. A study of the chemical industry showed a decreasingly rapid pattern of innovation with many of the greatest products such as polyethylene now a generation old (Duncan 1982). Hydrocarbon raw materials were imported and rapidly rising in price. Even if electronics or space travel offered routes to the bright industrial future, these seemed to be dominated by the USA. Impressed as they were, the Germans analysed their own requirements from first principles in a 1970 report on *Erster Ergebnisbericht des ad hoc Ausschusses 'Neue Technologien'*.[8] This identified three types of need: fundamental requirements such as food or raw materials, infrastructural issues such as transport, and environmental concerns. In 1972 a programme of Biology and Technology was established in the new BMFT. Six areas were identified as priorities: security of food and feed supplies; reducing environmental pollution; pharmaceutical production; new routes to raw materials; chemicals and metal production; development of biotechnological processes; and basic research. The close association with the New Technologies Program was clear. At a time when increasing priority was being accorded to environmental issues, biotechnology could be shown to address them all. The Germans having already specified the human centred targets including foods and medicines which their research of all kinds would address, could highlight the central role of Biotechnologie.

The idea that chemistry and physics had yielded up previous industrial revolutions and now was the moment for biology was already well entrenched. It had been around indeed as a resource since the beginning of the twentieth century. The rapid if not yet revolutionary development of biology in general and microbiology in particular were factors in attracting science policy advisors to biology. In both Germany and Japan this anticipation was converted into government policy.

The biological approach was attractive to the chemical industry. It would at least show a sensitivity to the pressures for environmental concern when even this post-war money-spinner could not act as if political support would last for ever. The increasingly successful opposition to the Vietnam War highlighted the contrasting needs of society, the individual and the environment on the one hand and of the military–industrial establishment on the other. Large companies such as Mitsubishi and Bayer promoted biological solutions with considerable public profile. Both companies established special institutes that came to be well known. Similarly, the views of society on the German chemical industry were reflected in the enthusiastic *Report on Biotechnology* presented to BMFT by DECHEMA. Here was a minimum accommodation to the idealists' complaints couched in terms which even the most visionary could not contest.[9]

Germany was the prime but not the only centre of enthusiasm for biotechnology in Europe. In Britain during the late 1960s and early 1970s, the government through the NRDC invested in such biotechnologies as enzyme manufacture, supporting the forming of Whatman biochemicals, and the early development of the single-cell protein product that came to be known as 'Quorn™'. Of course the British had been much influenced by the Germans. They had joined the German-inspired European Federation of Biotechnology and British definitions that emerged in the late 1970s were derivative from the German.

The DECHEMA initiative also informed a European Community initiative to foretell the industrial future. Three major changes were foreseen, as technical changes had a social impact: most immediately information technology, in the medium term biotechnology and in the long term a changing attitude to work itself. The concept of the Biosociety neatly complemented the idea of the 'Information Society' which was expected to have a greater short-term impact. The molecular biology explosion was hardly yet upon us, and it was not that which had prompted the concept. Instead it was the constellation of new environmental and human issues which seemed so important and so directly addressed by the category that made the biosociety a long-term transformer of society. The Biosociety proposal cites a 1978 definition of biotechnology: 'it incorporates fermentation and enzyme technology, water and waste treatment, and some aspects of food technology'. The editor did add 'the *modification* of such micro-organisms can be added to this list'.[10] The EEC was awash with agricultural produce, oil was apparently increasingly expensive and the environmentalists were an increasingly important constraint on the activities of the chemical industry. Working towards a Biosociety, argues the document, 'is the most promising basis for a sustainable, secure and prosperous future'. Its progenitor, Mark Cantley, the new head of the FAST Biosociety group, characteristically turned to DECHEMA as the key institution. DECHEMA and FAST in turn inspired Salomon Wald at OECD, who obtained a report on the subject from authors who were neither molecular biologists nor Americans, but instead were British microbiologists (Bull *et al.* 1982).

Thus, if the new biology was seen as a major resource by the European planners, it was but one factor in enabling a technology which had been invented out of need. The diagram entitled 'Outline of the "Bio-society"' as shown in Figure 14.1 is a dramatic demonstration of the sort of analysis which was being conducted across Europe. The promotion of biotechnology could therefore be seen as a resolution to the extreme tensions between opponents of traditional industry including environmental campaigners, guardians of the status quo, particularly the chemical industry seeking new defences against rising public disquiet and

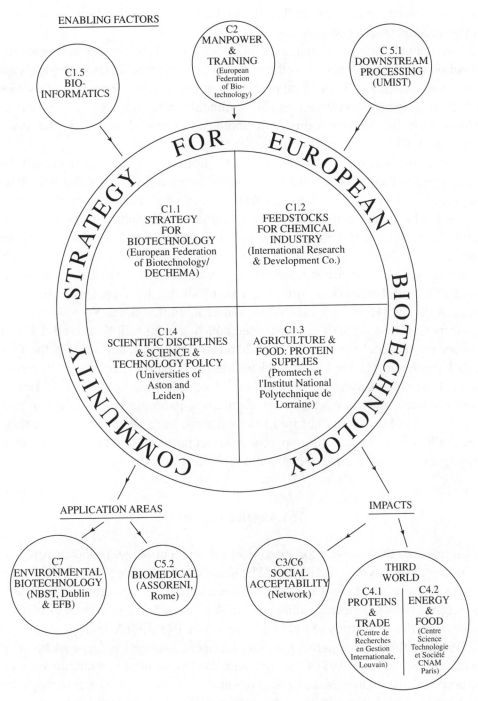

ENABLING FACTORS

C1.5
BIO-
INFORMATICS

C2
MANPOWER
&
TRAINING
(European
Federation
of Bio-
technology)

C 5.1
DOWNSTREAM
PROCESSING
(UMIST)

STRATEGY FOR EUROPEAN COMMUNITY BIOTECHNOLOGY

C1.1
STRATEGY
FOR
BIOTECHNOLOGY
(European Federation
of Biotechnology/
DECHEMA)

C1.2
FEEDSTOCKS
FOR CHEMICAL
INDUSTRY
(International Research
& Development Co.)

C1.4
SCIENTIFIC DISCIPLINES
& SCIENCE &
TECHNOLOGY POLICY
(Universities of
Aston and
Leiden)

C1.3
AGRICULTURE &
FOOD: PROTEIN
SUPPLIES
(Promtech et
l'Institut National
Polytechnique de
Lorraine)

APPLICATION AREAS

IMPACTS

C7
ENVIRONMENTAL
BIOTECHNOLOGY
(NBST, Dublin
& EFB)

C5.2
BIOMEDICAL
(ASSORENI,
Rome)

C3/C6
SOCIAL
ACCEPTABILITY
(Network)

THIRD
WORLD

C4.1
PROTEINS
&
TRADE
(Centre de
Recherches
en Gestion
Internationale,
Louvain)

C4.2
ENERGY
&
FOOD
(Centre
Science
Technologie
et Société
CNAM
Paris)

Figure 14.1. Biotechnology as interpreted by the European Commission in 1980.
Source: Biosociety report of the FAST project, 1980, EUR 7105.

resource scarcity, policy makers looking for new industrial standard bearers, and a public hungry for good news.

Ironically the main technological hopes of the 1970s – single-cell proteins, gasohol, self-nitrating plants – all came to pose considerable technological and economic problems. They disappointed regulators and chemical companies which had come to see them as the imminently available new generation of products for the chemical industry. In some countries such as Britain the result was a somewhat confused interpretation of the subject.

There, the Science Research Council continued to give a high priority to supporting fermentation engineering applied to microorganisms and industrial applications, while the Medical Research Council emphasized the application of molecular biological techniques. Like the Germans, the Science Research Council tended to manage it as a branch of engineering, and the head of its biotechnology directorate had come from the Engineering Board, whereas the Medical Research Council, like the Americans, treated it as the application of the new biology. In Germany and Denmark a more consistent distinction was preserved and Biotechnik and Gentechnik kept rather different connotations.

According to a 1990 Eurobarometer poll taken for CUBE, overall 15% of Europeans felt genetic engineering would make life worse, while only 7% thought that biotechnology would have a deleterious effect. Equally, genetic engineering yielded an optimism score of 0.45 and biotechnology 0.66 (INRA 1991). Governments, and that has meant regulators, particularly in Germany and Denmark and latterly the European Commission, have been restrictive in their regulations for experimenting on genetically engineered organisms, while they have been very supportive of conventional biotechnology.

The American model

If Europe was the seed-bed for the concept of biotechnology, it was in America that the subject, as it normally would be understood, was born. Of course, by the 1980s its meaning there was rather different from in Europe and no distinction between genetic engineering and biotechnology is made, even in such important surveys as the OTA study of public perception in 1988 (OTA 1987).

Certainly there were parallels to the European movement for a sustainable economy. By the early 1970s the environmental movement had already become so powerful that it gave rise to strong laws and agencies. Gasohol was perhaps the best symbol of a new biological product responding to changed economic conditions. The cultural and economic pressure on the chemical companies, however, was not yet as great as in Europe (Spitz 1988).

At the same time, perhaps more than in Europe, concerns about the consequences of the new biology had been widespread since the 1960s. The popularity of such movies as *The Andromeda Strain* highlighted the fear of alien organisms. Following the achievement of practical recombinant DNA techniques, biologists voluntarily imposed a moratorium and discussed it at the 1975 Asilomar conference. It was felt that without visible regulation, the public would wish to repress all such work. In 1976, the National Institutes of Health accordingly introduced regulations on the kind of work permitted in various laboratories. It defined what would be considered environmentally safe practices, preventing the escape of ecologically threatening organisms. This system reassured scientists who began work once more, but it did not cover either large-scale or industrial work. In Congress, Senator Edward Kennedy led a campaign for close supervision of industrial biotechnology during the 1970s. During 1976–7 sixteen bills to bring industrial-scale production were debated in Congress.[11] In the face of increasing enthusiasm for the economic potential of biotechnology, however, the pressure for legislation vanished. Instead, as Sheila Jasanoff shows (Chapter 15), it was to be through the courts and existing environmental controls that biotechnology would be challenged.

For almost twenty years Jeremy Rifkin has led the strongest of the challenges to the biotechnology industry, accusing it of unnaturalness and immorality. As in the case of the environmental campaigners, regulations were called for to protect America's natural heritage. Through the mid-1980s Rifkin fought the release of the bacterium known as ice-minus intended to increase the frost resistance of strawberry plants (Davidson 1988). He argued that an environmental impact statement should have been filed. More fundamentally the naturalness of nature seemed to be at risk. The former was answerable by regulators: after all an environmental impact statement was their own construct. Worries about nature have been left to evaporate, or to fester. Krimsky has explored the divergent rationalities in conflict over the release of the ice-minus bacterium (Krimsky & Plough 1988). He contrasts the instrumental rationality of regulators and scientists with the cultural rationality of protestors. The environmentalism of the government agencies, although originally proclaimed as a challenge to unfettered science, was in fact much closer to the practicality of the promoters than to the cultural unease of the protestors.

The scientists had not been against regulation either. The Asilomar meeting of 1975 had highlighted the belief that scientists themselves, with more or less enthusiasm, could reach a sensible consensus.[12] Within a year, however, most dramatically at the 1976 Cambridge hearings, this vision of a rational, single-minded community had broken down. Many of the scientific community became enthusiastic supporters of the fast exploitation of scientific results. Increasingly closely associated with the business world, many scientists would become part of an enthusiasts' lobby. Some, however, still felt that regulation would be required to provide public reassurance. The Office of Technology Assessment in the late 1970s has been described as 'Mount Olympus' by the geneticist Zsolt Harsanyi, who was leading a project to evaluate the impact of applied genetics.[13] When the analyst Nelson Schneider sought to bring the advantages of genetic research to the investing public in 1979 it was to OTA that he turned for expert testimony. His September workshop would be the seminal event in the enthusing of the investing public. Attracting 500 participants, it brought biotechnology's exciting future to Wall Street (Teitelman 1989).

The arguments put to Wall Street had not been originally developed for that purpose, but rather to counter the power of the sceptics in Congress. Irving Johnston of Eli Lilly, testifying to Congress a year earlier, before most of the subsequently famous small companies were born, had put all the arguments that would become typical later. Biotechnology had enormous potential. It was based on American science and would bring wealth to America. Any environmental hazards could be reduced to minimal significance by strict regulation.[14]

The arguments expressed during the 1978 hearings had roots that could be found from the early 1970s. The dreams being articulated by European contemporaries were picked up and recombinant DNA attached to them. As early as September 1974, at roughly the same time that the German DECHEMA report was appearing and shortly after his letter to Science, Paul Berg spoke in a televised debate at London's Royal Institution . There he appealed for a better appreciation of the potential of his subject: 'Probably because of exaggerated and misleading claims by scientists and the press, the words genetic engineering evoke terror as well as excitement at its prospects' (BBC 1974, p. 2). He then explained the new process of recombinant DNA, citing its scientific fruits and also its 'great practical significance'.

Of course Berg was responding to a scepticism very different in origin from the opposition to the German chemical industry's opposition. His concerns were with those who felt that the new biology threatened our concepts of life itself. In other words the rhetoric of advocacy was framed first in a regulatory context and was then transferred to commerce. When Harsanyi's report Impacts of Applied Genetics finally was published by the OTA in 1981, it duly opposed speculative benefits to speculative risks and formally confirmed the definition of biotechnology in terms of its unlimited and widespread prospects (OTA 1981).

Berg's syllogism, that the wonders of biotechnology would become possible not through the new disciplinary synergy foreseen in Germany but through the application of the basic science of molecular biology, made particular political sense within American industrial policy. Federal governments in America have long had to cope with the tradition that they would not intervene directly in the economy in the way that more corporatist societies have favoured (Etzkowitz 1990).

From the late 1960s, however, there was a wish to exploit the burgeoning medical research enterprise more successfully. Science was losing the automatic support of the public. Senator Proxmire, the anti-Vietnam War campaigner moved to criticizing inappropriate science funding, and annually announced his Golden Fleece award to the apparently most absurd beneficiary of a government grant.

As in Europe, regulators sought to overcome popular resistance to existing industries, so in America it was the taxpayers' resistance to science that proved a challenge. In the face of cancer's stubborn reluctance to be overcome, and the pressure on government research budgets, there was continuing congressional pressure for other tangible benefits from science. Industrial renewal would be such a key reward. The National Science Board's periodic publication *Science Indicators*, linking declining patenting to declining proportions of world science, was itself an indicator of this campaign (Elkana *et al.* 1978). More vividly, when at the end of the decade Peter Farley, President of Cetus, said that his company would be the next IBM, he was exploiting the American yearning for such another company, and his compatriots' belief that applied science would be its source.[15]

A science-based industry would be an ideal target for an American government well used to supporting science, and seeking to justify its already enormous investment in NIH (National Institutes of Health) but inhibited from formally supporting industry. Here was a third generation industry that would be the result of the application of American science. In biotechnology, an emphasis on basic science could be explained by an historical model according to which biotechnology was moving along a 'trajectory of innovation' from basic research through generic applied research to applied research (OTA 1984). Of course, the argument from basic science highlighted the role of molecular biology as a resource, and the economic necessity of giving scientists free rein. At one level the approach was remarkably successful: American legislators have been able to provide a fairly liberal regime. Sheila Jasanoff (Chapter 15) explores the formation of that policy in more depth. Yet it has made for distrust and anxiety, for the cultural threats associated with biotechnology have been overridden and repressed, perhaps resurfacing in the debates over abortion.

Conclusion

Biotechnology is more than an application of genetic engineering. The beliefs about its future potential, and of its dangers, have been central to its existence. Concerned with the mechanisms of industrial renewal, regulators sought to find a new kind of technology that was both novel and acceptable, and through regulations, promotions and objective reports define their image as that which was biotechnology. Their efforts succeeded in permitting the institutional survival of biotechnology and in rendering as a general truth what ten years earlier might have been an idea for enthusiasts only: that biotechnology was a technology of the future.

On the other hand, rarely have cultural concerns been answered in their own terms. Concepts important to laypeople and apparently threatened by bio-technology such as Nature, the natural, motherhood and feelings of danger have not been on the regulators' agenda. So, while successful in generally overcoming the cultural qualms of the public, regulators have been less successful in meeting them. As a result, I would suggest, new issues expressing those worries, whether they be the questions of patenting transgenic animals or the use of bST, keep recurring. They have become particularly acute in the case of the human genome project, and there they are being addressed directly. Yet if such issues reflect the new challenges scientific entrepreneurship throws up, they are also, partly, old issues in new guises. Current debates over the ethics and cultural consequences of modifying the human genome have potential implications for the definition of all biotechnology.

One can end with an allegory. It may well be argued that the founding event in the history of biotechnology was the publication of the novel 'Frankenstein'. Mary Shelley explored the implications of Dr Frankenstein's creation of a creature which in the end kills its maker. In one sense, of course, this is a gothic horror expressing public revulsion from the fantasies of the newly confident chemists. Yet we may reflect that, fortunately, every character in this story is imaginary, with one exception: Mary Shelley herself. In a sense she was acting as the regulator concerned with the potential implications of potential innovations. Typically, she gave her speculations a material outcome, the report, still familiar after two centuries. Her example stimulates the perverse thought that hitherto biotechnology may really have been constituted by its regulations, rather than by the actual products and processes so far realized. The attempt to cope with anxiety over technology has therefore not been merely a retarding force, rather it has helped to steer, to power and even, at first, to constitute its development.

Notes

1   Thomas Parke Hughes led the way in showing how the development of electrical technologies were interlaced with regulatory issues in Hughes (1983). More recently other scholars have explored late 19th century electrical engineering, such as Schaffer (1992). Space and accelerator technologies are being elucidated by scholars such as John Krige who are showing the integration of technical and policy issues.
2   See as examples of the growing literature, Morris (1989); on cancer research see, for example Studer & Chubin (1980) and Panem (1984); for scientific instruments, see Stine (1992).
3   The arguments of the 1970s' promoters are explored in greater length in Chapter 7 of Bud (1993).
4   US Congress, House of Representatives, Committee on Science and Technology 1986, p. 102.
5   See for instance Orsenigo (1989) which hardly mentions the control factors, while Krimsky (1991) deals mostly with those control factors.
6   'Statement of Carl Djerassi', US House of Representatives, Committee on Science and Technology, Subcommittee on Investigation and Oversight and Subcommittee on Science, Research and Technology 1981, p. 149.
7   The classic analysis of the history of German policy is by the secretary to the DECHEMA committee, Buchholz (1979). For the development of Japanese policy, see Brock (1989).
8   BMBW, (1971). All students of the development of German biotechnology policy owe a great debt to Klaus Buchholz, himself a key participant, for his article (Buchholz 1979).
9   The German 'Green' Benny Haerlin suggested to me that this cooption of traditionally radical values by the chemical industry had given a special anger to later campaigners.
10 Commission of the European Communities, FAST Subprogramme C 1979, p. 3.
11 For a discussion of the atmosphere at this time, see Bud 1993, pp. 178–82.
12 For an evocative image of the meeting see Goodfield (1977).
13 Zsolt Harsanyi, personal communication.
14 US Congress, House of Representatives, Committee on Science and Technology, Subcommittee on Science Research and Technology (1977).
15 This is cited in Teitelman (1989).

References

BBC (1974).   Certain types of genetic research should be suspended. Broadcast BBC2, 16 September, transcript.

BMBW (1971).   *Erster Ergebnisbericht des ad hoc Ausschusses 'Neue Technologien' des beratenden Ausschusses für Forschungspolitik.* Schriftenreihe Forschungsplanung 6. Bonn: BMBW.

BOSSEL, H. (1978).   Die vergessenen Werte. In *Der Grüne Protest. Herausforderung durch die Umweltparteien*, ed. R. Brun, pp. 7–17. Frankfurt: Fischer.

BROCK, M. (1989).   *Biotechnology in Japan.* London: Routledge.

BUCHHOLZ, K. (1979).   Die gezielte Förderung und Entwicklung der Biotechnologie. In *geplante Forschung*, ed. W. van den Daele, W. Krohn and P. Weingart, pp. 64–116. Frankfurt: Suhrkamp.

BUD, R. (1993).   *The uses of life: a history of biotechnology.* Cambridge: Cambridge University Press.

BULL, A. T., HOLT, T. G. & LILLY, M. D. (1982).   *Biotechnology: international trends and perspectives.* Paris: OECD.

COMMISSION OF THE EUROPEAN COMMUNITIES, FAST SUBPROGRAMME C (1979). *Biosociety*. FAST/ACPM/79/14–3E. Brussels: Commission of the European Communities.

DAVIDSON, D. K. (1988). Out of the strawberry patch: biotechnology's search for legitimacy. Dissertation Abstracts 88–18894. Ann Arbor, Mich: University Microfilms International.

DUNCAN, W. B. (1982). Lesson from the past: challenges and opportunity. In *The chemical industry*, ed. D. H. Sharp and T. F. West, pp. 15–30. Chichester: Ellis Horwood.

ELKANA, Y. *et al.* (ed.) (1978). *Towards a metric of science: the advent of science indicators*. New York: John Wiley.

ETZKOWITZ, H. (1990). The capitalization of knowledge. *Theory and Society* 19, 107–21.

EFB, EUROPEAN FEDERATION OF BIOTECHNOLOGY AND DECHEMA (1982). *Biotechnology in Europe*. FAST Occasional Papers No 59. Brussels: EEC.

GOODFIELD, J. (1977). *Playing God. Genetic engineering and the manipulation of life*. London: Hutchinson.

GRANT, W. (1989). *Government and industry: a comparative analysis of the US, Canada and the UK*. Aldershot, UK: Elgar.

HOUNSHELL, D. & SMITH, J. K. (1988). *Science and corporate strategy: Du Pont R & D, 1902–1980*. Cambridge: Cambridge University Press.

HUGHES, T. P. (1983). *Networks of power: electrification in Western society, 1880–1930*. Baltimore: Johns Hopkins University Press.

INRA (EUROPE) SA/NV FOR 'CUBE' (1991). *Opinions of Europeans on Biotechnology in 1991*, Eurobarometer 35.1.

JAPAN, SCIENCE AND TECHNOLOGY AGENCY (1977). *Outline of the White Paper on science and technology*. Aimed at making Technological Innovations in Social Development, February. (translated by Foreign Press Centre).

JASANOFF, S. (1985). Technological innovation in a corporatist state: the case of biotechnology in the Federal Republic of Germany. *Research Policy* 14, 23–38.

KRIMSKY, S. (1991). *Biotechnics and society: the rise of industrial genetics*. New York: Praeger.

KRIMSKY, S. & PLOUGH, A. (1988). *Environmental hazards: communicating risks as a social process*. Dover, Mass: Auburn Homer.

MORRIS, P. (1989). *The American synthetic rubber research program*. Philadelphia: University of Pennsylvania Press.

ORGANIZATION FOR ECONOMIC COOPERATION AND DEVELOPMENT (1988). *Biotechnology and the changing role of government*. Paris: OECD.

OFFICE OF TECHNOLOGY ASSESSMENT (1981). *Impacts of applied genetics: micro-organisms, plants and animals*. OTA-HR-132. US Congress, Washington DC: Government Printing Office.

OFFICE OF TECHNOLOGY ASSESSMENT (1984). *Commercial biotechnology, an international analysis*. OTA-BA-218. US Congress, Washington DC: Government Printing Office.

OFFICE OF TECHNOLOGY ASSESSMENT (1987). *New developments in biotechnology: public perceptions of biotechnology*. Background paper, OTA-BP-BA-45. US Congress, Washington DC: Government Printing Office.

ORSENIGO, L. (1989).   *The emergence of biotechnology: institutions and markets in industrial innovation.* London: Frances Pinter.

PANEM, S. (1984).   *The interferon crusade.* Washington DC: Brookings Institute.

ROOBEEK, A. J. M. (1990).   *Beyond the technology race: an analysis of technology policy in seven industrial countries.* Amsterdam: Elsevier.

ROTHWELL, R. & ZEGVELD, W. (1981).   *Industrial innovation and public policy: preparing for the 1980's and 1990's.* London: Frances Pinter.

SCHAFFER, S. (1992).   Late Victorian metrology and its instrumentation: a manufactury of Ohms. In *Invisible connections: instruments, institutions and science,* ed. R. Bud and S. E. Cozzens, pp. 23–56. Bellingham, Wash: SPIE.

SPITZ, P. H. (1988).   *Petrochemicals: the rise of an industry.* New York: John Wiley.

STINE, J. K. (1992).   Scientific instrumentation as an element of U.S. science policy: National Science Foundation of Chemistry instrumentation. In *Invisible connections: instruments, institutions and science,* ed. R. Bud and S. E. Cozzens, pp. 264–76. Bellingham, Wash: SPIE.

STRICKLAND, S. P. (1972).   *Politics, science and dread disease: a short history of United States medical research policy.* Cambridge, Mass: Harvard University Press.

STUDER, K. E. & CHUBIN, D. E. (1980).   *The cancer mission: social contexts of biomedical research.* London: Sage.

TEITELMAN, R. (1989).   *Gene dreams. Wall Street, academia and the rise of biotechnology.* New York: Basic Books.

US CONGRESS, HOUSE OF REPRESENTATIVES, COMMITTEE ON SCIENCE AND TECHNOLOGY, SUBCOMMITTEE ON SCIENCE RESEARCH AND TECHNOLOGY (1977).   Research with Genetic Recombinations Generates Promise and Controversy. In *Science policy implications of DNA recombinant molecule research,* 95th Congress, 1st session, 474–9.

US CONGRESS, HOUSE OF REPRESENTATIVES, COMMITTEE ON SCIENCE AND TECHNOLOGY, SUBCOMMITTEE ON INVESTIGATION AND OVERSIGHT AND SUBCOMMITTEE ON SCIENCE, RESEARCH AND TECHNOLOGY (1981).   *Commercialization of academic biomedical research,* 97th Congress, 1st Session, 8–9 June.

US CONGRESS, HOUSE OF REPRESENTATIVES, COMMITTEE ON SCIENCE AND TECHNOLOGY, (1986).   *The biotechnology science coordination Act of 1986,* 99th Congress, 2nd Session, 4–5 June.

WILLIAMS, R. (1973).   *European technology: the politics of collaboration.* London: Croom Helm.

WRIGHT, S. (1993).   The social warp of science: writing the history of genetic engineering policy. *Science, Technology and Human Values* **18**, 79–101.

15

Product, process, or programme: three cultures and the regulation of biotechnology

SHEILA JASANOFF

Introduction

The development of a multinational regulatory framework for biotechnology during the past twenty years provides an unparalleled opportunity to study the processes by which technological advances overcome public resistance and are incorporated into a receptive social context. Through the vehicle of regulation, states provide assurance that the risks of new technologies can be contained within manageable bounds. Procedures are devised to limit uncertainty, channel the flow of future public resistance, and define the permissible modalities of dissent. Regulation, in these respects, becomes integral to the shaping of technology. A regulated technology encompasses more than simply the 'knowledge of how to fulfill certain human purposes in a specifiable and reproducible way.'[1] Regulation transmutes such instrumental knowledge into a cultural resource; it is a kind of social contract that specifies the terms under which state and society agree to accept the costs, risks and benefits of a given technological enterprise.

The passage of biotechnology from moratorium[2] to market in just twenty years exemplifies this process of social accommodation. During this period, biotechnology moved from a research programme that aroused misgivings even among its most ardent advocates to a flourishing industry promising revolutionary benefits in return for negligible and easily controlled risks. The transformation occurred almost simultaneously and with remarkable speed throughout Europe and North America. To facilitate the commercialization of biotechnology, the United States, and the European Community and several of its member states, adopted laws and regulations to control not only laboratory research with genetically engineered organisms but also their purposeful release into the environment.[3] Risks that once were considered speculative and wholly unmanageable[4] came to be regarded as amenable to rational assessment in accordance with sound scientific principles. Apocalyptic visions and the rhetoric of science fiction yielded to the weightier discourse of expert advice and bureaucratic practice. The research community coalesced to persuade the public that the risks of biotechnology could be assessed in a reasonable way and that earlier fears of ecological disaster were mostly unfounded.

These changes in the status of biotechnology were all the more noteworthy because, as of the early 1990s, the risks of genetic manipulation remained largely hypothetical. Scientists and industrialists confidently proclaimed that no serious harm would befall ecosystems or human health if our daily bread were baked with genetically engineered, quick-rising yeast, if economically significant crop plants were fitted out with herbicide resistance genes, or if fruit farmers sprayed their orchards with gene-deleted bacteria designed to prevent frost formation. Unlike toxic chemicals, however, the products of the new biotechnology have not been around long enough to display their whole range of beneficial and adverse effects. Despite repeated allusions to Bhopal and Chernobyl by opponents of bio-technology, there is no reservoir of precedents into which one can readily dip for historical parallels to the production and use of laboratory-crafted living organisms – products not of nature but of human invention.

Nonetheless, as regulators in different countries approve new uses of biotechnology and reassure their publics that the risks are manageable, they are obliged to place believable outer limits on the technology's potentially harmful impacts. An important question for students of technology to ask is whether the resulting accounts of risk have diverged cross-nationally, conditioned by varying socio-political influences, as predicted by the social studies of science and as previously documented in studies of environmental regulation and risk man-agement.[5] Were there observable differences in national regulatory responses to biotechnology and, if so, could they be traced to differences in national traditions of legal and administrative decision making? How, in turn, did the process of constructing the risks of biotechnology for regulatory purposes affect the opportunities for public participation and protest?

This chapter is based on a focused comparison of the way governmental authorities in Britain, Germany, and the United States conceptualized bio-technology as a regulatory problem in the specific context of releasing genetically modified organisms (GMOs) into the environment. Looking primarily at events in the decade from 1980 to 1990, I describe how public resistance and state response initially led to quite different understandings about risk in each national context, and hence to divergent characterizations of biotechnology as a policy issue. In all three countries, however, the dominant conception enabled regulators to devise strategies for managing uncertainty and neutralizing the most common forms of organized opposition. Although their techniques varied – legislation, bureaucratic reorganization and expert advice were differentially employed – regulators in each nation succeeded in rearranging a potentially limitless expanse of scientific unknowns into familiar paradigms of assessment and control. I conclude with some observations about what this analysis implies for mobilization against risk in advanced industrial societies.

Paradigms of control

In order to approve the deliberate environmental release of GMOs, regulators in the United States, Britain and Germany had to persuade their respective political constituencies that the risks of biotechnology, although novel, lay sufficiently close to their prior experience of technological risks to permit effective public control. Although the ultimate goal was the same everywhere, the strategy of public reassurance adopted in the three countries varied, especially in the willingness to admit that biotechnology poses novel or special risks to human well-being. 'Specialness' as it relates to the adverse impacts of biotechnology had been understood on at least three different levels since the 1970s. First, opponents of the technology argued that human intervention through genetic engineering would produce *physical risks* to health and the environment that were different in kind and magnitude from risks created by 'natural' processes of genetic combination and recombination. Secondly, some observers were persuaded that the widespread application of biotechnology in agriculture would create a variety of *social risks*, ranging from the commodification of nature to the elimination of family farms in the West and to severe economic dislocations in developing countries. Thirdly, the esoteric technical content of biotechnology was considered likely to increase the distance between expert decision makers and the lay public, thereby exacerbating the *political risk* – increasingly troubling in modern industrial societies – of excluding citizens from meaningful control over technologies that could transform their lives. As we shall see below, these three dimensions of risk, each entailing its own discourses of protest and legitimation, were emphasized to different degrees in the regulatory politics of the United States, Britain and Germany.

United States – a product-based approach

The first applications for conducting deliberate release experiments caught regulatory agencies in the United States without appropriate institutional mechanisms in place for conducting persuasive safety evaluations. The only supervisory body that researchers could turn to at the outset was the National Institutes of Health (NIH), which had been regulating laboratory experiments involving recombinant DNA (rDNA) molecules since the mid-1970s. Pursuant to guidelines first adopted in 1976 and substantially relaxed in 1978, all federally funded rDNA experiments had to be approved by NIH's Recombinant DNA Advisory Committee (RAC). Governmental control, in other words, was tied to the sponsorship of research, a scheme that proved increasingly vulnerable as biotechnology headed out of the laboratory toward commercial application.

The insufficiency of the NIH review process was dramatically exposed when two University of California scientists, Steven Lindow and Nickolas Panopoulos, sought permission to carry out a field test using the 'Ice-Minus' bacterium, a member of the *Pseudomonas* family that had been genetically engineered to increase the frost resistance of plants. The scientists advising the NIH reviewed the application, requested some modifications, and decided unanimously on the second round of review that the experiment was safe. Their conclusion, however, was set aside by a federal court of appeals, which blocked the experiment on the ground that NIH had not carried out a proper environmental impact assessment, as required by the US National Environmental Policy Act (NEPA). In *Foundation on Economic Trends v. Heckler*,[6] the court especially deplored NIH's failure to explain why a type of experiment that had been considered too risky to undertake under the 1976 guidelines could now be permitted to go forward with so little explicit consideration of its risks. The scientific community predictably saw this call for greater public accountability as an insupportable intrusion into safety evaluation by a 'technically illiterate' judiciary. All the same, *Heckler* threw into relief the fact that NIH's research-funding mission did not sit well with creating an appropriate institutional forum for airing lay concerns about the risks of commercial biotechnology.

The Ice-Minus episode among others forced the US government to regularize its procedures for controlling the commercial applications of biotechnology. In 1986 the president's Office of Science and Technology Policy (OSTP) published a *Coordinated Framework for the Regulation of Biotechnology*, identifying the responsibilities of the three agencies with most extensive jurisdiction over the new technology – the Environmental Protection Agency (EPA), the Food and Drug Administration (FDA), and the US Department of Agriculture (USDA). A Biotechnology Science Coordinating Committee (BSCC) was established to develop a common inter-agency approach to issues governed by the Coordinated Framework. In addition, each of the lead regulatory agencies developed new institutional capabilities for dealing with biotechnology. For example, EPA established a Biotechnology Science Advisory Committee (BSAC) to give advice on the scientific aspects of regulation.

These institutional arrangements reflected in the first instance a consensus across the US government that the authority contained in existing laws, aimed largely at controlling physical risks, was sufficient to regulate any novel problems associated with biotechnology. OSTP and the agencies participating in the Coordinated Framework persuaded Congress that regulations issued under the old laws would adequately clarify concepts and eliminate possible jurisdictional overlaps. This approach was consistent with the views of many scientists in research and industry that the risks of biotechnology were not in any sense special or unique, and that biotechnological products – pesticides, drugs, foods, and food additives – should not be treated any differently from similar products created by traditional biological or chemical processes.

While denying the need for new legal authority, the Coordinated Framework happily accepted the institutionalization of new scientific authority. The creation of an expert advisory committee, BSAC, at the individual agency level and a coordinating committee, BSCC, at the inter-agency level indicated that federal regulators viewed the task ahead primarily in scientific terms and were prepared to strengthen their institutional capabilities accordingly. OSTP's central role in developing the Coordinated Framework reinforced the view that regulating biotechnology was not a matter for broad participatory politics but for expert policy making at the highest levels of the executive branch. The object at every turn seemed to be to demonstrate that the mainstream forces of science – not activists like Jeremy Rifkin nor the assorted nay-sayers of the environmental movement – were in the driver's seat with respect to managing the emergent technology.

An influential report published by the National Research Council (NRC) in 1989 lent support to the US government's evolving position that commercial biotechnology should not be regarded as a specially risky enterprise in relation to human health and the environment.[7] On each of three issues where splits had developed among federal regulatory agencies,[8] the NRC report sided with the agencies that took the more benign view of biotechnology's hazards. Specifically, the NRC report concluded that

> (i) the *product* of genetic modification and selection constitutes the primary basis for decisions ... and not the *process* by which the product was obtained; (ii) although knowledge about the process used to produce a genetically modified organism is important ... the nature of the process is not useful for determining the amount of oversight; and (iii) organisms modified by modern molecular and cellular methods are governed by the same physical and biological laws as are organisms produced by classical methods.[9]

The message was obvious: mere use of biotechnological techniques did not make a harmless product dangerous; nor, conversely, were organisms produced by 'classical methods' safe simply because they were not genetically engineered. The report as a whole helped crystallize the conclusion that, for policy purposes, biotechnology was to be regarded as a supplier of familiar classes of products – not as a novel technological process threatening mysterious and incalculable harm to social well-being.

Elaborating on the theme of 'no special hazards', the NRC report on the whole belittled the possibility that GMOs would introduce uncontrollable risks into the environment. With respect to genetically modified plants, for example, the NRC committee concluded, first, that the potential for enhanced weediness was the most significant environmental threat. The committee then determined that this risk was likely to be low for a variety of reasons – for example, that the analogy between genetically modified crop plants and 'exotics' was 'tenuous' and that 'genetically modified crops are not known to have become weedy through the addition of traits such as herbicide and pest resistance.'[10]

As the last sentence suggests, the committee's emphasis throughout the report was on what was already known about genetic engineering and environmental release rather than what still remained unknown. For example, the report took pains to point out that molecular methods, whether used on plants or microorganisms, are highly precise and lead to modifications that can be fully characterized and understood.[11] This precision, the committee felt, provided sufficient safeguards against unpredictable behaviour by the resulting organisms. Assessing the social or political risks of biotechnology would have been out of place in a report that self-consciously disciplined uncertainty through technical language; indeed, no explicit discussion of social or political issues contaminated the apparent specificity of NRC's scientific analysis.

Debates concerning the 'scope' of regulation gave further evidence of US policymakers' reluctance to treat the risks of biotechnology as different in kind from those of more traditional biological manipulation. The 1986 Coordinated Framework, for instance, proposed two definitions for organisms requiring review: intergeneric organisms (that is, organisms formed by combining genetic material from sources in different genera), and pathogens.[12] During public comment, these proposals were severely criticized on the ground that they focused – inappropriately in the view of many scientists – on the process by which an organism was produced rather than on the probable riskiness of the product.

Arguments about the scope of regulation continued to divide official opinion for several years, with EPA's staff and scientists favouring a different approach from that of FDA and USDA. In 1990, EPA's biotechnology advisory committee proposed a quite inclusive and process-based definition of scope ('organisms deliberately modified by the introduction into or manipulation of genetic materials in their genomes'), from which it proposed to exclude all organisms that did not raise new risk assessment issues. The BSAC felt that this approach was broad enough to address potential risks, yet flexible enough to cover future developments in biotechnology. Critics complained, however, that EPA's formulation still displayed an excessive tilt toward process over product as the framing concept for regulation and that this stance contravened the recommendations of the NRC report.[13]

The existence of the NRC report allowed EPA's critics to legitimate their attacks on EPA's scope proposal through an appeal to scientific consensus. But 'science', as socially constructed in US regulatory debates, is often a double-edged sword, and it served as the discourse of choice for EPA's supporters as well. In particular, BSCC, the expert inter-agency coordinating committee that many saw as hostile to EPA, was itself attacked for straying beyond its charter, holding closed meetings, and impeding EPA's scientific inquiry. At the committee's December 1989 meeting, Margaret Mellon of the National Wildlife Federation expressed scepticism based on 'the composition of the BSCC – all high-level administrators, not scientists'.[14] Others accused the committee of unlawfully and heavy-handedly appropriating the review functions of the Office of Management and Budget (OMB), whose own intervention into issues of regulatory science had become a matter of considerable notoriety during the Reagan administration. By late 1990, these challenges led OSTP to rename the BSCC as the Biotechnology Research Subcommittee of the Committee on Life Sciences and to scale down its involvement in policy making.[15]

Confusion in regulatory circles, and associated boundary disputes over expertise and authority, rekindled interest in a legislative solution to managing biotechnology, but political pressure was insufficient to overcome a settled congressional reluctance to do anything that might endanger the US industry's competitive position. Instead, actions by the FDA and the White House, acting through OSTP, consolidated the policy position that only the characteristics of specific products were legitimate objects of regulatory assessment. Labelling theirs a 'risk-based' or a 'science-based' strategy of safety evaluation, these agencies continued to harp on the theme that any negative consequences of biotechnology could be adequately controlled product by product, without creating barriers against 'useful innovation'.[16]

The courts, which in the American political context might have provided an independent spur to a broader public debate on biotechnology, proved unusually quiescent throughout the period of policy development. In *Diamond v. Chakrabarty*,[17] the US Supreme Court held by a narrow five-to-four majority that biologically modified microorganisms could be patented under an existing law whose operational language had been drafted 200 years before the advent of biotechnology. The decision on its face dealt with a narrowly legal question: whether living things constituted patentable subject matter under the Patent Act. Researchers and industry, however, found more grounds for rejoicing in the decision's subtext, for by relying on existing law the Court implicitly rejected the argument that the risks of biotechnology were so novel as to require special legislative attention. Even the *Heckler* decision, which some had taken to be a sign of awakening judicial activism in matters of biotechnology regulation, refused to require a programmatic evaluation of all deliberate releases, and it proved in any event to be an anomaly rather than a trend-setter with respect to later judicial decisions. Most subsequent challenges to proposals for environmental release from groups like Jeremy Rifkin's were curtly dismissed for lack of standing to sue.

Britain – biotechnology as process

Events in Britain suggested that the government was prepared to take a somewhat more expansive view of biotechnology's risks than were federal policy makers in the United States. Since 1978, laboratory work involving 'genetic manipulation' had been controlled through regulations issued under the Health and Safety at Work Act of 1974. Applications to conduct such activities had to be approved by the Genetic Manipulation Advisory Group (GMAG), replaced in 1984 by the Advisory Committee on Genetic Manipulation (ACGM)[18] to the Health and Safety Commission (HSC), Britain's lead agency for worker protection. Biotechnological work with environmental implications was further reviewed by the Department of the Environment, which obtained expert advice from its own interim Advisory Committee on Introductions. By the late 1980s, however, it became clear to British authorities that many biotechnological activities, including large-scale industrial production and deliberate releases into the environment, could not properly be controlled through the existing regulatory structure for occupational safety.[19]

Developments within the European Community provided additional impetus for Britain's decision to enact more formal statutory controls. In April 1990, the Community adopted two directives relating to biotechnology: one on contained experiments and one on deliberate release of GMOs. Recognizing the need for new legal authority to implement the latter directive, the British government introduced into the Environmental Protection Act of 1990 (the so-called Green Bill) a new Part VI governing GMOs. Meanwhile, environmental and health and safety authorities decided to replace their existing expert committees with a single new committee to review applications for releasing GMOs into the environment. The resulting interdepartmental Advisory Committee on Releases to the Environment (ACRE) held its first meetings in July 1990.

Debate on the Green Bill provided a focal point for environmentalists to demand more public participation in decisions about GMOs, and the government responded by agreeing to include an environmental representative on its new advisory committee on environmental release. The first person selected for this position was Julie Hill, a member of the Green Alliance, an environmental lobbying group spun off from the Liberal Party that had been particularly active in commenting on the Green Bill. Within Britain's normally closed and consensual policy culture, Hill's appointment marked at once a blow to tradition and a concession to long-standing regulatory practice. Asking an environmentalist to sit on ACRE affirmed the state's acceptance of the lay public's interest in biotechnology as significant enough to be represented in future negotiations over safety, but after the appointment, as before, the power to make decisions remained closely held within an expert advisory body.

Broadening the range of participation on ACRE appeared on the surface to be more responsive to the special social and political risks of biotechnology than comparable actions of the US government. It was almost as radical a move in the British context, according to one observer, as inviting Jeremy Rifkin to give advice on biotechnology might have been in America. Sceptics note, however, that the new committee was formed under the aegis of the HSC, the most participatory of Britain's regulatory agencies; under the Health and Safety at Work Act, HSC and its various operating committees are required to be constituted as 'tripartite' bodies, representing industry, labour, and local governments. Given this tradition of participation, it was perhaps easier for ACRE to accommodate a new interest (environmentalism) than it would have been for less broad-gauged scientific committees, such as those attached to the Ministry of Agriculture, Fisheries and Food.[20] Further, the move came at a time when the conservative government was seeking to expand its ties among moderate environmentalists. For British government and industry, the Green Alliance may well have represented environmentalism with a human face – a voice of reasoned dissent that could be internalized without seriously jeopardizing the evolution of technology. In constructing an appropriate advisory committee on deliberate releases, then, the government simultaneously constructed an official form of green participation that regulatory authorities were prepared to live with.

In Britain as in the United States, a well-timed report by a prestigious expert body helped reinforce the government's efforts to sort out its legal and institutional arrangements for dealing with biotechnology. The Royal Commission on Environmental Pollution (RCEP), a standing body charged with advising the government on environmental matters, decided that the time was ripe for a thorough evaluation of deliberate release, looking both at the possible consequences of releasing GMOs and at procedures for identifying, assessing and mitigating their risks.[21]

Issued in 1989, like its US counterpart, the Royal Commission's report was both more expansive in its treatment of impacts and more open in admitting uncertainty than the corresponding US document. Thus, instead of dwelling on benign past experiences and the precision of molecular techniques, the British experts emphasized how much was still unknown – and hence how little could be predicted with assurance about the likely behaviour of GMOs in the environment. With respect to genetically modified plants, for example, the RCEP report considered a broader range of possible risks than the NRC and seemed unwilling to dismiss any of these risk scenarios as wholly improbable. Thus, the RCEP felt that the historical experience with exotics could be highly relevant if a GMO were released into an environment where it was not native.[22] With respect to herbicide resistance, the Commission considered not only the possibility that the resistant gene might spread to weedy species, but also that the genetic engineering of plants resistant to herbicides could lead to greater use of environmentally damaging herbicides.[23]

The RCEP's stance, acknowledging the unpredictability of nature, was echoed in official British policy. In its guidance note on environmental release,[24] the ACGM subcommittee on releases spoke of possible differences between natural evolutionary processes and results obtained through genetic manipulation, noting for example that the release of a novel organism could involve the introduction of larger numbers than in the case of natural mutations. In sum, the subcommittee concluded as late as January 1990 that 'the deliberate release of novel types to foreign habitats could occasionally disturb the natural equilibrium of those habitats'.[25]

British authorities seemed to accept without question the Royal Commission's recommendation that all GMO releases, to start with, should be subject to regulatory scrutiny. Put differently, this amounted to accepting the principle (denied in America) that the process of genetic modification was an appropriate basis for defining the scope of policy action. Officials at both DoE and HSE acknowledged that risk categories might eventually be established that would either exempt some products from evaluation or subject them to reduced oversight.[26] But they indicated that any such relaxation would have to be based on actual experience, that is, on empirically observed data from earlier releases. These views were seconded by Dr John Beringer, the first chairman of ACRE, who thought that all GMOs should in principle be subject to review, although it might eventually be possible to move to a two-tier system of clearances for new

GMOs – a 'fast track' for relatively familiar organisms and a slower track for all others.[27]

Having agreed to a case-by-case approach, British regulators were most concerned to ensure that the approval process would flow as smoothly as possible from the standpoint of the applicant. The creation of a single 'postbox'[28] in the form of ACRE bypassed the possibility of inter-agency differences of the kind that arose in America. This committee was to review all applications for release regardless of whether the product was a food, drug, pesticide or crop plant. Moreover, the risk assessment guidelines and notification procedures adopted by the ACGM subcommittee on deliberate release, ACRE's predecessor, were to serve as the blueprint for new interdepartmental regulations.[29] In particular, the guidance note outlined a risk assessment procedure, spelling out what information applicants should provide on an interdepartmental form to facilitate unified submissions. The instructions accompanying the form were symptomatic of the extent to which deliberate release in Britain had been redefined from an exercise in assessing uncertainty to a matter of following bureaucratic routine: 'Continuation sheets should be used wherever necessary. These should be in A4 format and clearly marked with the number of the item to which they relate.'[30]

Additional steps toward normalizing the regulatory treatment of biotechnology were taken with the publication of the Royal Commission's report on 'GENHAZ', a systematic approach to evaluating proposals for environmental releases of GMOs.[31] The Commission acknowledged that each release was likely to be unique, and hence that blanket exemptions were not warranted for any products of genetic modification. Nevertheless, the risk assessment procedure the Commission outlined provided reassurance on at least two levels. First, the proposed analytic approach was based on a method already in use in the chemical industry, a fact that tended to make biotechnology look more like another, less novel form of hazardous activity. Secondly, the procedure assumed that an experienced, interdisciplinary team of experts would be able to imagine the possible hazards of release, and hence to guard against potentially unacceptable consequences. This presumption essentially negated the possibility of significant hazards lying beyond the imaginative reach of the trained scientific mind.[32]

Germany – a programmatic view

The three major dimensions of biotechnology's risks – physical, social, and political – were perhaps most fully deconstructed, or thematized, in the German case, although public debate was slower to take shape in Germany than in Britain or the United States. The regulatory history of genetic engineering in Germany began in the early 1960s with a top-down decision by the federal government to target biology as an area for state-supported R & D. The biotechnology programme received a further boost with the creation in 1972 of the Federal Ministry for Research and Technology (BMFT), whose central mission was to channel funding toward designated 'key technologies'. Paralleling the work of NIH in the United States, BMFT supervised the German response to the Asilomar conference, where researchers first expressed concern about the risks of genetic manipulation. Guidelines closely modelled on NIH's were issued by a restricted, *ad hoc* committee of experts, including at first neither labour nor industry, though these interests were later represented in a twelve-member implementing commission.[33] Through the early 1980s, the strategy of containing regulatory debate within carefully structured expert committees ensured a relatively narrow focus on the physical risks of rDNA research and correspondingly muted attention to the social and political consequences of the new technology.

The rise of new social movements and the waning of previously controversial issues such as nuclear power opened the way for a more participatory politics of biotechnology by the mid-1980s.[34] The Green Party was first elected to the Bundestag in 1984 and soon created a working group on genetic technology. In the same year, an alliance between the Greens and the Social Democrats led to the formation of a parliamentary Commission of Inquiry (*Enquete-Kommission*) to examine the opportunities and risks associated with developments in genetic engineering. As the state's policy on biotechnology was subjected for the first time to systematic, institutionalized criticism, two views emerged concerning the novelty of the problem confronting policy makers. The Greens and the Social Democrats argued that the risks of biotechnology were sufficiently unsettling – uncertain, potentially catastrophic, perhaps irreversible – to require a new political order for their management and control. Key to this new order would be a more pronounced voice for the public, institutionalized through new forms of public participation. The Christian Democrats insisted, to the contrary, that biotechnology was amenable to control through established forms of assessment by technically trained experts.

Green opposition to biotechnology led in due course to litigation. In an unusual lawsuit against Hoechst chemical company, German environmentalists in Hessen challenged a planned facility for the production of genetically engineered insulin on the ground that the state had not sufficiently guaranteed the safety of biotechnology. Existing laws, they argued, could not be construed as providing an adequate basis for controlling risks whose unique characteristics required explicit legislative authorization, just as nuclear power had done a decade earlier. The administrative court of Hessen accepted this representation of uniqueness and, in a move that went beyond the actions of any US court, ordered the cessation of

Table 15.1. *Thematization of risk*

	Physical	Social	Political
US product	High	Low	Low
UK process	Medium	Medium	Medium
Germany programme	Medium	Medium	High

industrial biotechnological activity until a suitable legal framework was in place. Within a year, however, the German parliament set aside this inconvenient roadblock by passing the 1990 Genetic Engineering Law, a statute that critics denounced for repudiating the inroads made by participatory politics on the government's insulated, bureaucratic–technocratic structures of control.

By combining the functions of protection ('*Schutz*') and promotion ('*Forderung*') within a single law, the legislature affirmed the state's presumed capacity to undertake these potentially conflicting tasks without compromising the values or rights of its citizens, but early implementation of the law raised questions as to whether this optimism was justified. As a partial concession to public concerns, the law opened up participation on the government's key advisory committee and created a new public hearing process for deliberate release applications. These procedural innovations seemed responsive to the theme of political risk articulated during the controversy preceding the law's enactment. In practice, however, the first public hearings deteriorated into administrative wrangles and rhetorical stand-offs that led the government in 1993 to rescind the hard-won right to a hearing. The environmentalists' position on the safety evaluation committee, too, appeared likely to become bureaucratized, as the Greens, unable to pay for their representatives, considered replacing them with sympathetic government officials.[35]

The political construction of risk and resistance

I have argued thus far that the risks of biotechnology, particularly as regards their novelty, were construed in fundamentally different ways within the regulatory frameworks of three advanced industrial nations – the United States, Britain and Germany. The divergences during the 1980s are most strikingly apparent if one looks in retrospect at the dominant characterization of biotechnology as a regulatory problem in each country and the impact of this problem definition on later debates about risk. See Table 15.1 for a two-dimensional, and hence necessarily oversimplified, representation of the cross-national differences in the thematization of risk.

Table 15.2. *Resistance and response*

	Forms of resistance				State responses			
	Scientific debate	Legislative debate	Litigation	Party politics	Expert committees	Administra-tive rules	Legis-lation	Judicial action
US	yes	some	yes	no	new	yes	no	yes (pro-develop-ment)
UK	some	some	no	no	new/ expanded	yes	expanded	no
Germany	no	yes	yes	yes	expanded	yes	new	yes (anti-develop-ment)

The focus in the United States was increasingly on the *products* coming into the market-place and the physical risks they may pose to human health *or* the environment. In Britain, regulators appeared initially more prepared to accept the *process* of genetic modification as the frame for policy making, with concurrent attention to the physical and social dimensions of risk. But this acknowledgment of the technique's specialness was undercut to some degree by a bureaucratized hazard evaluation procedure that stressed routine and internalized possible opposition from environmentalists. German political debate on biotechnology was unique in taking as its domain the entire *programmatic* relationship between technology and society, as mediated by the state, a position that led to a full-blown discussion of risks. Eventually, parliamentary action, in the form of a special law on genetic engineering, confirmed that the state's programme of promoting and regulating biotechnology was sufficiently novel to require explicit legislative licence. (See Table 15.2 for a summary of the main forms of resistance in each country and the associated variations in the state's responses to public challenge.)

In the remainder of this chapter, I will argue, first, that these cross-national variations were consistent with previously noted features of each country's political culture and regulatory style; secondly, I will suggest that the divergent forms of political accommodation worked out in each country were similar in result – in each case, the selected policy initiative blocked significant avenues of public dissent and smoothed the way for a relatively untroubled further development of biotechnology.

The US case illustrates the well-known national preference for according science a central role in public decision making. US regulators have generally been more inclined to justify their actions with appeals to objective knowledge than their European counterparts. Extensive scientific records, mathematical modelling of risk and uncertainty, and detailed procedures for peer review and quality control, all bear witness to the US decision maker's need to enlist the impartial authority of science in support of costly and controversial policy decisions. Confronted with scientific uncertainty, American agencies are reluctant simply to admit ignorance and exercise subjective judgment. If an extrapolation must be made from limited data, it has to be according to prestated rules of decision that spell out technical methods for dealing with uncertainty.[36] More generally, science in the US frequently serves as a resource with which political adversaries seek to trump their opponents in the regulatory arena. Scientific disputes thus become a surrogate for unstated ethical or economic conflicts.

Not surprisingly, then, every major US player with a stake in biotechnology policy stated publicly that decisions in this area should be based on sound science. Competition among these actors to justify their positions in scientific terms underscored the power of science as a legitimating rhetoric in politics. EPA, the most risk-averse of the US agencies (and, in the Reagan–Bush years, also the most politically vulnerable), created a new scientific advisory committee, BSAC, to shore up its credibility in the politics of regulation. When the White House tried to seize control of biotechnology policy, it created the BSCC, ostensibly to provide *scientific* coordination across the government, but in practice to serve as a counterweight to possibly recalcitrant regulatory agencies. BSCC, in turn, relied on the National Research Council for a still more authoritative exposition of the scientific principles that should govern the regulation of biotechnology. In due course, the NRC report provided scientific ammunition for OSTP scientists, Vice-President Dan Quayle's Competitiveness Council, and others who wished to challenge EPA's cautious regulatory approach.

Scientific pluralism, the result of scientific claims being produced by parties with competing claims to authority, is inevitably a feature of American regulatory politics, showing that the effort to tame uncertainty through technical discourse does not necessarily resolve conflicts. The multiplicity of agencies (EPA, FDA, USDA, NIH) and committees (BSCC, BSAC, NRC study committee) with an active interest in biotechnology virtually guaranteed that multiple technical accounts of risk would proliferate in the public domain once decision making was narrowed to questions of physical risk and safety. The protracted battle over the scope of regulation was but one example of the fracture lines that arise when American political actors draw upon 'scientific principles' to justify their agendas with respect to risk.

The British style of policy making, in contrast to the American, tends to be informal, cooperative, and closed to all but a select inner circle of participants. Disputes are resolved as far as possible through negotiation within this socially bounded space, and the power of the judiciary is seldom invoked even for enforcement purposes. These differences have had an impact on the production and use of regulatory science (science used as a basis for policy),[37] which tends in Britain to be less diverse and less admitting of uncertainty than in the United States (Wynne and Mayer, 1993). Early attempts to manage the deliberate release of GMOs, however, showed British scientists and regulators as apparently more receptive than their US counterparts to admitting the special status of bio-technology and to recognizing a broad range of possible hazards, from the ecological to the social and (to a lesser degree) political.

This finding seems inconsistent at first blush with observations previously made in the area of chemical regulation, where British experts consistently represented the risks as less severe than their counterparts in the United States. While American regulators often banned substances based on animal evidence alone, British health and safety authorities refrained from aggressive action except in cases where there was observable harm to human health. At a deeper level, however, Britain's seemingly higher tolerance for chemical risks and lower tolerance for biotechnological risks can be traced to similar underlying views about what constitutes acceptable evidence for political action. The British policy maker's classic preference for empirical proofs, attested to by credible communities of experts, explains why so few of the risks of biotechnology were initially ruled out as improbable, just as it explains why chemicals were so often exonerated when they only damaged the health of test animals but showed no effect on humans. British caution over biotechnology proceeded from the fact that no one had yet had the opportunity to *see* how gene-altered organisms might behave in the environment, removed from the physical containment of laboratories. In the absence of direct evidence, it was easy for all sides to agree that experience alone could guide the making of regulations, including the establishment of risk criteria and classes of exemptions. Biotechnology thus classically lent itself to the case-by-case regulatory style favoured by policy makers in Britain; it was a style well suited in this instance to permitting incremental adjustments to the new technology.

Britain's sensitivity to the need for broader political representation in biotechnology policy was also consistent with that country's established practices for managing risks to health and safety. The framework of tripartite decision making in the field of worker protection was easily adapted to include a representative of the environmental community. Giving the 'greens' a formal role in ACRE at least temporarily neutralized the threat of public discontent. At the same time the move, which left the state in charge of choosing its environmental partner, seemed unlikely to upset the science–government–industry consensus that normally drives policy in Britain. Many observers of the British regulatory scene saw the expansion of ACRE as yet another instance of successful political cooptation whereby a potentially troublesome 'outsider' voice is brought into – and contained within – the channels of closed, consensual, and expert-dominated decision making.

Relations among science, technology and the state have historically been less transparent in Germany than in the other two countries, and public disputes among experts are something of a rarity in the regulatory arena. Yet, the German environmental movement scored early and relatively pronounced political success, winning representation in parliament at a time when British environmentalists were hardly visible as a national political force. Confrontations over technological risk in Germany have been intensely political, even violent at times, as in the case of anti-nuclear protests in the late 1970s and early 1980s. Again, these dynamics reproduced themselves with reasonable accuracy in the context of biotechnology. The German policy debate was most directly tied to the agendas of the major political parties. Perhaps in consequence, it was also most successful in forcing an open public discussion of the social and political ramifications of biotechnology, avoiding the strictly scientific framing that accounted for so much of the American discourse on risk. In a society where expertise is normally the prerogative of the few, insistence on the value implications of biotechnology (rather than exclusively on its technical uncertainties) powerfully legitimated citizens' claims that they should be accorded a wider role in the direction of the new technology. Yet, by enacting a comprehensive regulatory law, the state in the end re-established the very bureaucratic culture of risk management that had initially aroused public protest. The 1990 law permitted technology to develop without substantial fear of widespread citizen mobilization.

Conclusion

I have devoted much of this essay to the theme that political and regulatory culture counted in the way that members of three technological societies imagined, characterized, delimited, and controlled the products of their scientific ingenuity. In each country, an early phase of protest seemed at first to expand the vocabulary of resistance to a new and fearful technology. Contingent and culturally specific accountings of risk led in the 1980s to divergent national conceptualizations of the problem facing regulatory authorities. Cultural influences surfaced most strikingly in the science-centred definition of risk in the United States, in the political adaptation of existing expert bodies in Britain, and in the comprehensive legislative response to citizen mobilization in Germany.

The final twist to the story, however, becomes apparent only when we ask what

these preliminary characterizations of risk meant in terms of the future of
biotechnology. It is difficult to avoid the conclusion that all three countries,
despite their culturally conditioned ways of constructing biotechnology as a
policy issue, converged in their willingness to make the technology possible. In
each country, the dominant political framing appeared to rule out one or more of
the expected forms of public resistance, thereby ensuring that scientific uncertainty would not spill over into social and political unrest. Thus, in the United
States, congressional and judicial inaction left the discussion of biotechnology's
risks within a bureaucratic framework where the issue was most likely to be
analysed in the relatively narrow terms of physical hazards. Moreover, the
absence of legislation foreclosed new opportunities for judicial review and sharply
restricted the dissenting public's least constraining avenue of access. Similarly, in
Britain, despite an initially more expansive reading of biotechnology's uncertain
consequences, decision making was soon channeled into a framework of carefully
structured expert committees that provided assurance by internalizing dissent.
Finally, legislation in Germany re-established a working state–industry partnership that formally bowed to citizen concerns but closed down the kind of open-
ended political debate that had preceded the enactment of the genetic engineering
law. In all three national settings, then, historical contingencies and political
culture proved equally amenable to accommodating the determined thrust of
biotechnology's forward movement. Explanations for this ultimate convergence
lie in all probability in the theatre of international relations, where national
protest politics confronted, and eventually succumbed to, the rhetoric and politics
of global competitiveness.

Notes

1 Harvey Brooks, "Technology, Evolution, and Purpose," *Daedalus* 109 (1980), p. 66.
2 In the mid-1970s leading molecular biologists declared a moratorium on research with
   recombinant DNA until the risks were properly explored and regulated. A scientific meeting at
   Asilomar in 1976 laid the conceptual basis for research safety guidelines that were formally
   adopted by the National Institutes of Health (NIH).
3 European countries that have legislated on this issue include Denmark, Germany and Britain.
   In the Netherlands, regulations governing deliberate releases were developed under the
   Environmentally Hazardous Substances Act. As noted below, on 23 April 1990, the European
   Community adopted a directive on the deliberate release of genetically modified organisms.
   Member states were required to implement the directive by the end of 1991.
4 The 1976 NIH guidelines prohibited deliberate release experiments. Just two years later, NIH
   decided that the prohibition could be waived on a case-by-case basis.
5 See, for example, Sheila Jasanoff, *Risk Management and Political Culture* (New York: Russell Sage
   Foundation, 1986).
6 *Foundation on Economic Trends v. Heckler*, 756 F.2d 143 (D.C. Cir. 1985).
7 National Research Council, *Field Testing Genetically Modified Organisms – Framework for*

Decisions (hereafter referred to as *Field Testing*) (Washington, DC: National Academy Press, 1989).

8 Among the agencies participating in the Coordinated Framework, EPA's positions tended consistently to diverge from those of FDA and USDA.

9 Henry I. Miller, Robert H. Burris, Anne K. Vidaver, Nelson A. Wivel, 'Risk-Based Oversight of Experiments in the Environment,' *Science* 250 (1990), p. 490.

10 NRC, *Field Testing*, p. 52.

11 See, for example, NRC, *Field Testing*, Executive Summary, pp. 3–4.

12 *Federal Register*, 26 June 1986.

13 EPA Memorandum, p. 2.

14 'BSCC Urged to Hold More Meetings, Open Process', *Pesticide and Toxic Chemical News*, 27 December 1989, p. 7.

15 Jeffrey Merves, Congress and Administration Closer to Regulating U.S. Biotech Industry, *The Scientist*, 4 (22) (1990), p. 12.

16 See, for example, Office of Science and Technology Policy, 'Exercise of Federal Oversight Within Scope of Statutory Authority: Planned Introduction of Biotechnology Products into the Environment,' *Federal Register*, February 27, 1992, pp. 6753–6762. See also, David A. Kessler et al., 'The Safety of Foods,' *Science* 256 (1992), pp. 1747–49 and 1832.

17 *Diamond v. Chakrabarty*, 447 U.S. 303 (1980).

18 To deal with the issue of deliberate release, ACGM had created an Intentional Introductions Sub-Committee (IISC). ACGM's own name subsequently was changed to Advisory Committee on Genetic Modification.

19 Existing regulations were deemed defective not only because of their limited scope but because they referred to the no longer existent GMAG. See *The Impact of New and Impending Regulations on UK Biotechnology*, report of a meeting sponsored by the Department of the Environment, the Health and Safety Executive and the Bioindustry Association (hereafter cited as *Impact*) (Cambridge: Cambridge Biomedical Consultants, 1990, p. 12) (remarks of Richard Clifton, Health and Safety Executive).

20 I am indebted to Les Levidow for calling my attention to this point.

21 Royal Commission on Environmental Pollution, *The Release of Genetically Engineered Organisms to the Environment* (hereafter referred to as *Release of GEOs*), Thirteenth Report (London: HMSO, 1989). Although the Commission spoke of genetically engineered organisms (GEOs), the term genetically modified organism (GMO) eventually took over as the international standard term for organisms produced by genetic engineering. In this chapter, I follow the international usage.

22 RCEP, *Release of GEOs*, p. 21.

23 RCEP, *Release of GEOs*, p. 20.

24 ACGM, 'The Intentional Introduction of Genetically Manipulated Organisms into the Environment,' Guidelines for risk assessment and for the notification of proposals for such work, HSE Guidance Note 3 (revised), January 1990.

25 HSE Guidance Note 3, p. 5.

26 See comments of Richard Clifton and Douglas Bryce in *Impact*, pp. 15, p. 24.

27 Interview with John Beringer, London, July 1990.

28 Comments of Richard Clifton, *Impact*, p. 15.

29 HSE Guidance Note 3.

30 HSE Guidance Note 3, Interdepartmental Proposal Form, p. 1.

31 Royal Commission on Environmental Pollution, *GENHAZ* (London: HMSO, 1991).

32 Although GENHAZ looked like a further attempt to routinize biotechnology regulation, British industrialists were less than enthusiastic about this labour-intensive, cautiously empirical approach to safety assessment. As of the summer of 1993, it looked as though this Commission

proposal would probably remain on the drawing board except for isolated trial applications. The European Commission's efforts to streamline regulation, somewhat on the U.S. model, seemed likely to swamp any distinctively national assessment efforts.

33 Sheila Jasanoff, 'Technological Innovation in a Corporatist State: The Case of Biotechnology in the Federal Republic of Germany,' *Research Policy* 14 (1985), pp. 23–38.

34 See, generally, Herbert Gottweis, 'German Politics of Genetic Engineering and Its Deconstruction,' *Social Studies of Science* (in press).

35 Interview with Jens Katzek, German Bundestag, Bonn, July 1993.

36 Thus, in regulating chemical carcinogens, U.S. regulatory agencies have developed complex principles and mathematical models for extrapolating human risk estimates from animal data. British regulators have never adopted comparable analytical methods.

37 For an extended discussion of the properties of regulatory science, see Sheila Jasanoff, *The Fifth Branch: Science Advisers as Policymakers* (Cambridge, MA: Harvard University Press, 1990).

References

ACGM, ADVISORY COMMITTEE ON GENETIC MANIPULATION (1990).   *The intentional introduction of genetically manipulated organisms into the environment. Guidelines for risk assessment and for the notification of proposals for such work.* HSE Guidance Note 3 (revised), January 1990.

BROOKS, H. (1980).   Technology, evolution and purpose. *Daedalus* **109**, 66.

GOTTWEIS, H. (1995).   German politics of genetic engineering and its deconstruction. *Social Studies of Science* (in press).

JASANOFF, S. (1985).   Technological innovation in a corporatist state: the case of biotechnology in the Federal Republic of Germany. *Research Policy* **14**, 23–38.

JASANOFF, S. (1986).   *Risk management and political culture.* New York: Russell Sage Foundation.

JASANOFF, S. (1990).   *The fifth branch: science advisers as policymakers.* Cambridge, Mass: Harvard University Press.

KESSLER, D. A. *et al.* (1992).   The safety of foods. *Science* **256**, 1747–9 and 1832.

'IMPACT' (1990).   *The impact of new and impending regulations on UK biotechnology.* Report of a meeting sponsored by the Department of the Environment, the Health and Safety Executive and the Bioindustry Association. Cambridge: Cambridge Biomedical Consultants.

MERVES, J. (1990).   *Congress and administration closer to regulating U.S. biotech industry. The Scientist* **4**(22), 12 November, p. 12.

MILLER, H. I., BURRIS, R. H., VIDAVER, A. K. & WIVEL, N. A. (1990).   Risk-based oversight of experiments in the environment. *Science* **250**, 490.

NRC, NATIONAL RESEARCH COUNCIL (1989).   *Field testing genetically modified organisms – framework for decisions.* Washington, DC: National Academy Press.

OSTP, OFFICE OF SCIENCE AND TECHNOLOGY POLICY (1992).   Exercise of federal oversight within scope of statutory authority: planned introduction of biotechnology products into the environment. *Federal Register*, 27 February, pp. 6753–62.

RCEP, ROYAL COMMISSION ON ENVIRONMENTAL POLLUTION (1989).   *The release of genetically engineered organisms to the environment.* Thirteenth Report. London: HMSO.

RCEP, ROYAL COMMISSION ON ENVIRONMENTAL POLLUTION (1991). *GENHAZ*. London: HMSO.

WYNNE, B. & MAYER, S. (1993). How science fails the environment. *New Scientist*, 5 June, pp. 33–5.

PART IV
Comparisons of different technologies

16

Learning from Chernobyl for the fight against genetics? Stages and stimuli of German protest movements – a comparative synopsis

JOACHIM RADKAU

The assertion of some close affinity, or even 'striking analogy' (Harold Green: Radkau 1988a, p. 347), between atomic physics on the one hand and genetics on the other, already has a long history, starting even before Hiroshima and dating back to at least 1916. In that year, Hermann Joseph Muller, one of the founding fathers of modern genetics, pointed out this affinity in order to underline the immense potential of both disciplines (Roth 1985, p. 132). Hiroshima and the emergence of nuclear power made the close parallels even more exciting, but more ambiguous too. The deep conviction that this analogy really existed seems to have been a powerful driving force behind genetic engineering as well as behind the sharpest criticism of genetic engineering; plenty of evidence is to be found in the United States and also, later on, in West Germany. Erwin Chargaff, first a pioneer and afterwards one of the most prophetic critics of genetic engineering, exclaimed in 1977: 'The two greatest deeds – and probably misdeeds – in my time have been the splitting of the atom and the discovery of a way to manipulate the genetic apparatus. When Dr Hahn made his tragic discovery, he is reported to have exclaimed: "God cannot have wanted that!" Well, maybe it was the devil' (Radkau 1988a, p. 336). In America, first and foremost, an analogy was drawn between genetics and the atomic bomb. This analogy had its social background. The Union of Concerned Scientists and the Bulletin of the Atomic Scientists became an important public forum for criticism of genetics, as they had been for criticism of nuclear armament earlier on. The affinity of the two great discussions has influenced the terminology. Until today, the human genome project, which established big science in genetics, bears the nickname 'Manhattan Project of genetic engineering' (and sometimes the nickname 'Apollo Project' too). In Germany, the main emphasis lay in the analogy between genetics and civil nuclear energy. The widespread and passionate movement against genetic engineering would hardly have been imaginable without the conviction of this parallel.

What shall we think of this conviction? Generally speaking, analogy has a rather bad reputation in historiography. And with regard to such different types of technologies as nuclear fission and recombinant DNA, there seems to be good reason for a sound scepticism versus any rhetorical attempts at stressing the spectacular analogy. Certainly, the first duty of the historian consists of identifying the **individual** characteristics of technological systems. But after having done this task – and I did it for many years (Radkau 1983, 1987) – one may eventually discover that the analogy is not as artificial as it might appear at first sight. In order to check the relationship between nuclear technology and genetic engineering, let us first distinguish between different types and levels of analogy. Thereafter, I will give a comparative survey of the historical stages which one could distinguish during the development of the two great debates on new technologies. I will conclude with some remarks on the question of whether one can really learn by analogy, or whether analogy might have distorted the public approach to genetic engineering in Germany.

What kind of analogy?

1. From the time of H. G. Muller, an analogy has frequently been asserted at the level of the basic relation between science and nature. There have been numerous statements of the following type: just as atomic physics meant the scientific penetration into the basic elements of the inorganic world, so genetic engineering meant the penetration into the basic elements of organic nature, or to express it in a metaphorical manner, it meant handling the 'blueprints of life'. Indeed, both disciplines brought about a revolution in science. Therefore, one could ask: Is it not logical to conclude, that they would bring about a revolution in technology, too, or even a revolution in industry, moreover: a new period of history? In fact, many observers drew exactly that conclusion; in the 1950s they proclaimed the 'atomic age' (Radkau 1983, pp. 78–96), and later on they proclaimed the 'age of biotechnics'. However, from the view of history, these historical epochs do not really exist, and the deduction of an industrial revolution from a scientific revolution does not appear logical. Science is often mixed with technology (as in the name 'Science Museum'!); but the history of science is not at all identical with the history of technology nor of industry, not even in modern times. Nuclear energy again and again has been called a 'scientific technology'; but the assumption that nuclear power stations could be designed on the basis of scientific

research turned out to be a fundamental error (Radkau 1984). Recently, we often hear the assertion that modern society is shaped by science; but I do not find much factual evidence for this statement. The world remains chaotic; politics has not become scientific, neither has economics. Until today, there is also little evidence for a new industrial revolution effected by nuclear energy or by genetic engineering.

2. Of course, opponents of both technologies had their own kinds of analogy. They stressed the analogy of the risk produced exactly by the revolutionary character of these disciplines, namely by the penetration into the basic elements of nature. As Dorothy Nelkin pointed out, both technologies evoked 'archetypical fears' reaching back into pre-industrial times. She wonders that electronics, though causing a real revolution in our time and intervening deeply into every-day life, does not arouse similar fears. But in former decades, robots and other automatons used to play a disturbing role in science fiction. Nowadays, when the robots have become reality, they have lost much of their stimulating effect upon fantasy. Today, the old fears of human automatons have been transferred to the perspectives of genetic engineering.

Science meant safety for its adherents, but danger for its opponents. It was the scientific origin of the technologies themselves, their remoteness from everyday experience, which became an element of high risk in the eyes of many critics. While for the protagonists of these technologies, the scientific origin appeared as a guarantee of rational control, their opponents pointed out that there is no safe and perfectly controlled transition from the laboratory to the complex world (Bonss, Hohlfeld & Kollek 1992).

But also on the side of the critics, it is important to distinguish between different types of risk. Several critics, referring to nuclear energy, looked at Hiroshima or – later on – at Chernobyl, and, turning towards genetic engineering, they searched for a similar kind of danger. In time they seemed to become a little disappointed. But with regard to potential military applications of genetics, the analogy with Hiroshima was not wholly absurd. On the other hand, several critics argued that the most disturbing dangers are not the risks we know but those we do not know: the hypothetical risks or even the risks about which we have no hypothesis at all. Indeed, the most important analogy between both technologies might consist of the wider range of unknown risks. On the other hand, protagonists reply that it is useless to talk about risks we do not know: that it is perhaps the most critical focal point of the great controversies about the new technologies of our time (Radkau 1988a, p. 356). As Harvey Brooks and Rollin B. Johnson state, 'the biggest obstacle to consensus – both between scientists and the public and between scientists from different disciplines – is likely to continue to be the issue of where the burden of proof should lie in the face of inevitable scientific uncertainty. Should scientists be asked to prove the impossibility of various hypothetical scenarios? Or should those who express concern be obliged to demonstrate real though low risks?' (Brooks & Johnson 1991, p. 280). While in the controversy on genetic engineering, there seems today to be no compromise in sight, the German nuclear controversy was in this regard – at least in principle – already at an early stage one step ahead. As early as 1973 Wolf Häfele, the former head of the breeder project and an opinion leader within the German nuclear community, acknowledged that in the case of the high-risk technologies even hypothetical accidents had to be considered, and that in this regard nuclear energy could play a 'pathfinder role' (Häfele 1973). But in fact it was practically impossible to put all hypothetical risks into earnest consideration. The problem remains largely open to this day.

3. At least for the historian, the most striking analogy exists at the level of the **perception** of both technologies by the public: the perception of chances and risks, connected with the extraordinarily difficult character of the task of evaluating the pros and cons of these technologies. Perhaps we cannot decide today whether nuclear and genetic technology are similar with respect to their inherent character, but apparently they have been connected with similar hopes and fears; and these hopes and fears are part of their history. And one can say that this similarity was no mere coincidence, but was produced by the historical situation and exerted real influence upon the development of these technologies. The rise of both new technologies has been pushed by high ambitions and powerful interests; and it would be useless to discuss the perspectives of both technological developments without considering these motive forces. With regard to the factual historical situation, the protest movements were not without inner logic and consistency.

Stages of the nuclear controversy in West Germany: an attempt at periodization

Proto-history

The proto-history of the nuclear controversy, which lasted until the early 1970s, was characterized by isolated local protest, sometimes remarkably well informed about the risks of nuclear power, but without mass mobilization and without successful appeal to a wider public. Opinion polls reveal the surprising fact that during the 1950s, at the height of public optimism about the potential of nuclear energy, there already existed a silent majority filled with pessimism (Dube 1988), but this majority did not find public voice for a long time. The early opposition to the atom had in several cases a rather old-fashioned outlook; the students involved in the revolutions of 1968 had no idea of the dangers of civilian nuclear power. On the contrary, the old philosopher Ernst Bloch, friend of the famous student leader Rudi Dutschke, sharply blamed Western capitalism for not pushing the peaceful atom quickly enough (Radkau 1983, p. 81). Compared with the United States, opposition to civilian nuclear power did not rise immediately out of the protest movement against atomic armament and test explosions (Radkau 1983, pp. 434–8). On the contrary, the 'Göttinger Manifest' (1957) of German atomic physicists against nuclear armament of the Bundeswehr left the impression that the peaceful atom was a counterforce against the atomic bomb, and atomic physicists were highly esteemed by the German Left. The whole political constellation is very surprising from the viewpoint of the 1970s and 1980s. Early criticism of nuclear power in general was characterized by a fundamental disquiet about the noxious effect of radiation upon human inheritance. In post-Nazi Germany this was a rather conservative type of fear with even a smell of racism. The most remarkable work of that early stage, Friedrich Wagner's *Die Wissenschaft und die gefährdete Welt* (1964) – an astonishingly well-informed book for its time! – was alarmed by the rise of atomic physics and also by the rise of genetics, long before the development of DNA recombination technology. Fear of human mutations could be provoked by both nuclear and genetic technology. Theoretically, a combined criticism of both technologies could have grown out of these beginnings. But such opposition lacked any social basis within the forces of movement of that time.

Take-off

The great take-off of the mass protest happened in the years 1974/5. Within quite a short time, opposition in Germany to nuclear power grew stronger than anywhere else in the world. What was the reason? Or perhaps one should first ask: by which method, which paradigm, should the historian describe this deep and surprising change? Until today, nobody has succeeded in explaining this development by the usual means of social history. Surely, anti-nuclear protest was a middle-class movement; but this is a feature common to many other movements. Opposition to nuclear power was no social movement of the traditional type; but also the fashionable type of the 'new social movement' does not explain very much. Later on, the advocates of the atom frequently asserted that anti-nuclear protest had been stirred up by the mass media; but, as an inquiry ordered by the Federal Research Ministry had pointed out in 1974, of about 20,000 press articles on nuclear energy between 1970 and 1974, only 123 mentioned any criticism (Radkau 1983, p. 417).

At least in part, it seems to have been the rise of nuclear power itself – or, more precisely: the German path of nuclear development – which generated the wave of criticism. The German path was the adoption of the American light water reactor type, combined with the attempt to go into densely populated areas thereby neglecting the American distance rules (Radkau 1983, pp. 376–83); and it was the enforcement of nuclear energy in a land which lacked a military atomic complex and possessed of rich coal resources. One can point out that in Germany the discussion on nuclear safety spread to the public domain at exactly the moment when the immense hypothetical risk became apparent, and when the atomic industry and science proved themselves unable to discuss the risk dimension adequately. Therefore, I agree with Kristine Bruland (Chapter 6) that such controversies can be considered as a logical outcome and legitimate element of the rise of new technologies which need controversial public discussion. Without this assumption, the origin of the German protest movement against atomic power would remain more or less a mystery.

Action and militant escalation (1975–8)

From the perspective of the mass media, the real start of the anti-nuclear movement was the occupation of the Wyhl power station site in February 1975. From this time on, the protest against atomic power continually filled the main headlines of the press. The original motives of the Wyhl action had a rather traditional outlook, characterized by local animosity against the spread of industrialization into agrarian areas. This motive reached far back into the past; but the style of action – site occupation – appeared new and revolutionary and exerted its fascination upon the Left, who until that time had not been much concerned with nuclear energy. Apparently, the rapidly rising dynamic force of the anti-nuclear opposition originated partly from the growing influx of leftist intellectuals. From this stage onwards, the protest movement adapted the methods of the New Left: mass demonstrations, sit-ins and even more militant modes of action. Temporarily, the so-called 'K groups' (anarcho-communist and Maoist groups) took the lead – to the discomfort of other anti-nuclear groups. The broad alliance of protest against atomic power arose from unity of action, not from a commonly shared ideological basis, and it was endangered by a too militant style of action. Since the time of Wyhl, many leftists made strong efforts to justify their protest in Marxist terms; this was not easy because Marxism had until then usually been linked with faith in technological progress.

The new style of action was in some way a challenge for reflection on nuclear energy: the thinking had to become more precise in order to achieve the concrete targets needed for action. Now, the protest movement discovered the peculiar risks of the fast breeder and the reprocessing technology, justified until then by the harmless title of 'closing the fuel cycle'. The main enemies of the German protest movement of the 1970s were the state, the power industry and the capitalist system, not primarily science itself. In this regard, one can discern a sharp difference in comparison with the criticism of genetic engineering during the 1980s. During the 1970s, it seems that the traditionally high opinion of science by the German Left was still alive.

From today's perspective, it is a strange fact that the famous Asilomar conference, which started the public debate on genetic engineering in the United States and which took place exactly at the same time as the occupation of the Wyhl site (February 1975), did not arouse any interest in the anti-nuclear public in Germany. But these people (me included) were at that time totally occupied with discovering the mysteries of nuclear technology. They were on the way toward a **concrete** approach to the risks of new technologies; therefore they had little feeling for an **abstract** approach which would have been necessary for the transfer of concern from the atom to the genes.

The formative stage of the 'green' culture, 1978/9

After some violent and sometimes bloody clashes between protest groups and the police which had shocked many people, one could somehow feel a change in atmosphere. It can be characterized as a change from the terminology of fight towards a terminology of life, from a mere *contra* towards a *pro*: from the fight 'against nuclear power' to the agreement 'for a soft energy path'. 'Alternative' became the watchword of that time. The culminating point of this stage was the great movement against the reprocessing project of Gorleben, gathered under the slogan '*Gorleben soll leben*' ('Gorleben shall live'); the reprocessing plant was a controversial project even within the atomic industry, and criticism could depend on insider knowledge. Fighting this project, the anti-nuclear movement succeeded in leaving its former isolation. It gained an impressive political success when the Gorleben project was stopped in spring 1979 by the government of Lower Saxony, at exactly the time of the Harrisburg accident. This Gorleben decision, decreed even by a Christian Democratic government, was the first great defeat of the nuclear party in Germany. It was the peak, but – in retrospect – also the beginning of the decline of the 'classical' anti-nuclear movement in Germany. From 1979 onwards, the Federal parliament took over the controversy on nuclear energy by establishing the enquiry commission on 'future nuclear energy policy' while the anti-nuclear protest was increasingly absorbed by the new peace movement.

The changing atmosphere of the anti-nuclear protest perhaps can be illustrated by the following episode. At the day of penance in November 1977, the teacher Hartmut Gründler, speaker of an anti-nuclear group, burnt himself at the entrance of the Petri Church at Hamburg in order to protest against the Federal atomic policy (Rucht 1985, p. 246). But the name of Hartmut Gründler – in contrast to that of the East German Pfarrer Brüsewitz, who burnt himself in order to protest against Communist church policy – was soon forgotten by the German public; the memory of this terrible act of self-sacrifice was not cultivated by the anti-nuclear movement for a long time. In the period of 'Gorleben soll leben', most anti-nuclear people did not like heroes and martyrs, because they increasingly represented life and the soft path.

Anti-nuclear opposition in the context of the new peace movement (early 1980s)

Among the anti-nuclear activities of this time, the main action was aimed at stopping the new reprocessing project at Wackersdorf (Bavaria), the substitute for the former Gorleben project. During the 1970s the German protest movement had on the whole neglected the military dimension of atomic technology to a remarkable degree; but now, under the influence of the peace movement, which became the new driving force, the opposite tendency began to dominate, and the fighters against the Wackersdorf project tended to overestimate the presumed military motives of German atomic policy (Radkau 1988c). Mass demonstrations seem to have acquired more and more the character of ritual and symbolic actions. Chernobyl came as a terrible surprise even for many anti-nuclear people. Dorothy Nelkin and Michael Pollak wrote of the German scene in 1980: 'In fact the gap between ecologists and the organized left remains profound' (Nelkin & Pollak 1982, p. 197). Surely, one could get this impression from the outside. But since the late 1970s, a growing discontent with nuclear energy arose within the Social Democratic Party; only the strong rule of the nuclear hard liner Helmut Schmidt held back this increasing criticism from public appearance. However, when the Social Democratic Party was forced to leave the Federal government in autumn 1982, this party increasingly took an anti-nuclear position: the nuclear controversy became a political conflict along the traditional front line between right and left.

The time of Chernobyl (1986)

The first impact in Germany of the Chernobyl catastrophe was tremendous: the whole world seemed to have changed. Many people felt it was the deepest shock they had had since the Second World War, and in an initial apocalyptic mood, there rose a widespread feeling that a new era with a new awareness was emerging (Radkau 1987, pp. 329–31). The antagonism to nuclear energy became the dominant attitude in Germany – even among engineers! – and ceased to be the distinctive mark of an opposing culture: for the anti-nuclear movement, Chernobyl meant victory, but also loss of identity. Women played an outstanding role in the post-Chernobyl rallies. In the 1970s women had already became very numerous in the ranks of anti-nuclear protest; but now, in the age of the new feminist movement, their role became more visible in public. Not all of the old anti-nuclear fighters of the 1970s were completely happy with the post-Chernobyl form of protest: some called the new trend the 'Becquerel movement', grumbling that these anti-nuclear mothers were more interested in what to eat than how to fight. So widespread and vehement was the post-Chernobyl protest, yet so short-lived it turned out to be; it was no longer confronted with a real enemy. At that time, criticism of genetic engineering attracted many of the former anti-nuclear forces and received growing interest.

After all, one can conclude that the history of the German anti-nuclear movement was a success story, though many fighters felt themselves to be underdogs and helpless victims when they were confronted with the power of the police at the sites of nuclear power stations. What were the causes of this unexpected success in the end? Three points seem to be especially important, even more so as many anti-nuclear fighters did not fully realize their importance.

1. In the early 1970s there existed a real and urgent demand for critical public discussion on the risks of nuclear technology, considered even from the interests of the atomic industry, which itself was too little informed about the extent of the risks in its own business (Radkau 1983, pp. 421–2, 476–7). Even if one tries to explain the anti-nuclear movement partly by sociological methods, one should nevertheless not underestimate the stimulating force of the technological problems themselves, which were revealed gradually in the course of public debate. This process of revelation gave to the anti-nuclear opposition the charm and the intellectual excitement of detecting a hidden truth. In retrospect one could ask whether the positive and the negative fascination of the atom belonged together. Fascination often includes an element of fear; this element may become autonomous: in this way, faith in technology may create its own heresy. Until the 1970s, it was membership of the nuclear 'community' which gave intellectual excitement, self-consciousness and group identity; from then on it was instead the debunking of atomic energy which had all these attractions. It was the lasting achievement of the anti-nuclear movement that it invented a new type of collective identity – based on common fear and on the debunking of a technology with high symbolic value – a kind of group consciousness which proved to be fairly attractive. The movement against genetic engineering would be hardly possible without this psychological experience.

2. For a long time unperceived by the public, there existed a concealed convergence between the anti-nuclear movement on the one hand and the

interests of a section of the German power companies on the other. Until the late 1960s, German electricity producers were the least enthusiastic about nuclear energy, because West Germany had no urgent economic need for this new risky energy resource. Moreover, during the 1970s it turned out that nuclear energy was much more expensive than had been assumed in the 1960s. In particular, the fast breeder and the reprocessing project, which became the main targets of the opposition in about 1977, lacked any real economic foundation at that time. Therefore, at the peak of the nuclear controversy just those projects which constituted the 'magic' quality of nuclear energy – the promise of energy for millenia – were not defended in a really impressive and convincing way.

3. In general the protest movement did not influence the state, the bureaucracy and the jurisdiction very much; nevertheless, one can say that it succeeded in mobilizing the inherent braking force of bureaucracy. Already at a time when politicians were still paying lip service to nuclear energy, the bureaucracy combined with the administrative jurisdiction hindered the speed of nuclear development with interminable administrative regulations. While the public and the mass media mainly observed the scene of the great demonstrations and spectacular happenings, there existed another less spectacular scene which was at least equally important: the juridical scene occupied with a mass of local objections. Many complainants felt themselves powerless; but in comparison with the centralized French system, the more decentralized German system of licensing nuclear power plants gave some opportunities for the complainants. At least, the objections achieved the effect of retardation. If one does not take nuclear energy as being reasonable, one could conclude that the German administrative system had an inherent tacit rationality, whereas the pro-nuclear declarations of the government did not.

Stages of the discussion on genetic engineering

In the United States of the 1970s, the debate on recombinant DNA for some years formed a coherent story centred around the Asilomar conference of 1975: a story first presented by the molecular biologists themselves, who however lost control of the public debate for some years, but regained it during the late 1970s in a sort of happy ending (Watson & Tooze 1981; Krimsky 1982; Radkau 1988a). Today, in contrast, it seems to be an open-ended story. In West Germany, the controversy on genetic engineering started a decade later, with a dramatic appearance during the mid-1980s mainly in connection with the parliamentary enquiry commission 'chances and risks of genetic technology' (1984–7). As its chairman, the Social Democratic politician Wolf-Michael Catenhusen, remarked, this commission 'became the first parliamentary committee in the world to try to define and evaluate the opportunities and risks associated with genetic engineering and to make recommendations to Parliament for political action' (Catenhusen 1989, p. 117). West Germany's new leading role in the public discussion on genetics stems apparently from its long experience with the nuclear controversy. The whole debate had, whether intended or not, a distinct hereditary relationship to the preceding atomic conflict. Therefore, writing the history of the German genetic controversy up to the present time, one can distinguish three stages: the first, when discussion on genetics was overshadowed by the nuclear controversy; the second, when it was organized before the background of the nuclear debate; and the third, when the paradigm of the atomic conflict lost its efficiency.

Prehistory

Until the early 1980s, the critical German public seemed to have been too much absorbed by the nuclear question to become involved in the problems of genetic engineering, which in any case did not appear to be urgent as German industry on the whole still behaved hesitantly regarding biotechnology. Even critics sometimes wonder how slowly Germany went into genetic engineering. 'The reservedness of the West German chemical and pharmaceutical industry with regard to biotechnological research and development has a long history, without which one could not explain the delayed approach to genetic engineering' (Dolata 1992, p. 183). The strong German tradition in chemistry hampered its engagement in biotechnology, which rivalled traditional chemical methods. In contrast with the early period of civil nuclear energy, there never was an era of genetic enthusiasm in Germany, nor did the gene become an element of popular culture, as Dorothy Nelkin pointed out for the USA. But so long as genetic engineering had no enthusiastic adherents, it provoked no bitter enemies. As to the environmentalist movement, at first there was some uncertainty about whether one should blame modern biotechnology, or whether one should blame instead industrial chemistry for being too slow with it, as the path toward biotechnics had great advantages from an energy-saving point of view: this view was dominant among German environmentalists around 1980. The ZKBS (Central Commission on Biological Safety), founded in 1978 as an agency of supervision with regard to safety issues, was said to be a product of Asilomar, but was established at a time when in the United States the tide had already turned in favour of genetics, and its chief function was merely as an approval authority for the molecular biologist community (Radkau 1988a, p. 340). In 1979, the Federal government planned to enact a law on gene technology; but the combined lobby of industry and science already halted the first steps towards this act without any public intervention (Theisen 1991, pp. 43–4). A first hearing on 'chances and dangers' ('dangers', not 'risks'!) of genetic research, held by the federal Ministry for Research in September 1979 and published in 1980 (*Chancen und Gefahren* 1980), did not attract wide public attention.

Dramatic stage

From 1984 onwards, there was for some years a widespread feeling that genetic engineering was the new great technological issue, following the waning of nuclear energy. A deep split of opinion developed on the question of how to define the relation *vis-a-vis* the nuclear problem. On the part of industry and politics, there was a strong tendency to argue that, because Germany had suffered such a failure with the atom, the more important it would be to gain a striking success with the new biotechnics. Frequently one even gets the impression that industrial engagement in genetic engineering was based not so much upon clear-headed economic calculations as upon a kind of categorical attitude: the conviction that German industry urgently needed a great new innovative technology which would be able to substitute for the lost innovative force of the atom and the lost innovative force of traditional chemistry, too. On the opposite side, genetic engineering became the great target of protest in an historical situation, when the fight against nuclear power had lost a great deal of its former attractiveness. From an international point of view, it is interesting to note that '*Gentechnik*', not '*Biotechnik*', became the current term for the matter and entered also the title of the parliamentary commission: this definition stressed the peculiar character of genetic engineering *vis-à-vis* traditional biotechnology. Later on, German industry, getting on the defensive, tried to replace '*Gentechnik*' by '*Biotechnik*', but up to now, in Germany usually one talks about '*Gentechnik*'.

The parliamentary enquiry commission 'chances and risks of genetic technology' got its name in analogy with the risk-and-chance-pattern of the nuclear hearings. However, its chairman Catenhusen, formerly a critic of atomic policy, now started from the principle that genetics was a topic completely different from nuclear energy – rather comparable with electronics – and that in this case it would be impossible simply to say No. Instead of merely criticizing genetic engineering, Catenhusen argued, in this case there was a chance to direct technological development, because the political debate on biotechnics had started early in the formative stage of this technology, whereas the debate on nuclear energy had come too late (Theisen 1991, p. 66). However, the enquiry commission in the end proved hardly able to achieve that leading role, though it gained considerable public attention, even more than the former commission on future nuclear energy policy, which had made a greater attempt to correct the way things went.

Why did a sudden increase of public attention occur in about 1984? Around this time, the German federal government started to support genetic engineering on a large scale (Thurau 1990, p. 27). Similarly, as with nuclear energy during the early 1970s, by 1984 there existed a demand for public discussion on genetics, the more so as the question of deliberate release of manipulated organisms into the environment arose at that time, first in the United States (Wheale & McNally 1988, pp. 183–5; Krimsky 1991, pp. 133–51). But the origin of the German debate cannot be explained in every respect from the state of things in genetic engineering. Critical attention did not concentrate primarily upon the topic of release; more frequently at the centre of the debate stood the ethical problems of artificial insemination and of manipulating human embryos. Politics reacted to this situation: at the time when the parliamentary inquiry commission started to work, the ministries of research and of justice established the so-called 'Benda Commission' on 'in-vitro fertilization, genome analysis and genetic therapy'.

It is interesting to note some parallels between the American and the German developments in these years. In both countries, the moral, even religious accents grew strongly in the debate on genetics during the 1980s. In the United States, Rifkin with his quasi-religious rhetoric became – as Krimsky called him – the 'Czar' of the movement against genetic engineering. The German movement never acquired such a leading figure; but a New Age atmosphere became characteristic of at least part of the German Green movement during the 1980s. It is still an open and sometimes controversial question how important is the spiritual motive within the Green Party in general and within the protest against genetics in particular, and whether this spiritual motive is an element of power or an element of dissipation and diversion within the Green movement (Hesse & Wiebe 1988). In each case, spiritual motives are much stronger in the movement against genetics than in the former anti-nuclear movement. During the 1970s, the old leftist contempt of religion was still alive in the German left; in the 1980s, many leftists were even attracted by spiritual groups. In addition, the influence of the new feminism was much stronger in the 1980s than the 1970s; and, of course, the ethical problems of modern genetics posed a special challenge for feminist groups, in Germany as in the United States. From an industrial point of view, targets of this kind were fairly harmless, at least during the mid-1980s. At that time, there seemed to have been a consensus in German industry not to engage in experiments with human embryos; only a minority of scientists questioned this consensus. As for artificial insemination, the speakers for molecular biology pointed out – not without reason – that this was not identical with genetic engineering. The discussion of genetics did not lead immediately to the formation of a sharp front-line between 'pro' and 'con'. When Hoechst AG, a leading corporation of German chemistry which engaged itself in genetic engineering, celebrated its centenary in 1984, it invited the old ecology-orientated philosopher Hans Jonas, a German refugee like Erwin Chargaff, to give an address in which he warned against experiments with human inheritance; and even the Federal minister of research, Riesenhuber, welcomed this warning (Radkau 1988a, p. 342). By that time, a substantial economic interest of German chemistry in genetic engineering existed merely in the field of drug production. When during the following years there appeared a growing desire to expand genetic experiments to include human embryos, the main driving force seems to have originated in science, not in industry. For many critics of genetic engineering, science was the most dangerous opponent, even more than industry, and probably they were right. This front-line was different from that in the nuclear conflict.

A conference on the 'unclear danger potentials of genetic technology', held at Heidelberg in 1986 and sponsored by several environmental foundations, became a milestone for the public discussion of biotechnical risks in Germany; the problem of the 'Restrisiko', the immense uncontrolled risk – the most controversial issue of nuclear technology – now appeared in the discussion of genetic engineering, too. In 1987, the comprehensive report of the parliamentary inquiry commission appeared, which looked in large part like a manual of applied molecular biology. Its most remarkable advice was a five-year moratorium for free-land experiments. From today's view, German industry and science appeared at that time rather compromising, probably because they were afraid that the debate on genetic engineering would otherwise escalate like the conflict on nuclear energy. The report was widely discussed in German newspapers, particularly as it contained a detailed and well-founded minority statement by the Green members of the commission who advocated a strongly restrictive policy against genetic engineering. In retrospect, this discussion seems to have been the peak of public interest in genetic engineering in Germany. The last rise in public attention was in 1988/9, when the Federal Act on genetic technology passed through commissions and advisory boards (Ruhrmann 1992, p. 179). However, it is remarkable that *Der Spiegel*, since the late 1970s in the forefront against nuclear power, remained rather reserved during the controversy on genetic engineering. In 1990, the Act was passed by a Bundestag plenary with nearly empty benches.

From 1990 on to the present (1993)

Perhaps one should characterize this most recent period as the years of dissipation and of counter-offensive. To be sure, during these last years there has continued a widespread discomfort with genetic engineering; but at the same time one can observe a dissipation of criticism, a lack of a coherent protest movement and also a lack of concentration upon particular targets. Some critics of genetic engineering became uncertain whether it was correct to stress the analogies between nuclear energy and genetics. Molecular biology did not produce big power plants; in sharp contrast with the anti-nuclear movement, the critics of genetic technology had no plans for great demonstrations: a fatal handicap in the view of the demonstration culture of the German left!

Above all, after 1989 the whole German scene had changed. The German public was occupied with reunification; Western environmental problems were overshadowed by the terrible state of affairs in East Germany; the risks of genetic engineering seemed relatively harmless beside the destroyed landscape resulting from East German chemistry and brown coal. The Green Party, disturbed by reunification, searched for a new identity by advocating a liberal immigration and asylum policy; one can feel the atmospheric distance between the philosophy of natural barriers, based on opposition to genetic engineering, and the philosophy of open frontiers, based on the fight for a liberal asylum policy.

One may consider it as an historical paradox that exactly in this situation, industry and science are increasingly sounding the alarm asserting that genetic engineering in Germany is hampered, to a threatening extent, by the over-whelming power of the opposition. Protagonists of genetic research usually give the impression that nowhere in the world is resistance stronger than in Germany; but their colleagues in other countries used to make similar complaints against their own peoples. In recent times, German industry likes to blame its critics for the slow development of industrial genetics in Germany; but I believe that in the first instance industrial traditions, technological difficulties and economic uncertainty are responsible for the slow speed, and that industry is seeking culprits on the opposite side in order to turn attention from its own difficulties. When industry is planning genetic research departments abroad, it justifies this with domestic resistance; but one may doubt whether this resistance is so influential as to determine industrial investment strategies.

Probably, what has changed during the last years is the basic attitude of leading industrial corporations. Whereas during the late 1980s they widely accepted, at least to some extent, that genetic engineering is a matter of politics and of public control – as had always been accepted with regard to nuclear energy – in recent years a growing endeavour to fight any state intervention into genetic research has reappeared. In 1992, an '*Initiative Pro Gentechnik*', sponsored by the chemical industry presented itself to the public by an alarming advertisement campaign with the headline: '*Die Bürokratie droht die Gentechnik zu Tode zu verwalten. Die Folgen für Deutschland wären katastrophal.*' (This could perhaps be translated as: 'There is the imminent danger that bureaucracy kills genetic engineering by over-administration. The consequences for Germany would be catastrophic.') It is difficult to perceive a topical cause for this campaign. Ironically, it seems to have had a boomerang effect by stirring up anew critical discussion of genetic engineering.

Conclusion: what to do with analogy?

Sometimes, looking back at the nuclear controversy, one could be seduced by the philosophical thinking of Jürgen Habermas based upon the assumption that free public communication is able by itself and by its inherent logic to become a driving force leading to the victory of reason. Evidently the whole history of nuclear discussion indeed contained a kind of inherent logic; it cannot merely be reduced to group interests. Therefore, one should conclude that this controversy, which proved to be relatively successful in the end – though not only because of its own merits! – could become a useful paradigm for the debate on genetics, too. In fact, this debate has been pushed forward, at least for some years, by the paradigm of the nuclear conflict, and I suppose this pushing effect has on the whole been fruitful. There existed more similarities between these two problem areas than could be realized at first sight, so that the analogy could lead to some discoveries. Moreover, some questions which had been discussed only superficially in the course of the nuclear controversy were debated more intensely in the context of genetic engineering. This holds true especially for the question of whether 'nature' can be used as a criterion for the assessment of certain technological developments. Speakers on both sides of the genetic controversy referred to 'nature'; critics appealed to the 'wisdom of nature'; protagonists questioned that inherent wisdom (Radkau 1988b, p. 60). However, one may doubt whether the debate on 'nature' constitutes real progress towards clear thinking. The discussion may never end, and occasionally 'nature' appears as a confusing and misleading term: perhaps 'human health and happiness' is a better criterion than 'nature'! Does an inherent force of free public com-munication, in the sense of Habermas, really exist and lead to reasonable action? Public discussion on molecular biology seems to be more difficult than the public debate on nuclear energy; today, the nuclear question appears comparatively simple – but one should not forget: twenty years ago it appeared much more complicated! Nowadays, one can sometimes perceive the following dilemma: when discussion spreads to a wider public, it tends to focus on popular themes, but these themes are not necessarily the really important issues. For example, when discussion in Germany centred chiefly around ethics and the potential power to design human beings, I am not sure whether these old Frankenstein fears, repeatedly revived by science fiction, really concern the most important problems of genetic engineering.

Among German leftist and liberal intellectuals, considerable attention is usually paid to supposed Nazi or Nazi-like tendencies; therefore, when discussing biotechnics, they tend to make eugenics an important theme; but again I am not sure whether the most urgent problems of genetic engineering have much to do with eugenics. It has been often asserted that genetic research, by preventing the birth of handicapped babies, would lessen the tolerance of society to disabled persons. But this argument does not seem to be really logical from an historical point of view. On the contrary, one could argue that Nazi eugenics originated from the fear that the number of disabled persons would continue to increase in the future, and that therefore the abortion of handicapped embryos would contribute to reduce the whole problem for society.

In discussions on technology, we are frequently confronted with the problem of Snow's 'two cultures', with the culture of science and technology on the one hand and the culture of humanities and social sciences on the other. In Germany, the critical debate on technology is often dominated by the latter, as on the side of the scientific and technological communities there are few dissenters who risk making their views known to the public. This does not mean that there is total conformity of opinion within these communities; but public conflict usually produces public conformity on the side of the communities, at least with regard to the most controversial issues. Therefore, the division of 'pro' and 'con' tends to develop along the front-line of the two cultures.

One should acknowledge that in the course of the nuclear debate, the cultural gap has been bridged here and there, because many intellectuals educated in social sciences learned that a certain appreciation of technological detail is necessary in order to reach the core of the problem, though how deeply one had to go into technical matters remained an open question. The first and perhaps the most useful manual of arguments against nuclear power was published by a group of scientists (partly physicists) of the leftist Bremen University in 1975 (Bätjer et al. 1975); it took to pieces, point by point, a propaganda booklet of the atomic industry, which had already been a reply to critical questions. In this way, the nuclear controversy had from the beginning the charm of a ping-pong game, much more than the later dispute on genetic engineering. The Bremen book of 1975 became the pattern for a recent critical analysis of the Integrated Services Digital Network (ISDN): again, a brilliant manual of arguments which became respected even by its opponents (Kubizek & Berger 1990). The nuclear conflict produced patterns for the discussion of other new technologies. Not in every respect have these patterns been taken up until now by the genetic controversy.

In the case of genetic engineering, the number of critical German biologists remained very small; the core of the critics consisted mainly of a small group of young women without established positions in science, who were mostly forced to earn a living within the social sciences. So the discussion was partly dominated by intellectuals from the humanities who had great difficulties with the scientific details of molecular biology. In Germany, there is a long and prominent tradition of discussing 'die Technik' on a rather generalistic level of cultural philosophy without detailed knowledge of technical facts; sometimes there is a danger that public discussion of genetics moves into this type of cultural criticism – the more so, as many intellectuals consider a high level of philosophical generalization to be not a cul-de-sac but a progress of deep cognition!

Moreover, a critical approach to genetic engineering might learn from nuclear experiences especially with regard to the following three aspects.

1. At present, the topic of technological risk, which dominated the nuclear conflict, seems too much neglected in connection with genetic engineering. The very important question of whether a kind of 'biological containment' is possible in DNA recombination – a question raised by the American NIH guidelines of 1976 following the Asilomar recommendations (Wheale & McNally 1988, pp. 47–50) and a problem similar to the much-neglected question of 'inherent safety' in nuclear power stations (Radkau 1988d, pp. 111–12) – seems to have been too often postponed if not forgotten, though some aspects of the problem of 'biological containment' – raised anew by the German inquiry commission – remain exciting to this day (Bonss et al. 1992, pp. 160–2). One might be reminded of the state of the nuclear debate during the early 1980s when the issue of risk had become a little old-fashioned and when a rather vague topic like that of societal compatibility ('Sozialverträglichkeit') stood in the centre of discussion. Chernobyl demonstrated that the question of risk was in no way passé – surprising at that time even for some critics of the nuclear complex.

As to genetic engineering, probably many opponents were waiting during the early years of controversy for an accident comparable with Harrisburg or Chernobyl, or they suspected that with the advent of AIDS the biotechnical equivalent of Chernobyl had already happened. However, until now, the practical significance of several hypothetical risks of genetic engineering has not become clear. Perhaps it is the wrong way to look for a risk of the Chernobyl type – the search for an over-exact analogy may be misleading! – and perhaps it may be erroneous, too, to look for the kind of risk which is specific to genetic engineering alone. But one could learn from the nuclear experience that the question of risk deserves to be carefully observed over a long period, even if a spectacular accident will not happen in the near future. As to the tryptophan affair, making headlines in 1989, which caused the death of at least 27 people and illness in well over a thousand, scientists still believed in 1992 that 'it could take several more years to find an answer' to the question of 'whether genetic engineering played a role in the epidemic' (Swinbanks & Anderson 1992, p. 96). Risk consciousness needs a *'longue durée'* with regard to some new technologies. And a long memory is needed, too, in order to evaluate whether new technologies fulfil their promises.

2. One could learn from the nuclear conflict, that not only the risks, but also the **chances** deserve critical attention. Considering this, one may be surprised that the alleged manifold and immense chances of genetic engineering have not been thoroughly examined, even by many critics. Instead, opponents of genetic technology prefer to state that these chances themselves are the worst dangers for society because of their effects upon nature and society. Adherents and critics of genetic engineering frequently mix facts with future chances, which in reality are very different things, as every historian knows. In comparison with these gigantic perspectives, it is surprising how narrowly limited the real economic use of genetic engineering remains. Nearly everybody speaks of it as a new 'key technology', like electronics, but so far there is no real sign of it – not at all another industrial revolution, nor even an important new industrial structure (Teitelman 1989, p. 307). Most of the alleged chances of new biotechnics still exist only in the form of research programmes; the real industrial use is confined to certain limited areas of drug production, as was the state of affairs one decade ago. Even in world-wide drug production, products of genetic engineering do not account for more than a few per cent. It may be useful to remember that, at the end of the long nuclear controversy, it turned out that the real secret of nuclear technology was not – as many leftists had believed – the powerful capitalist interest, but on the contrary, its weak economic foundation. Could the same happen with genetic engineering? Certainly, today nobody knows; this open question could be the best basis for critical discussion (Radkau 1988a, pp. 351–4).

3. The history of anti-nuclear protest shows that a certain concentration of critical energy is useful for getting practical results. Criticism of genetic engineering has appeared during the last few years rather incoherent and diffuse. Certainly this has much to do with circumstances mentioned above. Presumably, it would be the best to accept those applications of genetics which do not necessarily need governmental regulation and to concentrate critical power upon the assumed high-risk sectors. It is true, admittedly, that one can construct high hypothetical risks everywhere, and that it is not easy to distinguish between those risks which have no practical significance, and those which have. Perhaps historical examples might give some orientation, and the experience with nuclear energy, the worst risk of which seemed to be merely hypothetical for a long time, could help to carry practical risk analysis a little further. After all, society itself has

to decide which hypothetical risks it will treat as practically important. In view of this, the best heritage from the nuclear conflict could be a fresh and confident feeling that society is free and able to choose between different technological paths, not compelled to be subdued under the authority of technological determinism.

References

BÄTJER, K. *et al.* (1975). *Zum richtigen Verständnis der Kernindustrie.* 66 Erwiderungen, Berlin: Oberbaumverlag.

BONSS W., HOHLFELD, R. & KOLLEK, R. (1992). *Zur Unsicherheit in der Gentechnologie, Technik und Gesellschaft,* Jahrbuch 6, pp. 141–74. Frankfurt: Campus.

BROOKS, H. & JOHNSON, R. B. (1991). Comments: public policy issues. In *The genetic revolution: scientific prospects and public perceptions,* ed. B. D. Davis, pp. 266–81. Baltimore: Johns Hopkins University Press.

CATENHUSEN, W.-M. (1989). Public Debate on Biotechnology: The Experience of the Bundestag Commission of Inquiry on the Opportunities and Risks of Genetic Engineering. In *Biotechnology in future society,* ed. E. Yoxen and V. di Martino, p. 117. Luxembourg: Aldershot.

CHANCEN UND GEFAHREN DER GENFORSCHUNG (1980). *Protokolle und Materialien zur Anhörung des Bundesministers für Forschung und Technologie.* Munich: Oldenbourg.

DOLATA, U. (1992). *Weltmarktorientierte Modernisierung. Die ökonomische Regulierung des wissenschaftlich–technischen Umbruchs in der Bundesrepublik.* Frankfurt: Campus.

DUBE, N. (1988). *Die öffentliche Meinung zur Kernenergie in der Bundesrepublik Deutschland 1955–1986. Eine Dokumentation.* Berlin: Wissenschaftszentrum Berlin für Sozialforschung.

HÄFELE, W. (1973). *Hypotheticality and the new challenges: the pathfinder role of nuclear energy.* Laxenburg: IIASA Research Report.

HESSE, G. & WIEBE, H.-H. (1988). *Die Grünen und die Religion.* Frankfurt: Athenäum.

KOLLEK, R., TAPPESER, B. & ALTNER, G. (ed.) (1986). *Die ungeklärten, Gefahrenpotentiale der Gentechnologie.* Munich: J. Schweitzer Verlag.

KRIMSKY, Sh. (1982). *Genetic alchemy: the social history of the recombinant DNA controversy.* Cambridge, Mass: MIT Press.

KRIMSKY, Sh. (1991). *Biotechnics and society: the rise of industrial genetics.* New York: Praeger.

KUBICEK, H. & BERGER, P. (1990). *Was bringt uns die Telekommunikation? ISDN – 66 kritische Antworten.* Frankfurt: Campus.

NELKIN, D. & POLLAK, M. (1982). *The atom besieged. Antinuclear movements in France and Germany.* Cambridge, Mass: MIT Press.

RADKAU, J. (1983). *Aufstieg und Krise der deutschen Atomwirtschaft 1945–1975. Verdrängte Alternativen in der Kerntechnik und der Ursprung der nuklearen Kontroverse.* Reinbek: Rowohlt.

RADKAU, J. (1984). Kerntechnik: Grenzen von Theorie und Erfahrung. *Spektrum der Wissenschaft* **12**, 74–90.

RADKAU, J. (1987). Die Kernkraft-Kontroverse im Spiegel der Literatur. Phasen und

Dimensionen einer neuen Aufklärung. In *Das Ende des Atomzeitalters?* ed. A. Hermann and R. Schumacher, pp. 307–34. Munich: Moos.

RADKAU, J. (1988a). Hiroshima und Asilomar. Die Inszenierung des Diskurses über die Gentechnik vor dem Hintergrund der Kernenergie-Kontroverse. *Geschichte und Gesellschaft* **14**, 329–63.

RADKAU, J. (1988b). Gentechnik: Die ungeklärten Risiken beim Übersehreiten einer naturgeschichtlichen Schwelle. *Geschichte lernen*, Heft 4, 55–61.

RADKAU, J. (1988c). Der Nebel wurde jäh zerrissen. *Der Spiegel*, No. 6, 95–104.

RADKAU, J. (1988d). Sicherheitsphilosophien in der Geschichte der bundesdeutschen Atomwirtschaft. *S + F, Vierteljahrsschrift für Sicherheit und Frieden*, Heft 3, 110–16.

ROTH, K.-H. (1985). Sozialer Fortschritt durch Menschenzüchtung? Der Genetiker und Eugeniker H. J. Muller (1890–1967). In *Gen-Technologie: Die neue soziale Waffe*, ed. F. Hansen and R. Kollek, pp. 120–51. Hamburg: Konkret Literatur Verlag.

RUCHT, D. (1985). *Von Wyhl nach Gorleben. Bürger gegen Atomprogramm und nukleare Entsorgung*. Munich: C. H. Beck.

RUHRMANN, G. (1992). Genetic engineering in the press: a review of research and results of a content analysis. In *Biotechnology in public*, ed. J. Durant, pp. 169–89. London: Science Museum.

SWINBANKS, D. & ANDERSON, C. (1992). Search for contaminant in EMS outbreak goes slowly. *Nature* **358**, 96.

TEITELMAN, R. (1989). *Gene dreams*. Wall Street, Academia, and the Rise of Biotechnology: New York.

THEISEN, H. (1991). *Bio- und Gentechnologie: Eine politische Herausforderung*. Stuttgart: Kohlhammer.

THURAU, M. (1990). *Gute Argumente: Gentechnologie?* Munich: C. H. Beck.

WAGNER, F. (1964). *Die Wissenschaft und die gefährdete Welt. Eine Wissenssoziologie der Atomphysik*. Munich: C. H. Beck.

WATSON, J. D. & TOOZE, J. (1981). *The DNA story. A documentary history of gene cloning*. San Francisco: W. H. Freeman.

WHEALE, P. & MCNALLY, R. (1988). *Genetic engineering: catastrophe or Utopia?* Hemel Hempstead: Harvester Wheatsheaf.

17

Individual and institutional impacts upon press coverage of sciences: the case of nuclear power and genetic engineering in Germany

HANS MATHIAS KEPPLINGER

Basic circumstances of new technologies

Criticism of technology is nothing new. For some years, however, this criticism has had a new quality. A brief look back demonstrates this. The development of road, rail and air traffic did not trigger any fundamental political debates, but rather occurred within the framework of administrative guidelines. Mass vaccination for numerous diseases, chlorination of drinking water and pasteurization of milk were introduced in a similar way, without any significant public debate concerning the advantages and disadvantages, which might have endangered these projects. In contrast, plans for the fluoridation of drinking water in the USA in the 1960s provoked intense public controversy, which already contained important elements of later disputes concerning nuclear power. This is true both for the structure of the conflict – the appearance of citizens' action groups who mustered their own experts, the turning of their actions into a current topic by the mass media, and the shifting of decisions to political institutions – and also for the type of arguments – the assertion that there was an invisible threat, that the whole population was in danger, and that there was a possibility of unrecognized long-term effects, etc. (Sapolsky 1968).

One fundamental reason for the changes mentioned may be the existence of social groups who intervene in politics in an unconventional manner and who find a platform for their criticism and demands in the mass media. The growth of these groups and the new role of the mass media can be traced back to changes in society. I would first like to present this in theory and then give empirical evidence using the discussions concerning nuclear power and genetic engineering.

Structures of society

Post-industrial society

According to their stage of development, we can distinguish between pre-industrial, industrial and post-industrial societies. Post-industrial societies are characterized by:

1. the transition from a goods-producing society to a service-orientated society,
2. the pre-eminence of a class of professionalized and technically-orientated professions,
3. the central role of theoretical knowledge as the source of innovations and the starting point for political programmes,
4. the control of technical progress and the evaluation of technology, as well as
5. the analysis of the interactions between various subsystems of society (economy, science, technology, culture, etc.).

The path to the post-industrial societies is characterized by contrary processes: on one hand, the diversification of society is growing and with it the specialization of its individual sections; on the other hand, mutual dependence is increasing, coupled with the demand for mutual exertion of influence and control (Bell 1973).

The USA was the first society which – in the 1960s and the 1970s – began the transition to a post-industrial society. Thus it was there that the growing dependencies of the various social subsystems were first seen. This became particularly clear in the discussion regarding the peaceful use of nuclear power.

Counter-elites

In developed industrial societies, the proportion of those among the total work-force having intellectual occupations is increasing. This is also true for the occupations that pass on the values and norms of societies. Included here are, above all, university professors, teachers, ministers, journalists, writers, etc. Like Helmut Schelsky (1975), we may characterize people in these occupations as 'reflective elites' or 'producers of meaning'. Conflicts of interest arise between the producers of meaning and the producers of goods. People belonging to the reflective elites, for example, urge for an expansion of the education sector in the widest sense, since this improves their basis of livelihood. In the end, the expansion has to be financed by the producers of goods by means of taxes and duties. Although acting in their own interests, as already mentioned, the reflective elites, for justification, make use of general values such as the right to education, the importance of which can hardly be denied.

The possibilities of influence for the reflective elites are based in particular on criticism of existing institutions, on their role as disruptive factors. Joseph A. Schumpeter (1950, pp. 235–51) already referred to these fundamental facts in his sociology of intellectuals. Similar approaches are found in C. P. Snow's two culture thesis (1967), which especially emphasises the contrast of scientific and technical intelligence on the one hand and literary intelligence on the other. In the discussion concerning nuclear power, the contrast is manifested by the growing importance of social scientists, who have become the critics' spokesmen. As a consequence, today two elites confront each other, basing their standpoints on differing assumptions: scientists and engineers on one side and sociologists and social philosophers on the other side (Short 1984).

Mentalities

The growing affluence in the industrial societies, together with the expansion in education and leisure time, has freed ever-increasing sections of the population from serious worries of preserving their livelihood. Because of this, people have become free to take an interest in what is happening outside the narrow sphere of their own experience. An increasing number of people are taking this sort of interest in an increasing number of subjects, of which they have no knowledge from their own direct experience. At the same time the feeling is growing that they must join in the discussion and decisions on such subjects.

Owing to the change outlined, in the population the sounding board is growing for the discussion of topics which previously attracted no attention even if, regarded objectively, they were considerably more urgent (Douglas & Wildavsky 1982). Examples of this in the Federal Republic of Germany are the number of motor vehicle fatalities, the contamination of rivers and the pollution of the air. In all these cases the interest and the concern increased, while the damage decreased (Kepplinger 1989). There were similar discrepancies in the USA (Mazur 1981).

For the reasons mentioned, the reduction of the negative side effects of technology – or the risks of technology – does not necessarily lead to a decrease in fears and concern. Instead, even small dangers become the occasion for great concern due to increased interest in remote events and potential incidents. It can therefore hardly be supposed that an increase in safety of nuclear power plants or genetic engineering automatically increases acceptance. Making increased safety a theme of topical interest would presumably rather add to the concern than reduce it, because it brings facts found to be threatening into people's consciousness without the population being able to understand the arguments.

Evidence for this supposition is given by experiments concerning the acceptance of the fluoridation of drinking water: the acceptance dropped due solely to the subject being made a theme of popular interest. This was still the case even if the arguments in favour of fluoridation were presented in a suitable way (Mueller 1968).

Roles of the professions

Members of the professions may be characterized by the following features: they use

1. specialist knowledge, which is
2. based on a theoretical foundation, that
3. was obtained during systematic training, the mastery of which
4. was examined in a special test, which
5. thus regulates entry to the profession. Members of the professions
6. may make use of a professional organization; they are
7. committed to their professional ethic, possess
8. a great personal responsibility and on the basis of their training and professional position
9. have relative autonomy in the sense of a freedom from control by lay people. Further, members of the professions carry out their work
10. in support of generally recognized social values (Kepplinger & Vohl 1976).

The most important aspect in the present context is the relative autonomy in the sense of independence from control by lay people. Its foundation is the peer group orientation of those belonging to the professions: the criterion of specialist qualification is exclusively the judgement by colleagues. Examples are the reviewing procedures in the funding of scientific projects, concerning publication in scholarly journals and appointment of scientists. The justification for these procedures is based on the specialist qualifications of the members of the professions, which were obtained by means of a systematic training and documented by means of special records of achievement. The crucial consequence of peer group orientation is the collective rejection of lay criticism, which is regarded as uninformed and unqualified and thus as possessing no professional relevance. This fundamental attitude is reflected in the claim by members of the professions that they have the right to make all significant professional decisions themselves. An example of this is the view of the medical profession concerning the choice of a therapy (Faden *et al.* 1981).

During the last two or three decades the basis of the professions' authority has changed. Previously, the basis of their authority was cutting themselves off from the outside world – pure orientation to colleagues, no discussion of professional problems with lay people, no justification of professional decisions to the general public, etc. Thus the limits and insufficiencies of acquiring knowledge and of the possibilities for action remained to a great extent unknown outside the professions. This lack of knowledge was an essential reason for lay people's trust in the competence of professionals. Spectacular errors, such as the approval of thalidomide, shattered this trust. Outsiders from various scientific disciplines have emerged and been established as counter-experts whose views are shared by many journalists. The development of investigative journalism has considerably reduced the possibility of covering up errors and keeping the public in ignorance. Thus cutting themselves off from the outside world can, for ethical and practical reasons, no longer be the basis of authority for the professions.

The changed basis of the professions' authority compels them to communicate with the public. This requires a new orientation: in addition to the colleagues in the profession, the lay people also have to accept the professionals' arguments. This means that the arguments have to take the lay people's expectations into consideration. Thus, among other things, the arguments have to take the interests and concerns of the lay people seriously – even if in some cases they seem to be unfounded from the professional viewpoint – and arguments are to be presented in language which the lay people can understand (Maier-Leibnitz 1986; Noelle-Neumann & Maier-Leibnitz 1987). Many scientists, probably the majority, have not yet learned that lesson and are not able to communicate effectively to lay people.

The role of the mass media

Journalists' conception of their role

In the last 30 years journalists' conception of their role has altered drastically. Until the mid-1960s they saw themselves above all as passive observers, whose news coverage remained as neutral as possible. Since then, they have seen themselves increasingly in the role of involved critics, who with their reporting actively intervene in what is happening. Modifying a quote by Karl Marx, they no longer wish only to portray society, they want to change it. The possibility for this is given by the journalists' influence on the definition of reality – the idea of what is important and right in a society. The change in the journalists' image of themselves was accompanied by a reinterpretation of the concept of criticism, which has its origin in the 'Critical Theory' of the Frankfurt school of thought. Originally, a journalist was regarded as critical if he or she published a report only when its correctness had been checked. Checking also required an answer to the question of whether criticism by others, which was to be reported on, was justified. Today, a journalist is regarded as critical if he or she actively criticizes others or respectively, reports critically on their criticism.

Examples of this development are to be found in the Swedish and German press. From the beginning of the century until the mid-1960s, about 7% of all contributions in the Swedish press contained references to criticism of persons, organizations or conditions. This changed dramatically in the second half of the 1960s. Within a short time, the share of contributions containing criticism rose by a quarter to nearly 30% (Westerstahl & Johansson 1986). In the German press the share of contributions containing criticism of technology rose between 1965 and 1986 from 15–20% to 30–50% (Kepplinger 1989, p. 143). In the news programmes of one German radio station (HR), the share of reports on negative events (economic crises, wars, unrest, accidents, etc.) rose from 27% to 41% between 1955 and 1985 (Kepplinger & Weissbecker 1991).

The change in the journalists' understanding of their role had four consequences in particular. First, criticism of persons, institutions and circumstances received more publicity. Secondly, the intensified publicity increased the readiness for public criticism. Thirdly, it became more likely that negative events, among them the unintentional results of science and technology, received news coverage. Fourthly, the subjective points of view of journalists increasingly shaped the contents of news coverage. This takes place, among other things, through 'instrumental actualization', that is, the preferred news coverage of events and opinions which confirm journalists' own views (Kepplinger, Brosius & Staab 1991a; Kepplinger 1992, pp. 201ff).

Organization of editorial departments

Newspaper and magazine staffs are usually divided into different departments. The individual departments are responsible for producing the different sections of the publications. A similar structure can be found in radio and television broadcasting stations. The individual departments are responsible for the presentation of certain subject areas, the economic department for economics, the scientific department for science, and so on. One department, the politics department, spreads into all other subject areas. The reason for this is that all topics can become politically relevant and then come into the sphere of this department's responsibility. The politics department has considerably more employees and a considerably greater influence with the editors than all other departments. In addition, the politics department has considerably more space at its disposal for news and commentaries.

When a theme becomes politically relevant, this has repercussions concerning the type and extent of the news coverage: The writers and editors in the politics department claim a monopoly on the news coverage. As a result, newspapers and magazines report on politically relevant aspects of technology mainly in their political sections. Owing to its great volume, news coverage in the political section has a formative influence on the overall news coverage.

The writers and editors in the various departments are – at least in the quality publications – specialists in certain areas. A high proportion of the economic editors, the scientific editors and the features editors have studied economics, sciences or the arts respectively. Others have acquired specialist knowledge in previous occupations or during their journalistic careers. Political journalists have had their academic education mostly in the humanities. Therefore, the two cultures mentioned above can be discerned in the differences between the academic background of writers in different editorial departments – the science departments on one hand and the political departments on the other.

The shift of news coverage, e.g. of science and technology, from the special departments to the politics department is, for the reasons mentioned, presumably connected with a loss of specialist competence and a change in perspectives: the writers in the science and technology departments present topics from their special areas with more expert knowledge and thus make judgements rather from the point of view of scientists and engineers than their colleagues from the politics department. In particular, they will quote recognized experts from the relevant professional disciplines. In contrast, their colleagues in the politics department will more frequently let counter-scientists have a say, whose opinions reflect their own opinions or those of the editorial line. Political angles thus displace scientific and technical aspects in the presentation of controversial topics. In the following, these themes will be discussed, using as examples the coverage of genetic engineering and nuclear power in the German press.

The case of genetic engineering

One of the basic questions is whether the views held by scientists are shared by science writers, and whether their views shape the coverage of the leading newspapers and magazines. To identify the views of the leading researchers in genetic engineering, we have interviewed 30 top ranking scientists of various universities, Max Planck institutes and research departments of companies in Germany. In addition, we have interviewed 30 science journalists who were among the most respected in their profession. The relevance of the answers is based on the expertise of the scientists and science journalists interviewed, not on their number. In addition, we have interviewed 30 editors responsible for the political sections of various regional and national newspapers. They are not representative for all journalists of that type but their answers may give some information about their opinions on genetic engineering. The scientists as well as

Table 17.1. *Opinions of journalists and scientists concerning the risk of genetic engineering*

Question: 'Various risks are being discussed in connection with genetic engineering. Please go through the following list: which of these risks are, in your estimation, to be taken very seriously, quite seriously, less seriously, or can be ignored completely?'
N, Number of people questioned who were of the opinion that a risk was 'to be taken very seriously' or was 'to be taken quite seriously'.

	Scientists (N = 30)	Scientific journalists (N = 30)	Political journalists (N = 30)
'Discrimination of individuals with unusual genetic features on the job market.'	15	23	18
'Decreasing willingness to accept the life of handicapped people.'	12	21	19
'Military use of organisms altered by genetic engineering.'	12	20	26
'No clear-cut transitions from the correction of defective genes to optimization of human hereditary factors.'	11	19	26
'State-enforced abortion of unborn babies with genetic defects.'	11	11	8
'Uncontrolled spread of manipulated organisms in outdoor experiments.'	10	22	27
'Increased use of chemical pesticides as a result of the development of herbicide-resistant plants.'	10	19	20
'Pressure for rationalization and accelerated structural change in agriculture.'	10	18	16
'State-enforced information on genetically-caused risks to one's own health.'	10	16	18
'Impairment of health in laboratory staff through contacts with new combinations of organisms'	5	18	21
'Unintended creation of new pathogenic organisms from originally non-pathogenic material.'	3	23	28
'Systematic "breeding of humans" by private or state organizations.'	3	13	16
'Environmental pollution due to accidents due to new combinations of organisms.'	2	18	27
Total	114	241	270

Sources: Kepplinger *et al.* (1991b); Kepplinger & Ehmig (1994).

Table 17.2. *Presumed causes of shortcomings in press coverage*

Question: 'The mass media's representation of science and technology is often criticized as being inaccurate. Regarding this I would like to read you an opinion. Please tell me whether you think that this opinion is correct, partly correct or incorrect.

The opinion is as follows: The shortcomings in the representation of science and technology are not based so much on the fact that the scientific journalists have had poor training and are badly informed. It is due much more to the fact that their articles hardly get a chance in day-to-day press coverage, because the political editors take every topic of general interest out of their hands.'

	Scientists N	Scientific journalists N	Political journalists N
'I agree completely with this opinion.'	6	4	1
'I agree in part with this opinion.'	12	14	11
'I do not agree with this opinion.'	8	8	17
'I don't know.'	1	—	—
No definite answer	3	4	1
Total	30	30	30

Sources: Kepplinger *et al.* (1991b); Kepplinger & Ehmig (1994).

the journalists were interviewed personally using a structured questionnaire (Ahlheim 1991; Kepplinger, Ehmig & Ahlheim 1991b).

A considerable number of journalists had a negative attitude toward genetic engineering (Kepplinger *et al.* 1991b, pp. 28–30). This is especially true for political editors, but is also the case for a section of the scientific journalists. The journalists' attitudes were evident both in their general opinions concerning genetic engineering as well as in their views concerning its risks (Table 17.1).

Scientific and political journalists had different opinions about the causes of the mass media's portrayal of science and technology. The majority of the science journalists agreed that shortcomings and errors were 'due to the fact that their [own] articles hardly get a chance in day-to-day press coverage'. The majority of the scientists shared this belief whereas half of the political journalists rejected it. This can be seen as an indicator of a latent conflict between science writers and political journalists, with the latter being more influential because of their number and of their position in the news rooms (Table 17.2).

Table 17.3. *Authors of the articles on genetic engineering*

	National daily newspapers (N = 302)%	Regional daily newspapers (N = 50)%	Weekly newspapers (N = 41)%	Magazines (N = 17)%	Popular science magazines (N = 114)%	Total (N = 524)%
Scientific journalists	20	12	22	6	13	17
Other journalists	20	14	7	6	3	15
News agencies	18	16	—	—	—	12
Other identifiable sources	2	2	4	—	5	3
Non-identifiable sources	38	28	66	29	72	47
No information	3	28	—	59	7	8
Total	101	100	99	100	100	102

Sources: Kepplinger *et al.* (1991b); Kepplinger & Ehmig (1994).

To analyse the impact of the different views upon the coverage of genetic engineering, we have carried out a quantitative content analysis of six national daily newspapers, four regional daily newspapers, two weekly papers, two weekly magazines and four popular science magazines. As a rule, the basis of the analysis is a random sample from one-sixth of all editions in the years 1987–9. The popular science magazines are an exception – from these we took account of half of all editions. In the editions investigated, the eighteen newspapers and magazines published a total of 524 articles with 4494 evaluative statements on genetic engineering. Projected to all editions in the period of investigation, this corresponds to 2688 articles with 22,024 statements (Kepplinger *et al.* 1991b).

Of all the articles analysed, only 17% were written by authors who were clearly proved to be scientific journalists or who were identifiable as such. The source of a relatively large number of articles could not be identified. Even if one supposes that half of these articles were written by scientific journalists – which is very improbable – it can be established that the image of genetic engineering in the press was determined in particular by journalists who have no specific expert knowledge. The role played by the news agencies was remarkable, especially for the news coverage in the daily newspapers (Table 17.3).

Contrary to a widespread assumption, most statements on genetic engineering did not appear in the science sections of the papers but in their political sections. The science sections took only second place. In the political sections, genetic engineering was characterized rather negatively, in the science sections definitely positively. It should be remembered that the political sections are given attention by many more readers than the science sections (Table 17.4).

In the science sections of the papers, members of organizations developing and applying genetic engineering were especially cited or referred to. However, in the political sections of the papers, it was primarily the members of organizations criticizing genetic engineering who had a say. Those who developed and made use of genetic engineering did not have a sufficient chance to express their points of view in the places where most of the coverage on genetic engineering appeared and where most of the readers informed themselves (Table 17.5).

The journalists' opinions about genetic engineering – as indicated by their own evaluative statements – and the experts' opinions quoted were not independent from each other. Instead, the journalists of the various media put into the spotlight experts who conformed to their own views: the more positive their opinions were, the more often they cited or referred to experts with similarly positive views and vice versa. That is what we call instrumental actualization: the journalists used the available opinions of experts to underline their own points of view. Thus, some papers (*Natur* – which is not identical with the English language magazine *Nature*, *Tageszeitung*, *Rheinzeitung*) created the false impression that – in

Table 17.4. *Statements on genetic engineering in different sections of newspapers and magazines*[a]

	Number of statements (N = 3259) %	Tendency of statements (N = 2867)[b] $\bar{x}$
Political sections	38	−0·14
Business sections	6	+0·32
Feature sections	4	−0·66
Local sections	5	−0·26
Science sections	27	+0·59
Other sections	19	−0·01
Total	99	+0·09

[a] 6 national dailies, 4 regional dailies, 2 weekly papers, 2 magazines.
[b] Means (± 3). Without statements on general political, economic and legal framework of genetic engineering.
Sources: Kepplinger *et al.* (1991b); Kepplinger & Ehmig (1994).

line with the journalists' opinions – most experts would criticize genetic engineering (Figure 17.1).

The case of nuclear energy

To identify long-term trends in the coverage of technology, we have analysed the coverage of nuclear energy, which has been a topic of the mass media for several decades. Assuming most of the coverage to appear in the political sections, we have concentrated on these sections of four nationally distributed dailies (*Frankfurter Allgemeine Zeitung, Süddeutsche Zeitung, Die Welt, Frankfurter Rundschau*) and three weeklies (*Die Zeit, Der Spiegel, Stern*). We have chosen these papers because they are held in high regard by German journalists and function as a frame of their coverage. In addition, the four dailies represent the political spectrum of the press in Germany. Therefore, their coverage can be regarded as an indicator of the coverage of most of the mass media in this country. The study is based on a representative sample including 13 editions per paper and per year. Thus 286 editions per paper or a total of 2002 editions were analysed. We have analysed the first four pages of the political sections of the newspapers and the whole political sections of the magazines (Kepplinger 1988, 1989, 1992).

From 1965 to 1986, the papers published 6046 evaluative statements on

Table 17.5. *Statements by various authors in the political and scientific sections of newspapers and magazines*[a]

	Political sections		Science sections	
	Number of statements ($N = 329$) %	Tendency of statements[b] ($N = 293$) $\bar{x}$	Number of statements ($N = 212$) %	Tendency of statements[b] ($N = 191$) $\bar{x}$
Academic research institutions, industry	43	+0·90	87	+0·56
Alternative research institutes, interest groups, churches, the Greens	57	−1·64	13	−1·86
Total	100	−0·52	100	+0·30

[a] 6 national dailies, 4 regional dailies, 2 weekly papers, 2 magazines.
[b] Means (± 3). Without statements on general political, economic and legal framework of genetic engineering.
Sources: Kepplinger *et al.* (1991b); Kepplinger & Ehmig (1994).

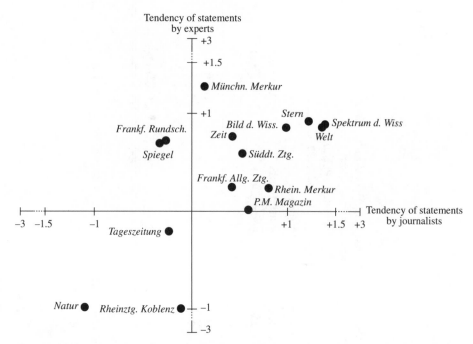

Figure 17.1. Instrumental actualization of experts. Basis: statements by journalists and scientists on genetic engineering. Mean values (± 3). Sources: Kepplinger *et al.* 1991; Kepplinger & Ehmig 1995).

nuclear energy. As evaluative statements we have regarded all explicit assessments of nuclear energy as well as any information on positive or negative events related to nuclear energy – independently of its correctness. In the late 1960s, the papers clearly presented nuclear energy in a positive light. By 1969, however, there were already indications of a negative turnabout. As early as 1972 the tenor of reporting was mainly negative. The following year it was definitely positive again, which can be seen as a reaction toward the oil crisis of 1973. Beginning in 1974, the papers almost continually presented nuclear energy in a negative light. Thus, the re-evaluation of nuclear energy in the papers investigated had taken place a long time before the reactor accidents in Harrisburg and Chernobyl. However, it was not noticed because the papers devoted only a small space to nuclear energy. Once the tenor of portrayal was negative overall, the intensity of coverage increased, creating the mistaken impression – especially in the late 1970s – that nuclear energy had suddenly been re-evaluated. The tenor and the intensity of reporting on nuclear energy over time suggest that the accidents in Harrisburg and in Chernobyl as well as a number of less important incidents, rather than resulting in a re-evaluation of nuclear energy, were taken as an occasion to intensify reporting based on prior re-evaluation. At least on a

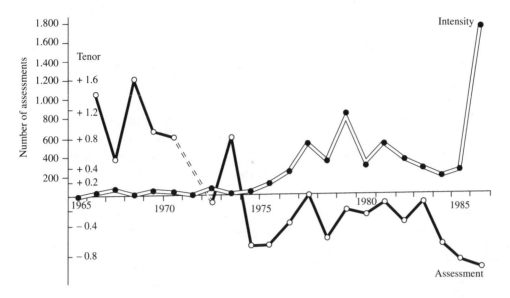

Figure 17.2. Intensity and tenor of reporting on nuclear energy over time. Source: Kepplinger 1988.

subjective level, the accidents confirmed the negative assessment of nuclear energy, which had already gained acceptance in some sectors of journalism and which made them so important (Figure 17.2).

Of the 6046 evaluative statements on nuclear energy, more than two-thirds originated from politicians (40%) and from journalists (30%). Other sources, such as scientists (5%), companies (5%) and environmentalists (4%) played only minor roles. The rest of the statements (16%) came from a broad variety of other sources. During the period under investigation, the relevance of the various sources had changed. From 1965 to 1975 – when the total number of statements was low and the tendency more or less positive – journalists were the most important sources (52% of all statements). From 1976 to 1986 – when the total number of statements was high and the tendency negative – politicians were the most important sources (42% of all statements). The relevance of journalists' opinions had dropped from 52 to 28% of all statements. The relevance of scientists had dropped from 10 to 5%, whereas the relevance of the companies and of the environmentalists remained more or less unchanged over time. It should be remembered that the total number of statements had remarkably increased, resulting in a higher visibility of all sources mentioned.

The different sources made very different statements on nuclear energy in the papers analysed. Three categories of sources can be distinguished. In the first category all sources are grouped which more or less stuck to their attitudes during the period analysed. Among these are companies (mainly those constructing and running nuclear power plants), including their spokespersons, who made positive statements, displaying strong fluctuations. Among these are also environmentalists, who assessed nuclear energy negatively, showing almost no fluctuations. Companies and environmentalists can thus be regarded as opposing camps in the conflict on nuclear energy. Both camps, however, entered the discussion relatively late, with the environmentalists preceding the companies by three years. This indicates that companies at first did not perceive, or did not take seriously, the criticism of nuclear energy, and as a consequence gave way to their opponents. Also, it is striking that companies assessed nuclear energy less decidedly and less extremely than environmentalists.

In the second category, all sources are grouped who have changed their attitudes toward nuclear energy over the years. Among these are journalists and politicians, who were quoted or referred to mostly with positive assessments up to 1972–4 and mostly negatively ever since. They changed camps, so to speak. The journalists did it much earlier than the politicians. Simultaneously, their views approached the views of the environmentalists. As politicians, at the beginning of the 1970s, still largely made positive statements, an increasing gap developed at that time between politicians and journalists. The politicians closed this gap in the following years by a sharp turn of their attitudes. In this, they followed a trend set by the journalists.

A third category of sources is formed by scientists. In the 1960s, they were mostly quoted or referred to with positive assessments. In the phase of the sharp re-evaluation of nuclear energy, hardly any statements by scientists were to be read in the media. Not before the re-evaluation was completed among journalists and politicians did they appear again, now alternating between negative and more neutral assessments. The low involvement of scientists during the crucial phase of the re-evaluation and an astonishing passivity of companies up to the mid-1970s might have had a considerable share in the rapid and radical change of public attitudes. Figure 17.3 shows the development in detail.

The long-term relationship between the portrayal of nuclear energy in the press and the views of the population on nuclear energy can be tested using a question presented by the Institut für Demoskopie Allensbach to representative samples of the population of the Federal Republic of Germany eight times since 1975 (Institut für Demoskopie 1987). The question reads: 'There are different views as to what the advantages and disadvantages of nuclear power plants are. Some of the things you hear people saying and which you read about nuclear power plants are included in these cards. Would you please look at them and lay aside those where you would say this is true, this applies to nuclear power plants?' The cards listed seven advantages and seven disadvantages, i.e. benefits and costs, which were also determined in the content analysis of reporting. This included,

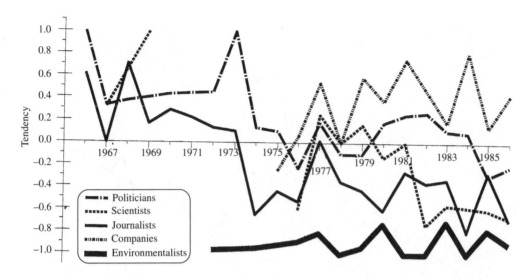

Figure 17.3. Tendency of statements on nuclear energy by journalists, politicians, scientists, companies and environmentalists. Mean values (± 3).

for example, the positive assessments 'Nuclear power plants reduce our dependence on foreign countries' and 'Nuclear power plants are more compatible with the environment than other power plants', as well as negative assessments such as 'River water heats up too much because of the cooling water draining off into it, thus becoming more polluted', 'Nuclear power plants create hazards resulting from nuclear waste' and 'The areas surrounding nuclear power plants are contaminated by radioactivity'. To analyse the relationship between reporting and the population's view, popular opinion on the seven positive and on the seven negative aspects was compiled as an index and compared with a corresponding compilation of assessments of the consequences, the quality, and the operation of nuclear power plants.

When the reporting of positive and negative aspects of nuclear energy in the press is compared with the population's views on the subject, a high degree of agreement is evident. Beginning in 1974, the papers mainly dealt with negative aspects of nuclear energy. The population, who initially took a mainly positive view of nuclear energy, began to appreciate its positive aspects less and less, beginning in 1975. By 1979 at the latest, mainly the negative aspects were noted. After 1979, reporting became less negative and the popular image of nuclear energy improved until a positive view of nuclear energy prevailed in 1985. The population's views on the subject then became extremely negative in the wake of reporting on Chernobyl (Figure 17.4).

The figure showing the relationship can be supplemented by a statistical analysis using time-lag correlations on an annual basis with the missing data

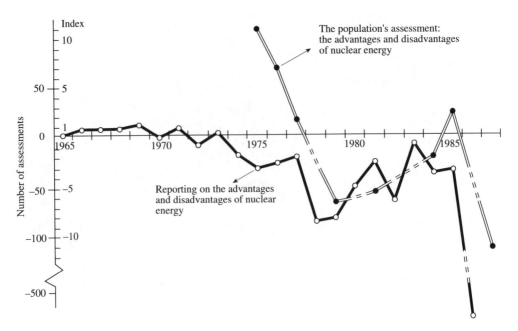

Figure 17.4. Reporting on the advantages and disadvantages of nuclear energy and the population's views on the subject. Source: Kepplinger 1988.

being interpolated. The time-lag correlations show that – taking the entire period from 1975 to 1986 – the coverage by *Stern* (0.66), *Der Spiegel* (0.71) and *Frankfurter Rundschau* (0.47) anticipated the views of the population by three, two or one years respectively. Reporting by *Frankfurter Allgemeine Zeitung* (0.45) and *Süddeutsche Zeitung* (0.48) paralleled popular opinion while the coverage by *Die Welt* (0.58) lagged behind by a period of two years. In contrast, *Die Zeit* (−0.45) responded to the dominant opinion trends with the opposite tendency at an interval of one year. Thus, it is mainly *Stern* and *Der Spiegel* which may be considered opinion leaders, leaving open the question whether these publications merely anticipated the views of the population (and of other publications), or initiated this way of thinking.

Based on the available data showing the relationship between reporting and popular opinion, we may rule out the idea that the coverage simply reflected the views of the majority of the population on nuclear energy. At best it reflected the views of a minority, but this cannot be proven on the basis of the data available. Focusing on the majority of the population, we may surmise that reporting by the mass media shaped the population's views on nuclear energy, or at least anticipated them. This does not exclude the possibility, however, that once the fears of the population had been induced, fears also resulted in expectations, exerting pressure to publish negative assessments, so that toward the end of the period studied a self-reinforcing process took place.

The results of the empirical studies

1. Scientists have different opinions on the benefits and risks of genetic engineering from science journalists and political editors, the science journalists taking a position between the other two groups.
2. Scientists and science journalists believe – contrary to political editors – that the shortcomings in science coverage is due to the fact that political editors dominate the coverage on science.
3. Science journalists played only a marginal role as authors of articles on genetic engineering in the German press. Most of the articles came from non-identifiable sources, other journalists and news agencies.
4. Most of the coverage on genetic engineering was published in the political sections of the newspapers and magazines, followed by the science sections. The coverage had a negative slant in the political sections and a positive slant in the science sections.
5. In the political sections primarily members of alternative research institutions were put into the spotlight presenting negative views on genetic engineering, whereas in the science sections primarily members of academic and commercial research institutions were referred to and cited. They presented positive views on genetic engineering.
6. There was a clear correlation between the tendency of evaluative statements made by journalists in the various media and the tendency of the evaluative statements of experts referred to and cited by these media. Thus, the experts seemed to confirm the views of the journalists of nearly all the media under investigation.
7. The coverage on nuclear energy in the German press was positive and of low intensity till the mid-1970s. Since then it has been negative. Therefore, the accident of Three Mile Island was not the cause of a re-evaluation of nuclear energy but an incident which made visible that there had been one.
8. Three different types of sources of evaluative statements can be identified in the German press: first, the companies and the environmentalists, who made positive the respective negative statements over the whole period under investigation; secondly, the journalists and the politicians, who changed from supporters to critics, starting with a re-evaluation of nuclear energy by journalists; and thirdly, the researchers, who did not appear in public debate when the issue became controversial.
9. The trend in media coverage on nuclear energy went about two years ahead of the trend in the opinions of the general public. Therefore, the coverage of the mass media did not primarily reflect but shaped the opinions of the majority.

From these data, one has to conclude that there was a conflict between
scientists and journalists as well as a conflict between science writers and political
editors. In this conflict, the scientists as well as the science writers were not very
successful in publicly presenting their points of view. This can be traced back to
the role perception of political editors and their position within the news
organizations – which in turn have to be seen in the context of general trends in
society.

References

AHLHEIM, C. (1991). Das Spannungsverhältnis von politischen Journalisten,
Wissenschaftsjournalisten und Wissenschaftlern am Beispiel der Gentechnik. Master's
thesis, Johannes Gutenberg University, Mainz.

BELL, D. (1973). *The coming of post-industrial society. A venture in social forecasting.* New
York: Basic Books.

DOUGLAS, M. & WILDAVSKY, A. (1982). *Risk and Culture. An Essay on the Selection of
Technological and Environmental Dangers.* Berkeley: University of California Press.

FADEN, R. R., BECKER, C., LEWIS, C., FREEMAN, J. & FADEN, A. E. (1981). Disclosure of
information to patients in medical care. *Medical Care* **19**, 718–33.

INSTITUT FÜR DEMOSKOPIE ALLENSBACH (1987). *Medienwirkung und
Technikakzeptanz.* Allensbacher Bericht über ein Forschungsprojekt des BMFT. Allensbach.

KEPPLINGER, H. M. (1988). Die Kernenergie in der Presse. Eine Analyse zum Einfluß
subjektiver Faktoren auf die Konstruktion von Realität. *Kölner Zeitschrift für Soziologie und
Sozialpsychologie* **40**, 659–83 and **41**, 413.

KEPPLINGER, H. M. (1989). *Künstliche Horizonte. Folgen, Darstellung und Akzeptanz von
Technik in der Bundesrepublik.* Frankfurt: Campus.

KEPPLINGER, H. M. (1992). Artificial horizons: how the press presented and how the
population received technology in Germany from 1965–1986. In *The mass media in
liberal democratic societies*, ed. S. Rothman, pp. 147–76. New York: Paragon House.

KEPPLINGER, H. M., BROSIUS, H.-B. & STAAB, J. F. (1991a). Instrumental actualization:
a theory of mediated conflicts. *European Journal of Communication* **6**, 263–90.

KEPPLINGER, H. M., EHMIG, S. C. & AHLHEIM, C. (1991b). *Gentechnik im Widerstreit.
Zum Verhältnis von Wissenschaft und Journalismus.* Frankfurt: Campus.

KEPPLINGER, H. M. & EHMIG, S. C. (1995). Press coverage on genetic engineering. Facts,
faults and causes. In *Modern biotechnology: legal, economic and social dimensions*, ed. D.
Brauer. Weinheim: VCH (in press).

KEPPLINGER, H. M. & VOHL, I. (1976). Professionalisierung des Journalismus?
Theoretische Probleme und empirische Befunde. *Rundfunk und Fernsehen* **24**, 309–43.

KEPPLINGER, H. M. & WEISSBECKER, H. (1991). Negativität als Nachrichtenideologie.
Publizistik **36**, 330–42.

MAIER-LEIBNITZ, H. (1986). *Lernschock Tschernobyl.* Zürich: Edition Interfrom.

MAZUR, A. (1981). *The dynamics of technical controversy.* Washington, DC:
Communications Press.

MUELLER, J. E. (1968). Fluoridation attitude change. *American Journal of Public Health* **58**,
1876–81.

NOELLE-NEUMANN, E. & MAIER-LEIBNITZ, H. (1987). *Zweifel am Verstand. Das Irrationale und die neue Moral.* Zürich: Edition Interform.

SAPOLSKY, H. M. (1968). Science, voters, and the fluoridation controversy. Conflict among perceived experts leads voters to act negatively on the fluoridation innovation. *Science* **162**, 427–33.

SCHELSKY, H. (1975). *Die Arbeit tun die anderen. Klassenkampf und Priesterherrschaft der Intellektuellen.* Opladen: Westdeutscher Verlag.

SCHUMPETER, J. A. (1950). *Kapitalismus, Sozialismus und Demokratie.* Bern: Francke.

SHORT, J. F., Jr (1984). The social fabric at risk: toward the social transformation of risk analysis. *American Sociological Review* **49**, 711–25.

SNOW, C. P. (1967). *Die zwei Kulturen.* Literarische und naturwissenschaftliche Intelligenz. Stuttgart: Klett. [Translation of *The two cultures and the Scientific revolution* (1959), Cambridge University Press.]

WESTERSTAHL, J. & JOHANSSON, F. (1986). News ideologies as moulders of domestic news. *European Journal of Communication* **1**, 133–49.

18

Forms of intrusion: comparing resistance to information technology and biotechnology in the USA[1]

DOROTHY NELKIN

Exploring the public resistance to technology in America, one is immediately struck with a paradox. Some technologies provoke organized opposition; others, no less invasive, no more benign, are welcomed, or, at the least, they are accepted with comparatively little debate. The contrast is rather extraordinary when we compare the response to two important technologies that have burgeoned over the past decade: information technology and biotechnology. These are both pervasive and rapidly expanding technologies, and both have their share of social costs as well as benefits. But they have evoked a very different public reaction.

In this chapter, I will first briefly remark on the diverse responses to these two technologies in the United States, and then explore these differences along several dimensions. Note that by resistance, I refer to overt opposition, not to the passive reluctance of individuals to use word processors or to buy bio-engineered products (see Bauer, Chapter 5). My purpose in the comparison is to shed light on the values and priorities that shape the public response to new technologies in America, and to highlight some fundamental contradictions between the rhetoric of support for science and technology and the reality of public attitudes as expressed in behaviour.

Responses to information technologies

Information technologies – from computers to communications – have obviously had an overwhelming social impact and their economic and social benefits hardly need explanation. But they have also intruded on our privacy, threatened our civil liberties, and imposed on many of our rights (Westin 1970). Computerized data banks empower bureaucratic authorities by providing easy access to personal information – about credit ratings, school performance, housing,

medical histories and tax status. And in the future, they will no doubt allow access
to genetic profiles, providing information about our predisposition to certain
behaviours or disease. Such information may be available to employers, insurers,
product advertisers, banks, school systems, university tenure committees, and
other institutions that exercise enormous control over our lives (Nelkin &
Tancredi 1994). Computerization might be called the 'cursor' of our time.[2] It has
enabled the relentless extension of advertising through sophisticated distribution
of mailing lists. Telephone propaganda and telemarketing solicitations shame-
lessly intrude on our home life, disturbing us at mealtimes with automated
messages that have got out of hand. Information technologies have displaced
people from jobs and turned many potentially skilled workers into low-level
computer technicians. Computers have, in many ways, facilitated our work as
scholars, but they have also turned us into typists; yet, from this most articulate
community, one hears hardly a complaint. They have turned the simple act of
buying a plane ticket into an endless manipulation over frequent flyers and fares,
but we welcome the so-called convenience. They have encouraged new forms of
crime and fraud, but we describe them with grudging admiration. They have
allowed new types of vicious weaponry, but we call them 'smart bombs'.

Perhaps most important, information technologies have extended the power of
the mass media, creating unprecedented possibilities for political manipulation
and changing the very nature of political life. The media creation of politicians
was obvious during the 1992 United States presidential campaign. But, also, the
use of electronic communication has reduced accountability, threatening one of
the most important ways we protect democratic values. And in many other ways,
they limit speech, restrict exchange, and challenge First Amendment Freedoms
(Solla Pool 1983).

Many years ago, George Orwell predicted that information technologies would
bring about an era of mind-control; but the symbolic year, 1984, came and went
as if his scenario were only a science fiction plot. While there have been many
critiques of information technologies, they mainly come from an elite –
sociologists, ethicists and others professionally concerned about the problematic
legal, social and political implications of electronic technologies. Humanists
worry about the blurring of image and reality brought about by telecommunications (Winner 1988). Sociologists worry about the effects of these
technologies on work (Garson 1988). Educators worry that computers in the
classroom may undermine the child's desire to read, reduce careful thinking to
impulse shopping, and turn dynamic problem solving into predigested programs.
Computers, as one educator put it, are 'the high-tech pacifiers of a vacuous
information age'.[3]

I could go on with examples, but my point is to suggest that reservations mainly come from scholars and their warnings have never gained a public following. There is nearly total absence of organized public concern about a set of technologies with profound and highly problematic social and political implications.

Responses to biotechnology

What a contrast to the response to technological advances in biotechnology! These advances have been the focus of persistent public opposition, and indeed biotechnology has replaced nuclear power as the symbol of 'technology-out-of-control'. This is a technology with many positive benefits, for example, for the development of more productive agriculture, the creation of disease-resistant crops, the development of new pharmaceutical products and therapeutic procedures, and the enhancement of biomedical research. But biotechnology is actively resisted on many grounds. Critics of biotechnology have mobilized to oppose the siting of research laboratories, the marketing of better-tasting tomatoes, the field testing of genetically engineered bacteria intended to inhibit frost damage, and the creation of transgenic animals designed to provide special breeds of laboratory animals useful in cancer research (Krimsky 1991). They have opposed the development of bovine growth hormones, the creation of genetically engineered food products, and, indeed, nearly every biotechnology application.

The opposition to biotechnology is not limited to a few articulate individuals such as Jeremy Rifkin. He is surely a ubiquitous gadfly, but he has an active and diverse following, willing to participate in public hearings and engage in public demonstrations and civil disobedience. Resistance has engaged the farm organizations representing small farm interests who view biotechnology developments as symbolizing the decline of the family farm. It has engaged animal rights groups concerned about the exploitation of animals (Jasper & Nelkin 1992). Religious groups have opposed biotechnology as a threat to 'natural' boundaries, and they anticipate with horror a future of genetic engineering that will violate the sanctity of life.[4] Chefs have agreed to boycott the 'Flavr Savr' Tomato. Environmentalists have organized against biotechnology applications as a violation of safety standards. Communities have blocked laboratory facilities as constituting 'biohazards'. And gene splicing techniques have brought forth anxieties about the role of biological weapons in warfare, evoking fears about the creation of virulent strains that may be used as weapons (Wright 1990).

The issues of concern

It is clear that both of these rapidly developing and pervasive technologies have costs as well as benefits. But, aside from occasional professional critiques and some concern about radiation exposure from computer screens, there has been no popular or organized resistance to the remarkable development and diffusion of information technologies. Indeed, they are viewed as the symbol of progress, the icon of ingenuity, and the test of American competitiveness in the economic marketplace. Biotechnologies, on the other hand, have become the focus of sustained and vocal public opposition. I believe that this striking difference reveals something about what matters in American society, about certain values that guide our response to science and technology; so let me try to explain the paradox by examining, in greater detail, the issues at stake.

In drawing out the contrast in the public response to these technologies, I have suggested that information technologies pose three types of problems; they intrude on personal privacy; they offer the means for institutions to control their clients; and they encourage practices that threaten certain democratic values. In the case of biotechnology, the stakes are very different. New biotechnologies can directly affect the economic interests of particular groups such as small farmers. They may be a source of environmental risks. And, they are seen as a moral threat, involving 'tampering' with fundamental aspects of nature. Let me look more closely at these six issues, exploring in each case the ways in which they affect, more generally, the public response to technological change.

Intrusion on individual rights

First, the potential intrusion on individual rights – in particular, the right to privacy. In the individualistic culture of America, resistance to technology is often cast in the rhetoric of rights. Animal advocates call for animal rights, anti-abortionists make claims for fetal rights, environmentalists advocate the rights of future generations, and the elderly claim the right to die. Rights talk has become the way that Americans express the fundamental and frequent tensions between individual expectations and social or community goals. Thus even technologies intended to improve public health, such as fluoridation, universal vaccination, or the automobile air bag, have all been resisted because they intruded on the rights of individuals to make their own choices.[5]

Rights, as defined by philosopher H. L. H. Hart, are 'moral justifications for limiting the freedom of another' (Hart 1955). Thus, rights claims are inevitably a source of conflict and contradiction. Perhaps nowhere is this more evident than in claims to the right of privacy where those who claim the right of access to information must necessarily confront those concerned about confidentiality, and who fear the abuse of information.

Privacy in America appears to be an important value. While not specifically mentioned in the constitution, the right to privacy is inferred from various provisions of the Bill of Rights such as the right of association and the protection against unreasonable searches and seizures and against self-incrimination. Rhetorical support for the right to privacy is extremely high. A survey in the 1970s found that 76% of the public believed that privacy should be added to life, liberty and the pursuit of happiness as a fundamental right. But, in fact, how deep is the commitment to privacy when it conflicts with other values? Another survey suggests that most people support measures that would require psychiatrists to report to the police a patient's expressed intention to commit a crime. And attitudes towards wiretapping are equivocal; political extremists and potential enemy sympathizers are considered fair game (McCloskey & Brill 1986).

In fact, observing the American scene, I would argue that most Americans seem to care little about privacy. We all know about data snoopers who, helped by sophisticated software, have become a veritable industry. Spying and surveillance gimmicks have made many millionaires.[6] But there is little public outcry in a society often willing to challenge industrial practices. A survey by the March of Dimes Birth Defects Foundation in October 1992 found that most Americans believe that genomic information (probably about others, not themselves) should be available – not only to directly affected relatives, but also to employers and insurers.

As a society, Americans tolerate an extraordinary amount of intrusive noise: people accept Musak in shopping malls, supermarkets and airports. They accept televised surveillance in department stores and other public places. Media audiences seem to relish the intrusions on personal privacy when the networks explore the sex lives of public figures. And to an amazing degree, people talk about their own personal problems in public. Thus, popular magazines and media talk shows are full of lurid and embarrassing personal confessions. The remarkably popular self-help movement is characterized by a confessional mode of discourse. The confessional style of Alcoholics Anonymous has been extended to deal with smoking, gambling, and overeating, suggesting that relinquishing privacy is seen as a way to solve personal problems. Far from demanding privacy, Americans let it 'all hang out' (Kaminer 1992). Perhaps this explains why, despite their obvious intrusion on privacy, information technologies have not been resisted.

Potential for social control
The second issue at stake in the development of information technology has to do with its potential for social control. The availability of computerized data on many

aspects of personal behaviour has enabled a striking level of institutional control over individuals. This has been the source of some professional and philosophical concern, but has not brought about significant public resistance. Computers and fax machines have been marketed as a means of empowering and liberating the individual: of expanding individual choice. Perhaps no industry has been more successful in turning the latest gimmick – the extra megabyte, the latest fax machine, call waiting and now the videophone – into dire necessity. For the middle class who form the core of most resistance movements, these are familiar and useful technologies that seem to give people more, not less, control.

Discriminatory abuses of computerized information, for example its use for surveillance or for denying insurance, have been examined in legislative debates, and in investigations by civil liberties groups such as the American Civil Liberties Union. But such inquiries seldom raise fundamental structural questions. While protests against biotechnology question how these technologies are developed and diffused, those concerned about information technologies focus on particular incidents and often treat them as aberrations. And some of the abuses – for example, computer crime – are admired as creative, clever, a way to 'beat the system'. And such issues seldom generate a popular outcry; for consumers who are affected by the abuse of personal information are dispersed and difficult to organize. There are few groups prepared to mobilize protest against such abuses. The Gay activists who have organized resistance to the flagrant abuses of information from HIV tests are an exception that proves the point.

Related to concerns about social control are the questions of trust that commonly underlie popular resistance to technology: will the inevitable corporate control over technological applications sacrifice public or individual interests to the imperatives of private profit? Recall that the computer industry was generated by commercial entrepreneurs, while biotechnology came directly out of academia. Yet it is biotechnology that evokes dour images of the military industrial complex and overt mistrust of commercial motives (Kenney 1986). For some reason, few seem to mind the tradeoffs between corporate efficiency and individual rights when people become digits in data banks.

Nor do we seem to care that along with the Global Village comes the risk of hegemonic control over the images and messages we receive from the media. We welcome the advances in information technology that have brought cable systems and multiple channels as 'pluralism'. But this pluralism, as one critic cynically suggests, may just be 'code for a corporate controlled mediasphere that isolates consumers into ever narrower pigeonholes of taste and cash flow'.[7] Today there are plans for digital broadcast satellite services offering no less than 1000 channels – truly technology out of control. Yet the most common popular response is that expressed in Bruce Springstein's song: '57 Channels and Nothing on'.

In contrast, genetic engineering evokes scenarios of social and corporate control that have become a part of popular culture. They appear, for example, as central themes in recent comic books with names such as *DNAgent*, or *Extinction Agenda*. In *DNAgent*, a company called Matrix Inc. exploits genetic engineering to expand its control. In *Extinction Agenda* scientists from a country called Genosha genetically engineer mutants as slaves. True, these are fantasies, but they are reflected in the rhetoric of resistance to biotechnology, and the sense that this is the technology that is out of control (Nelkin & Lindee 1995).

The threat to democratic values

Let me turn to the third and related issue, the threat to democratic values – an important theme in the history of resistance to technology. Controversies over power plant siting or the use of toxic substances in the work-place have often focused on the question of public control over technological decisions. Typically, opponents of a technology seek to participate in decisions that affect their interests. Challenging the authority of experts and questioning the motives of public officials, they seek to increase accountability. Thus, technical obfuscation and its limiting effect on public accountability has been an important issue in the resistance to many technologies. But the technical language of bits and bytes, of Dos and disks, of macros and mice, has entered the vernacular. To the middle class, the group most often engaged in resistance movements, information technologies appear to be decentralized, comprehensible and controllable.

This is to ignore, however, the capacity of electronic technologies to reduce the citizen's capacity for reflective engagement in politics, to substitute digitalized responses for active participatory exchange (Lyon 1988; Winner 1988). Thus, when the 1992 American presidential candidate, Ross Perot, proposed to revive the old and discredited idea of electronic democracy, no-one, even in the contentious climate of a political campaign, tried to debate its political implications. Advocates of electronic democracy fail to see the difference between the inundation of information and reflective political exchange. And computer advocates fail to see the broader issues of manipulation and loss of political accountability as problems; to them, the technology appears to enhance individual choice.

Interests affected and resistance

The fourth issue has to do with affected interests. Resistance to a technology can be maintained only if activists can count on a group of people who offer a base of political support, and who will become part of a social movement. These may be people who are directly affected by the siting of a noxious facility in their neighbourhood, or by the economic implications of a technology for their livelihood. Or resistance may be supported by those who share broad ideological or religious convictions. The resistance to biotechnology has been able to draw on groups of people who were already mobilized to protect their economic interests and moral concerns. Biotechnology was a ready-made issue for small farm organizations concerned about the decline of family farming and organized to oppose technological change; they felt directly threatened by new biotechnology applications that seemed designed for agribusiness. And for the powerful and well-organized animal rights movement, seeking ways to extend their popular cause, the creation of transgenic animals was a perfect target. Such groups were able to enforce a federally imposed moratorium on animal patenting from April 1988 to December 1992 when the Patent and Trademark Office issued the first new animal patents since the infamous Harvard onco-mouse. Meanwhile the office has a backup of several hundred applications, some pending for years.

In contrast to the specific interests of these groups, concerns about the invasion of privacy, the potential for social control, or the threat to democratic values, are vague and diffuse. These issues have no natural constituency, no organized group that will speak out in protest. Thus, resistance is expressed, less through organized protests than through the individualized procedures of the courts in response to specific abuses. And the legal system operates more to protect individuals than to challenge the development of the technology. Nor do information technologies evoke the negative images of corporate abuse – that have helped to mobilize popular protest against biotechnology. In the American mythology, the history of agribusiness calls forth quite different associations than Silicon Valley. And helpless animals, like besieged farmers or vulnerable fetuses, become easy lightning rods for social movements.

'Biohazards'

The fifth issue, the possibility of 'biohazards', is one of the critical sources of the protests against biotechnology. Concerns about endangering human health or damaging the environment have, of course, spawned many technological disputes. The fear of generating new and possibly dangerous organisms – an 'Andromeda strain' – was the basis of the first biotechnology controversy in the mid-1970s over the siting of a Recombinant DNA Laboratory in Cambridge, Mass (Krimsky 1982). Since then, citizens have organized to oppose specific projects, and to demand appropriate safety standards for releasing transgenic plants and micro-organisms into the environment. The potential risk of biotechnology applications attracts a constituency already mobilized through the environmental and especially the anti-nuclear movements of the 1970s. Since then, their fears have been enhanced by increased knowledge about other risks emanating from science, creating the suspicion that science, in a context of corporate greed, can in itself yield dangerous products. And suspicions are increased by the continued disputes among scientists over the nature and extent of many risks.

Over the past few decades, we have been deluged with warnings about invisible health hazards: from PCBs, freon, radiation, food additives, antibiotic-resistant bacteria, and even the possible risk of constant exposure to video display terminals. Indeed, this is one of the few contentious issues concerning information technology. Biotechnology raises many of the same problems: the hazards are invisible and there remains uncertainty about the health effects of low-level, long-term exposure. Like nuclear power, biotechnology evokes images of warfare and fantasies of monsters and mutations. In his analysis of the sources of nuclear fear, historian Spencer Weart observed the pervasiveness of these images and fantasies in popular culture, suggesting their importance in shaping public attitudes towards nuclear power (Weart 1988). Anti-nuclear sentiments were expressed in images of mad scientists and Frankenstein monsters. The horror film, *Dr Cyclops*, portrayed an irradiated scientist shrunk to 13 inches. And a series of amazing atomic creature films in the 1950s portrayed ants, spiders and scorpions growing to the size of 747 aircraft after straying into the path of atomic tests. These archetypal fears are also played out in the remarkably similar imagery of biotechnology mutations – shrunken scientists, oversized mutant cows, deformed transgenic pigs – that appear in biotechnology protests. Thus, unlike information technology, the response to biotechnology has been powerfully influenced by its association with risk.

Morality and tampering with nature

Finally, one of the most important issues underlying the resistance to biotechnology has been the implications of 'tampering' with nature, with 'violating' the sanctity of life. Embedded in this complex issue are concerns about authenticity, about tampering with 'natural' or God-given features of human life. Thus, biotechnology evokes Frankenstein images. The Harvard genetically engineered onco-mouse has been called Frankenstein. Genetically altered tomatoes are Frankenfruit. The current revival of Frankenstein films, the best selling books such as Robin Cook's *Mutation*, and hundreds of x-men mutant comics all feature images of mad scientists who engineer human souls.

I cannot resist noting, however, that information technologies present, perhaps, more of a challenge to authenticity. While not tampering with the body, they tamper with the mind, creating bizarre confusions between fact and fantasy, between the image and the real. What can be more intrusive than the distortion of mental images involved in the simulation of virtual reality? But this manipulation of mentality, for some reason, evokes little public dismay. The mind, it seems, can be sacrificed while the manipulation of diseased genes for therapeutic purposes or the creation of bio-genetic mice for research purposes becomes a serious moral dilemma.

Concerns about the morality of biotechnology, I believe, must be understood in terms of the strongly embedded fundamentalist tradition in American society where moral and religious agendas have extraordinary importance. Gallup polls suggest that some 90% of Americans profess their belief in God and 70% belong to a church. Every US President, including Clinton, has invoked God in his inaugural address, and surveys find that 63% of adult Americans would not vote for a President who did not believe in God.[8] Indeed, religious values seem to be increasingly important in shaping American attitudes, and in driving resistance movements. Note, for example, the moral opposition to fetal research that succeeded in stopping federal funds for medical research, the religious resistance to the teaching of evolution that generated the creation controversy, the anti-instrumental values that continue to drive the animal rights movement, and the beliefs about the sanctity of nature that motivate ecologists. I find it interesting that the American press paid considerable attention to the new catechism of the Roman Catholic Church specifically condemning genetic engineering as a violation of 'the personal dignity of the human being and his unique, unrepeatable identity'. Indeed, any technology that threatens to tamper with nature is bound to confront organized opposition.

As research in biology touches on human evolution and the nature of life it is increasingly subject to religious-based resistance, and scientists find themselves increasingly engaged in moral politics. The position of those who are driven by moral sentiments is absolute and does not countenance compromise or negotiation. Thus, arguments about the medical, economic and social benefits of biotechnology applications often fail to allay their concerns. There seems to be no parallel in the response to information technologies. The Bible, after all, is on line.

Patterns of resistance reveal the hierarchy of values

Both information technology and biotechnology promise enormous benefits to society, and both present certain risks. Comparing the public response to these technologies – laying out the issues of concern – suggests the hierarchy of values that more broadly shape public attitudes toward science and technology. It also exposes certain contradictions between rhetoric and reality. We give lip service to the importance of 'rights': the right to privacy, the freedom from social control, and the preservation of democratic values. But the issues that are most likely to generate resistance to a given technology have more to do with its potential risk to health, its impact on organized interests, and especially its effect on moral and religious agendas. It is these concerns that underlie the contrasting attitudes towards two of the most critical technologies in society today.

Notes

1 I would like to acknowledge the National Center for Human Genome Research of the National Institutes of Health, Grant 1R01HG0047–01 for supporting the research for this paper.
2 I credit my husband, Mark Nelkin, for this awful pun.
3 David Gelernter, 'Babes in Computer-land–Op-Ed', *New York Times*, 23 December 1992.
4 For discussion of these diverse interests, see Nelkin 1992–3, pp. 203–10.
5 For case studies see Nelkin 1992.
6 Roger Rosenblatt 'Who killed privacy?', *New York Times Magazine*, 31 January 1993, pp. 24–8.
7 Julian Dibbell, 'It's the end of TV as we know it', *Village Voice*, 22 December 1992.
8 *Time* magazine poll on religion, summarized in Nancy Gibbs, 'America's holy war', *Time*, 9 December 1991, p. 60.

References

GARSON, B. (1988).   *The electronic sweatshop*. New York: Simon and Schuster.

HART, H. L. (1955).   Are there any natural rights? *Philosophical Review* 64, 175–91.

JASPER, J. & NELKIN, D. (1992).   *The animal rights crusade*. New York: The Free Press.

KAMINER, W. (1992).   *I'm dysfunctional, you're dysfunctional*. New York: Addison Wesley.

KENNEY, M. (1986).   *Biotechnology: the university industrial complex*. New Haven, Conn: Yale University Press.

KRIMSKY, S. (1982).   *Genetic alchemy*. Cambridge, Mass: MIT Press.

KRIMSKY, S. (1991).   *Biotechnics and society: the rise of industrial genetics*. New York: Praeger.

LYON, D. (1988).   *The information society*. Cambridge: Polity Press.

MCCLOSKEY, H. J. & BRILL, A. (1986).   *Dimensions of tolerance: what Americans believe about civil liberties*. New York: Russell Sage Foundation.

NELKIN, D. & TANCREDI, L. (1994).   *Dangerous diagnostics: the social power of biological information*. Chicago: University of Chicago Press.

NELKIN, D. (ed.) (1992).   *Controversy: politics of technical decisions*, 3rd edn. Newbury Park, Calif: Sage Publications.

NELKIN, D. (1992–3).   Living inventions: animal patenting in the United States and Europe. *Stanford Law and Policy Review*, Winter, pp. 203–10.

NELKIN, D. & LINDEE, S. (1995).   *Sacred DNA: the gene as a cultural icon*. New York: W. H. Freeman.

SOLLA POOL, I. de (1983).   *Technologies of freedom*. Cambridge, Mass: Harvard University Press.

WEART, S. (1988).   *Nuclear fear. A history of images*. Cambridge, Mass: Harvard University Press.

WESTIN, A. (1970).   *Information technology in a democracy*. Cambridge, Mass: Harvard University Press.

WINNER, L. (1988).   *The whale and the reactor*. Chicago: University of Chicago Press.

WRIGHT, S. (ed.) (1990).   *Preventing a biological arms race*. Cambridge, Mass: MIT Press.

PART V
Afterword

19

Towards a functional analysis of resistance

MARTIN BAUER

In this chapter I develop two ideas about resistance in social processes in a speculative manner, with the help of a functional analogy: (a) resistance is primarily a functional event in social processes – dysfunctionality is possible but secondary; and (b) resistance is a contribution that urges consideration of whether to sustain a process, in analogy to 'acute pain'[1], and if so, how.

In whatever context, political, technological or economic, resistance is an action attribution, and as such the achievement of a communication system (Heidenscheder 1992). This analysis of resistance is mainly concerned with resistance in areas of present day technology, but makes use of ideas from other historical and political contexts. I explore a discursive schema with two main actors: an innovator and a resistant. Further differentiation is conceivable according to the various roles of the change agency (Ottaway 1983) and resistance (see Bauer, Chapter 1). The innovator proposes a project that is not acceptable and rejected tel-quel by the resistant part; in that mismatch mutually unexpected expectations meet. Concrete actors may change their roles in two ways. First, the innovator resists changes to the project; and resistance may become an initiator. Second, these parts of innovator and resistant are not scripted: they change as they are enacted.

Being interested in the function of resistance in a process, I focus on effects: how does resistance affect the process that is its target. At the start is the thesis that resistance is a **signal that things are going wrong** – an idea that is suggested in analogy to acute pain in terms of self-monitoring. What is 'right' or 'wrong' is a matter of communication; but the fact of irritation puts the matter on the public agenda. The effects of resistance depend on contexts and identify the way the system deals with challenge. The communication situation is characterized by the alarm signal addressed to the innovator, who needs to interpret and to accommodate future action accordingly.

Formally the problem may be seen to have the following logic. The verb 'to resist' is a predicate with as many as nine arguments:

$$R (A, B, S, TD, D, C, Eff, N, O)$$

(A) and (B) form a temporary communication system (S) that defines itself. (B) reacts to a technological design (TD) that is proposed by (A); (B) acts in a manner (D), in the context (C), with the effects (Eff) on (A) and (B) and therefore on (S); these events are narrated in form and content (N) from a perspective of either the outsider (O), the innovator (A), or the resistant (B).

For the innovator an obstacle is in the way; determination (A) encounters another will (B). To propose a change may be seen as a provocation. The distinction between the proposal (TD) and the innovator (A) is useful, because it allows us to distinguish resistance to technology from resistance to the innovator/promoter and his or her ways (see Chapter 1). (D) refers to the various forms that resistance can take; (C) are the resources and motives of resistance. (Eff) refers to the consequences of resistance (D) on communication and future action; this includes direct effects on (A), recursive effects on (B), or effects on the way (A) and (B) structure their communication (S). Events are narrated with form and content (N). A narrative of 'resistance' is produced by the innovator, the resistant himself, or a third observer. To a certain extent actors take the other's point of view. As the outsiders' point of view is taken up it becomes part of the communication system.

I regard this formalism as the logic of resistance – a framework within which to talk about resistance, once you have decided that it may be helpful to do so.

Multiple perspectives and foci of observations

This conflict situation can be analysed from at least three points of view that may have different foci. Table 19.1 shows three perspectives (innovator, resistant, outside observer) and three foci of observations (other, self, and system). Because any observation is done by an observer, we need to observe the observer. Resistance is an attribution of action, and as such a product of communication; in other words, a social system describes an aspect of itself. The segmentation of the stream of events into meaningful actions is an achievement of communication where meaning is created and maintained selectively. Resistance is attributed to: the other, as stable disposition mostly with negative connotations (techno-logic); or to oneself, mostly in positive terms (resistance logic) as action in a situation; or it appears in functional analysis as input **and** output, as process of dissenting (system logic).

Table 19.1 compares internal and external attributions as forms of explanations. We tend to attribute failures by external circumstances and preferably

Table 19.1. *Resistance perspectives and attribution of problems to solve*

Focus of observation	Perspectives		
	Innovater	Resistant	Observer
Other	Deficiency, bias irrationality resistance	Abuse of power	Observing observers
Self	'Rationality' self-observer	Resistance moral imperative	Observing self-observations
System	Control problem	Contradiction problem	Interaction; functions of resistance

blame the 'other' (external attribution, scapegoating); in contrast success tends to emanate from our own qualities (internal attribution). Social psychology suggests that this self-serving bias of attribution is correlated to cultural individualism (Moscovici & Hewstone 1983). Individualistically minded actors are often blind to their own conditions in order to avoid self-blame. This may be one of the factors that explain why in stories of innovators and technological projects the contribution of resistance to these achievements remains hidden; resistance takes the blame for failure, but not the credit for success of innovations (see Staudenmaier, Chapter 7).

The innovator perspective (techno-logic)

From the innovator's perspective resistance is commonly explained in deficiency terms: deficits in cognition and information processing (e.g. bias, irrationality), in personality (e.g. nuclear phobia, cyberphobia, rigidity, conservatism) or in the structure of the social system (e.g. bureaucracy). From the perspective of promotion, traditions are obstacles to progress; and society becomes an object of manipulations and marketing strategies (Evers & Nowotny 1987, pp. 272f). Resistance is seen as a relation among five variables or questions: actors (who?), means (how?), conditions (why?), target of resistance (to what?), and the motives (why?). Innovative actors often see themselves in possession of the one best way, the 'Rationality' writ large; this provides a source of self-confidence that may be necessary to carry an innovation through many obstacles.

Diffusion research is a tool for organizers of innovations to speed up the process (Rogers 1983; Mahajan & Peterson 1985). New things and ideas spread in a characteristic way: slowly at the beginning, at an increased rate in the middle, and more slowly again towards the end of the process, when the saturation level is approached. Resistance delays, slows, concentrates and limits the penetration of a social system by a new idea or product. Resistance means reluctance to take it up. It is assumed that the unit of diffusion, an idea or a device, remains identical throughout the process of diffusion. The number of adopters changes, but not the item of adoption itself.

Resistance is a controllable factor. It is explained in terms of risk perception. The idea of risk has three pragmatic implications. First, it takes the existence of a technology for granted and excludes its existence as an option. Secondly, bias in risk perception is seen as a cognitive deficit, and it abstracts from genuine motivational differences. Thirdly, it shifts our discourse from 'danger', which we want to avoid by changing our actions, to 'risk', which we seek and deliberately take in acting the way we do. The implicit demand for change in action is different in danger and under risk. Risk is defined as expected damage. This means that situations with low damage and high probability, and high damage and low probability, may be mathematically equal while psychologically they are not. Empirical risks (objective risks) are compared with perceived risks, and deviations between the two are diagnosed as deficits in non-expert reasoning: lack of knowledge, inadequate knowledge, biased experience, pathologies of inference, biased processing of information (Slovic, Fischhoff & Lichtenstein 1984). These deficits seem to be manageable through special training and de-biasing information (Fischhoff 1982), the provision of which is a profitable business (for a critique of this techno-logic approach see Berkeley & Humphreys 1982; Evers & Nowotny 1987).

More realistically, from another point of view, a network of ideas, people and machines grows by co-opting and accommodating 'resistances' (Latour 1988, p. 259). Technological design is never finished: it continues to change once it is put to use. In accounts ideas and devices are reconstructed *a posteriori* as a fixed identity such as the radio, the computer, or the nuclear power plant. Closer scrutiny reveals that each of these products of earlier times is not the same as today. They are homogenized as functional equivalents with variable structures: devices to receive radio waves, to calculate and to store information, and to produce electrical power. The functions may stay the same, but need not. The diffusion model is 'Whiggish history', as if a glorious idea or device remained the same through all encounters over time. However, the present is rarely the achievement of a single line of tradition; more likely it is the outcome of a messy interplay of initiatives, resistances, trial and error.

The transfer of technology from one context to another means accommodating the original design. New techniques are altered in design and implementation to fit the terms of a different locality. Local resistances to new techniques and ideas contribute to this adaptation. The diffusion model assigns no significance to resistance other than delaying, slowing, concentrating and limiting the modernization process of otherwise fixed ideas and devices. This assumption hides a fruitful paradox: on the one hand resistance is 'bad' because it delays the process contrary to expectations; and on the other hand it focuses accommodations so that the diffusion is possible at all.

Resistance perspective (resistance logic)

Resistance will often thematize actions that mark the misuse of power. In effect it limits the scope of the possible. It may, but need not, be a category of self-description. Resistance is often justified by appeal to moral and ethical principles (constitution, civil disobedience, 'the right to resist': *Widerstandsrecht*), and it may consider effects of resistance as part of the rationale to move people. However, resistance may be tacit and clandestine, often not conscious of itself as 'resistance', parochial in motives, but nevertheless effective in interfering with the intentions of the innovators (Scott 1985, pp. 241ff).

The richest literature concerns resistance against totalitarian regimes. The keyword 'resistance' leads to military studies on the conditions and strategies of resistance movements (Haltinger 1986). The view is that resistance has a natural advantage over initiative, other things being equal (Clausewitz 1832). Another positive view on resistance comes from studies on Fascism (e.g. Wippermann 1983; Schmädeke & Steinbach 1985; Balfour 1988). Actors, forms, conditions, legitimization and effects of resistance are extensively covered. A third source is civil disobedience and non-violence (Sharp 1973). We find philosophical reflections to express the resistance point of view (e.g. Saner 1988, 1994). In democracy the legitimacy of resistance poses a dilemma. The 'right to resist the law' is a contradiction at the core of the law. The legal system is a process, and resistance signals shortcomings in that process such as slow reactions, legal gaps, and the decay of democratic institutions under changing circumstances (Rhinow 1985). The judges judge for themselves whether there is such a shortcoming or not, but they may need a reason to convene.

Resistance is grounded in the ethos of revolt. Albert Camus' existential reassurance of the individual among others is achieved 'in revolt'. This is echoed in Touraine's variation of the Cartesian credo: '*je resiste, donc je suis*' (Touraine 1992, p. 318). Resistance creates the 'subject', a distinct notion that is based on the triple distinction between '*soi*' (= self), '*moi*' (= me) et '*je*' (= I). In becoming a modern subject we face a 'neither–nor' situation, rather than an 'either–or' choice. We are called simultaneously to resist the temptation of two evils, individual narcissism (the illusory '*soi*' = self) on the one hand, and collective self-denial (the internalized role '*moi*' = me) on the other hand. Between the delirium of style and the festival of grand causes and their depersonalized bureaucracy modern subjects find themselves in resistance as an existential condition. Both evils ultimately avoid recognizing the '*je* = I' as an autonomous subject (ibid., pp. 307ff). The 'I' gains strength and self-assurance only in equidistance between narcissism and the self-denial of the bureaucratic 'me'. The subjectified individual steers clear of both the Scylla of particularism and the Charybdis of universalism. This tension cannot be resolved; it can be endured only with lucidity and action. This condition of modernity, to which a dual progress towards bureaucracy/efficiency and subjectification has led, overcomes both the inefficiency in the sphere of production and the repression of collectivism. However, the danger persists as modern bureaucracy negates subjectivity, the second achievement of modernity. The social basis of resistance changes from one societal formation to another. In an entrepreneurial industrial society the basis of resistance is the old-style landed property (landed aristocracy) on the one hand, and small property or no property at all ('proles', trade unions, communities) on the other which at times form an alliance. In a post-industrial knowledge society the major line of social conflict may be over the control of science and technology that confronts managerial experts and non-experts. Resistance against the enclosing of public science in the private corporate interest may become a characteristic of post-industrial society (Evers & Nowotny 1987; Bauer, Durant & Evans 1994).

Resistance in social processes (communication logic)

The systemic approach urges us to observe observers: we observe how innovators and resistants act, think and communicate; and we understand how the actions are structured by communication. Cognition and communication serve at least two basic functions: they direct **and** legitimize actions. For the analysis of resistance the 'perspective' (who observes?) is a variable. An observer processes events as 'resistance', and this is done as observation of others or as self-observation.

Observing observers is useful for two reasons. First, human cognitions are symbolic – they are in conversations before they become individual thoughts; social by origin, they draw upon resources that are shared with other people; descriptions of acts converge and become acceptable forms for self-description and for descriptions of others. Such forms are called social representations (Farr & Moscovici 1984). Secondly, observing observers allows us to compare the pragmatics of accounts as guidelines and legitimation for future actions. I develop this process-logic of resistance in a functional analogy of resistance and acute pain. I discuss the notion of living systems and its double environment, and the function of communication for its operations. A scientific or technological project has characteristics of a living system that move in time and space by accommodating internal and external demands, challenges and irritations more or less successfully.

Social systems and functional differentiation

Social systems are seen as living phenomena to be explored on various levels of activity. Functional analogies compare the processes on several levels: from individual consciousness, to formal and informal, parallel and distributed communication (Boulding 1956; Luhmann 1984; Miller 1986; Cranach 1992). Coevolution is a characteristic of living systems that evolve together with their environment. Living systems are defined by a number of problems that need to be solved in order to continue 'living'. Failure to solve these problems adequately jeopardizes the future of the systems. Structures are temporary solutions for orientation, self-monitoring, perception, information storage, goal setting, regulation, evaluation or consumption of events (Cranach, Ochsenbein & Valach 1986). For the present purposes orientation and self-monitoring are in focus. Both processes explore the requirements for action, orientation for the outer environment, and self-monitoring for the inner environment such as needs and expectations. The comparative method identifies structures as functional analogues with reference to these problems; functions may vary for similar structures. This diversity is limited by concrete situations.

Functional differentiation of society is the result of historical development. A hierarchy of spheres becomes increasingly a heterarchic order of relatively autonomous social spheres of activity such as the economy, religion and culture, science and technology, politics and public opinion. Each subsystem builds images of the world according to its own logic, which includes the other systems as the environment. There is no privileged place from which to observe and to deal with the world, only such claims which can be observed (Luhmann 1984). The modern problem emerges of how to ensure the cooperation of these relatively autonomous spheres each following a particular logic. In democratic societies we resort to public opinion as a procedure to solve conflicts between various tendencies. In the following I suggest that public opinion has become an increasingly relevant environment for an otherwise autonomous science and technology.

Science and technology form a complex social system that consists of diverse activities, networks, projects and programmes, more or less institutionalized in places and disciplines which communicate with the specialized jargon of paradigms. This structure changes shape as they move through time and space; the definition of time and its refinement is a product of that system itself.[2] A time-lapse film would reveal how scientific networks start, grow, split and merge, decline and disappear, in a chaotic manner. Traditions and paradigms provide the rules upon which to evaluate individual contributions. Consistency is temporarily required (Leydesdorff 1993).

Social systems normally operate with a plurality of self-images. A unified identity is more an aspiration than an empirical reality, but may be, under difficult conditions, an operational requirement. Without a distinguished and relatively stable self-image a system dissolves into the environment and ceases to be a perceptual whole.[3]

Social systems face a double challenge: the outer and inner environments. They define their boundaries in communication (see 'boundary work': Gieryn 1983) and in doing so maintain relative autonomy in their self-image and from the perception by observers. Autonomy suggests that the origins of changes are to be found within the system. The environment is marked with a boundary set by the system itself. Environments vary both in meaning and shape. The texture but not the existence of the environment is a function of the system's operations; their basis and products are images (Boulding 1956). This environment is communicated and acted upon; communication and actions are informed by symbolic representations. Its features are constituted by gate keeping, selections and elaborations of meaning.

This forces us to understand systems in their own terms, and to observe self-observations. Social systems are open for matter, energy and information, but closed in their operations (Kueppers & Krohn 1992). Information 'enters' a system in the form of local irritations; their relevance and content are generated internally. Information is a difference that makes a difference **for** the system (Bateson 1972, p. 315).

Bi-directional accommodation: outer and inner environments

Social institutions such as science, the economy, politics, the legal system, education and culture confront each other as environments that communicate proposals and present irritations to each other. These communication systems observe each other in order to make adequate contributions and to extend their control by imposing processing criteria such as instrumental truth (science), money (economy), beauty and moral truth (culture) or decisions and power (politics) (Kueppers & Krohn 1992; Luhmann 1992) on other systems: products are true or false, profitable or useless, beautiful or ugly, depending on the observer's affiliation and code used. Institutions decide relatively autonomously how and what to produce. Autonomy is a problem that requires continuous efforts to sustain. Science and technology is one communication system in society among others (Luhmann 1992). Networks of science and technology decide which contributions to include and which to exclude. Science and technology are traditionally considered separate activities that are closely linked. Science solves problems of theory, technology solves material problems. Technology makes scientific observations possible, and technologies build upon scientific observations.

Like other social activities, scientific and technical projects deal simultaneously with inner and outer environments to accommodate activities viable for the future. The outer environment is everything that is not scientific or technical as defined by science and technology themselves. At times the system defends its autonomy against hegemonical influences; at times it expands its own hegemony. Universities, laboratories and research institutes are institutionalized projects and networks which offer publications to read, patents to exploit, conferences to attend, and problem solutions to apply. These are achievements for other social systems to use. In so doing unexpected reactions occur; the system will face irritations of all kinds: surprise, challenge and unexpected expectations from its environments. Science and technology face other social systems as environments on the same level; on a lower level they face individuals' consciousness. Persons are not parts of a social system such as science and technology; they only make certain contributions. The person as an autonomous subject is part of the system's environment. Personal consciousness, strengthened in multiple loyalties, may become a source of irritation for the communication system of science and technology.

Inner environment and internal structure

It is helpful to distinguish 'inner environment' from the 'internal structure' of a system (Willke 1991, p. 41). The inner environment comprises the resources – energetic, material, ideal and symbolic – to which activity has in principle access, in other words the potential of the system or its structure-for-action. A particular project is always a temporary mobilization of resources, the structure-in-action, accommodating the internal structure to the demands of both inner and outer environment in order to remain a viable process. This internal structure defines the social system in focus and its operations. They are complex, relatively independent from the involved individuals, and with variable structures and functions throughout history.

The inner environment of science and technology comprises the persons who are more or less integrated and identified with a particular institution. They make contributions to scientific and technological projects. Multiple loyalties are important to allow for critical distance. Individual consciousness, formed and strengthened by other institutions, constitutes an inner environment which is functionally equivalent to the outer environment. The ultimate reality test of scientific and technological activity comes in the form of negative feedback from the inner and outer environments. The scientists' and engineers' personalities form the inner environment in which the system needs to accommodate itself.

Public opinion, politics and the economy form the outer environments that bring challenges to a scientific and technical project. Resistance works as negative feedback on such projects, and adds to the directing of the process. Science and technology solve mental ('truth') and material problems ('solutions') as contributions that are variably appreciated by other systems, personal or social. Each environment judges science and technology by its own code of appreciation and priority: politics and public opinion appreciate in terms of power and decisions; culture and religion in terms of beauty and moral truth; economics in terms of money and profit; individual consciousness in any of these terms depending on identity and loyalties.

The question arises of how the scientific and technological system construes these challenges in order to deal with them. The definition of the problem has implications for action. What are the reactions of science and technology? How does the system take a challenge of which the language is not its own? Several irritations from different environments may compete simultaneously for attention and consideration. How does public opinion become a relevant environment for science and technology? If it becomes a relevant environment, how is 'public opinion' conceived (Neidhardt 1993; Wynne 1993)?

It seems that irritations, diverse and conflicting tendencies, and sensibilities are the conditions which provide adequate preparation for the future. We focused on the challenge which resistance vociferously puts to the scientific and technological projects of nuclear power, information technology and biotechnology. One can expect that when challenged, projects and activities turn rigid, and, reassured by successful traditions and habits, most inadequately in new situations, produce more of the same (stress-rigidity-hypothesis: Miller 1986); on the other hand, they may change their operations based on more reflexivity given the space and time for it. What makes resistance to technology effective to induce reflexivity? To answer that question we need to analyse empirically the effects of resistance. For example, the controversy on nuclear power has demonstrated how resistance is more effective as soon as the issue is defined by 'dissident' insiders with a double loyalty to science and to the public good. As soon as physicists started to voice doubts about the programme of civil nuclear power, the anti-nuclear movement gained strength and impact (Mazur 1975; Rüdig 1990).[4] Of course, any challenge from inside faces the old problem of 'heresy', of being shunned by the powers that be; in history heretics had to face death, but nowadays 'only' economic disadvantage.

Resistance: the 'pain' of the technological process

Looking at resistance within this framework of systemic actions allows us to specify its functions. The systemic view mediates a functional analogy between acute pain on the level of individual action, and resistance on the level of collective activity; this is expressed in the metaphor of 'organizational pain' (Bauer 1991, 1993).[5]

Resistance may be usefully analysed in analogy to acute pain or strain: this analogy applies particularly to its power to change activity for the benefit of future activity.[6] Acute pain is a signal that something is going wrong; it (a) focuses attention internally, (b) enhances reflective activity and self-images, (c) evaluates and (d) alters the course of action. Like pain, resistance has diagnostic and pragmatic value, and it is probably more usefully characterised by its effects than by its causes (Wall 1979).[7] Its pragmatic value lies in the fact that a change of action is urged from within the system as otherwise the health of the system is jeopardised. Used diagnostically resistance indicates the location of a problem in self-observation or to an observer; however, its diagnostic value is less reliable than its unspecific action implication. Ignoring and suppressing the sign is always an option, but a dangerous one, as it may jeopardize the future. However, neither pain nor resistance is unmediated: its significance is due to symbolic elaboration.

The pain metaphor is a device of persuasion; the pain analogy is a device for creative thinking. To elaborate the metaphor to analogy and hypotheses is a theoretical and empirical challenge (Bauer 1991, 1993). Analogies are neither true nor false, but more or less fruitful of new ideas.

To compare acute pain and resistance we need an abstraction to frame the relevant processes of both pain and resistance. For my purposes this is provided by the theory of self-monitoring (Cranach & Ochsenbein 1985), which suggests that acute pain and resistance serve similar functions of self-monitoring, albeit with different structures and on different levels of analysis. What pain does for individual action, resistance does for social activity.[8]

Table 19.2 compares the pain analogy across different levels of analysis that need to be separated (Miller 1986). On each level the sign process is mediated and elaborated by integration in other processes. Pain affects individual action and is mediated by cognitions; resistance affects informal interactions in social groups, and is mediated by informal communication; resistance movements affect institutionalized projects and are mediated by formalized structures such as negotiations, committees and mass media. Effects on one level are likely to affect other levels as well, though not necessarily in the same direction. In this volume we explored effects of resistance at the group level and at the institutional level.

This framework of social systems and functional analysis with reference to self-

Table 19.2. *Levels of analysis and functional analogy*

Structure	Form of mediation and elaboration	Area of effect
Pain	Conscious cognition	Individual action
Resistance	Informal communication	Network of interaction; organisations
Resistance movement	Formalized communication e.g. media, rituals,	Institutions; technological networks; society

monitoring makes it possible to achieve a relatively unbiased analysis of
resistance, and to integrate both the innovator's negative attributions and the
ethical imperative of 'resistance'. Pain and resistance have a double nature: they
are negatively connoted and trigger anxieties about a situation to be avoided and
about the future; but through this very quality they function as important
signals, as the bad news that calls for action. Information is a difference that
makes a difference for the system's operations; and only bad news seems to have
that quality; good news is nice but has no action implication. The capacity to
experience pain and to register resistance is 'life serving' and secures the future.
Any social system, a dyad, a small group or an organization, encounters
resistance in its movement, and in turn resists unwelcome influences. An
individual unable to experience pain loses his or her life sooner rather than later
when minute injuries develop into major complications (Melzack & Wall 1988).
The pain which we experience in action indicates our limits and forces a change
of pace and direction. Similarly, resistance makes the point of reality and shows
the limits of a project.[9]

Three basic functions of resistance

The self-monitoring approach shifts the problem from controlling resistance to
the analysis of its concrete effects in various contexts; resistance is part of
autonomous social processes. The particular question is: what difference does
resistance make for the scientific and technological project under various
circumstances? I have elaborated on three basic functions of resistance in the case
of a computer project in a bank environment (Bauer 1993 and Chapter 5). Here
I make an attempt to abstract from this context, which may allow us to explore
similar functions of resistance on the level of technological projects and processes.
Resistance is basically structure preserving. Three functions make it possible to

achieve this in various contexts: attention allocation, process evaluation and process alterations. Further research may suggest additional functions.[10]

Attention allocation and increased process awareness

Resistance directs attention to important features of the target process. The importance of such features is judged by operations of the project itself, i.e. by communication. The span of attention, or the number of topics that a communication system can elaborate on, are limited. The concrete themes that dominate communication in a science–technology project define the span of attention of the project. This may include the procurement of adequate funding, publications, public relations, manpower policy, political lobbying, and public reactions. The relative weight of these issues at any time defines the temporary attention structure. Resistance points to the importance of some goals and issues that have to be dealt with immediately.

To show the attention allocation function of resistance, one correlates the changes in the content of communication with the registration of resistance. Resistance in communication is a strong signal which changes the themes communicated in a science–technology project. Problem sensing is an important skill of project management. For the sensitive observer resistance is an early warning system about issues and expectations that are relevant in the inner and outer environment which the project has to accommodate to be viable (Raschke 1988, p. 385). Sensitivity is a quality of the social system that is expressed in the time-lag between the registration of resistance and the changes in the project. Past events influence the present sensitivity.

Resistance actions are often organized as happenings that may have news value. The media cover public happenings such as the sit-in at a new nuclear power plant or the occupation of a field where genetically altered plants are grown. Attention is reallocated when the fora of science and technology, their conferences, journals and committees put the issues symbolized by resistance on their agenda. Correlating the coverage of resistance in the media and the uptake of an issue in scientific networks is the way of studying this problem. Agenda setting suggests the power of the media to make an issue salient for public opinion (the weak form of agenda setting), or to change public opinion in line with the media source (the strong form of agenda setting). One would study the coverage of various media over the life cycle of an issue (Downs 1972), and compare the coverage of debates in specialist networks, as well as in popular TV, press and radio (see, for cyberphobia, Bauer, Chapter 5). Media coverage of scientific and technological issues seems to peak before they reach the political agenda, indicating its signal function for the political system and for the scientific and technological project (Mazur 1984); on the other hand, media coverage of issues is in turn put on the agenda by expert channels of communication (Strodthoff, Hawkins & Schoenfeld 1985; Kepplinger, Chapter 17).

Process evaluation

Resistance evaluates the ongoing activity in process and outcome. Resistant people reject a proposal that is made in the form of a particular technological project. Various elements of a project can be evaluated: a technology *per se* can be regarded as good or bad, acceptable or not; the people identified with the project may be mistrusted; the consequences of the technological change may be feared. Resistance evaluates elements of the process; this may be done diffusely at an initial stage, and in a more focused way at later stages when informed attitudes polarize (Bauer 1993). International comparisons of survey data show a tendency towards polarized public attitudes towards science and technology with raising levels of knowledge and economic development (Bauer, Durant & Evans 1994). The evaluation of a technological project is an input for the future course of action; it is part of an institutional learning process.

The mere fact of allocating attention may imply a negative evaluation of the technology (Mazur 1975: the strong version of the agenda setting theory). If things go smoothly, attention is not necessary – things can happen routinely. If this becomes inadequate because of an error or incident or because people's expectations change, attention is required which leads to an evaluation of the traditional routine. Evaluation processes are more or less formal. Formal examples are public hearings and inquiries. Resistance motivates people to learn about their object of resistance, which makes them a source of informed criticism of the technological design (see Bauer, 'technophobia', Chapter 5).

Process delay, alteration and reinvention

Resistance directs **and** motivates immediate changes to a new technology project. Changes may range from abandoning the project altogether, to making alterations on various levels of strategy, tactics or operations. The alterations are conditioned; reactions are of the system's own making within the context of culture and memories of past events.

A way of assessing these changes is to compare project plans at different stages of the project and relate these changes to concrete resistance (see Rucht, Chapter 13). Changes can be on the level of action or on the symbolic level. An example of symbolic changes is changing expectations; such changes can be typified as optimism or pessimism, for example in the context of nuclear energy. The expectations of the nuclear industry about its production capacity by the year 2000 have been reduced dramatically over the past 35 years (see Bauer, Chapter 1).

Learning is generally defined as adaptive structural change that persists. The technological process undergoes structural change in both the design and the people who support and use it. Resistance affects this institutional learning process. The accommodation of a project to resistance encountered differs according to contexts. Jasanoff (Chapter 15) shows how different legal traditions respond to similar public demands voiced in form of resistance to biotechnology.

The most obvious effect, which is cause for major anxieties for social systems in competition, is the delay of the project. Conflicts and self-reflective activity divert resources and energies which otherwise go directly into the project, with the effect that the pace of the project slows. Time targets may be missed.

Another type of alteration concerns resistance itself. Recursive effects influence the resistance itself. Such self-alterations may lead to progressive decay of the monitoring function of resistance in analogy to the decay of acute pain into a chronic and debilitating condition. Such decay can be due to inadequate counter-actions. For example, ignoring the resistance may lead to chronic conflict and violence, jeopardizing the entire project. To secure the functions of resistance, the right thing needs to be done.

How is the signal function of resistance constituted?

Resistance raises attention and directs self-observations;[11] it stimulates further thinking and communication. Effective resistance is always mediated. The features of mediation that make the efficient self-monitoring possible and enhance its flexibility are selection, schematic representation, increased self-awareness, and multiple processing. The symbolic presentation of needs and requirements that resistance produces are selective, schematic, and multiply processed. Symbolic presentations of resistance may be erroneous: events are transformed, enhanced or suppressed by elaboration.

The signal function establishes communication between the project and resistance. Resistance selects a certain message, the issue, for a certain target audience, the innovators, who need to make sense of this message by further interpretative selection (why us and why that particular message?).

I have argued that the warning function of resistance may be analysed into three processes: attention allocation, evaluation, and alteration of action. These three functions mediate the **transfer of knowledge** in emergency situations and to **solve conflicts**. This implies that resistance cannot be analysed in raw form; the ultimate effect depends on its elaboration in discourse, i.e. who observes resistance, where, and in what way?

Communicating resistance and its symbolic elaboration

Communication solves primarily three problems: it directs and motivates action in the face of obstacles, it solves conflicts, and it transfers knowledge to where it is needed. Secondarily, communication works on its own refinement through meta-communication and cultivation (Cranach 1992).

Pain that is not felt and symbolically associated has limited effects on action; equally, resistance that is not communicated and symbolically elaborated, has a limited effect. However, the problem that is indicated may persist and finally threaten the future of the system. Hence one may distinguish two empirical problems (Bauer 1993): resistance is more or less frequently a theme of communication; and if it is communicated, this is done in a certain manner. The first is a matter of quantitative content analysis of media channels (see Rucht, Chapter 13); the second is a problem of qualitative analysis of the elaborated meanings given to resistance (see, for example, Gamson & Modigliani 1989).[12]

Selection of particular events

Self-monitoring selects certain states or processes of the inner environment in the form of a message. The product is the presentation of this internal state. With the milieu continuously changing, any selection temporarily creates a foreground on a background. Selected processes have form and intensity (threshold), deviate from the normal state of affairs, require fast and energetic reactions, and demand reconsideration in information processing and action. Specialization to depict internal processes in a particular way characterizes different self-monitoring systems: a financial accounting system highlights different events from resistance in a project.

To study this selection process one would concentrate on the various foci of resistance and the issues that are indicated by them (see Bauer, Chapter 1). Resistance seems to focus on changes in the social structure that accompany technological innovations, and on detecting misuses of power in such projects.

Schematic representation

The messages are typified and presented in coded form. Schemata of events are stored in individual or collective memory and are triggered by key events. Schemata form a relatively fixed link between perception and action. Time and location of events are indicated which attract attention and redirect actions.

On the level of communication, schemata come in the form of theoretical and symbolic elaboration of what resistance is and how it works. In observing observers, we register any theory that explains resistance as the dependent variable as such a schema. This comprises ideas from the perspectives of the innovators of technology (techno-logic), as well as the reasoning about its legitimacy under social and legal constraints (resistance logic). To study these representations of resistance one analyses the various accounts, lay and expert, given by innovators and resistants, both about themselves and about each other (observing the observers).

Internal attention and increased project awareness

Resistance redirects attention to inner problems, and enhances the awareness of the inner environment; similarly pain redirects attention to the enhanced body image as a result of localized pain. The focus of attention is at the expense of everything else.

By allocating attention inside, an internal state of affairs becomes thematic that urges accommodation of action. As the span of attention is narrow, only a limited number of topics are processed at any one time. Any network of communication selects topics to be figures on ground, priority and marginal; messages from different self-monitoring processes may compete for attention and communication.

One effect of self-monitoring by resistance is increased attention to 'inner politics'. Resistance points to a problem in one's own constituencies, rather than a demand from another social system. As a consequence of such attention the project's inner workings are highlighted. A self-image of inner workings existed before, but this is likely to be revised in the process, and a new image emerges in response to resistance; hence we can talk about enhancing the self-awareness of the project by resistance.

Project analysis and evaluation in emergency situations are essential tools to enhance prospects for the future; such an analysis may be initiated by resistance as in the case of the software project (Bauer, 'technophobia', Chapter 5). The outcome is an elaborate description of the project to locate and identify problem areas. The project management is finally confronted with three options: to accept this image, to create an alternative, or to call upon an external observer to tip the balance between conflicting self-descriptions. The latter case is a lucrative field for project consultants. The self-image allows the location of problems geographically and socially.

In speculation one could argue that the social study of science and technology, as a scientific effort, gains momentum in response to the rising discontent and resistance to large technological projects after the Second World War, which changes the image of how science works and how science sees itself.

Multiple channel processing

Complex systems are likely to have many self-monitoring processes, which produce competing accounts of what is the case and what needs to be done. Conflicting accounts are unavoidable where actors and observers meet. The analysis of resistance needs to pay attention to a multitude of communication channels that operate in parallel, with variable intensity and different ways of elaborating the message.

Science and technology produce various accounts of their own operations. Several networks of reflexivity operate in parallel, with different results. Science policy, philosophy, history, scientometrics, and the sociology and psychology of science constitute diverse reflexive traditions, all talking about the same process with different foci.

However, self-observation is not unproblematic: it requires resources, takes time, and distracts temporarily from actions and necessary decision making, and it may produce unpleasant 'truths'. Hence, not surprisingly we find vigorously defended 'blind spots' about how science and technology 'really' work. The contested reception of Kuhnian history and the more recent sociology of scientific knowledge show the difficulty of changing collective self-images. Wynne's (1993) diagnosis of 'institutional neurosis' comes to mind; others may explain this difficult change of self-images with the peculiar manner by which social science theories spread: first, they are dismissed as mistaken, then they are stating the obvious, and finally they are taken for granted as common sense, and by then their origin is forgotten or reinvented.[13]

Public resistance emerges in various accounts, which may acknowledge, enhance, marginalize or suppress, and try to control the effects of resistance on the system. As resistance is communicated and elaborated in a network, past experience will inform the dealing with present conflicts. On the level of institutions, resistance may be channelled into formal negotiations, or be ignored. The integration of resistance into formal communication changes its character to become opposition (see Bauer, Chapter 1). Resistance is channelled into expectable forms of action.[14]

Formalized institutions for peaceful conflict resolution are important fora of communication in modern society. The way resistance is communicated, and channelled from informal to formal communication, characterizes a particular social era. Symbolic integration opens the experience to cultural refinement. Social arrangements, such as technological impact analysis and technology assessment, create a plurality of public arenas in between the individual and the state. These new arenas of public debate emerge in response to mounting public resistance and become an object of sociological analysis (e.g. Beck 1993). Such analyses may in turn increase the sensitivity of these institutions to resistance. In similarity to the nineteenth century debate on poverty and insurance, new institutional arrangements are an attempt to control the uncertainties of social life (Evers & Nowotny 1987).

Error and dysfunction

Symbolic elaboration of resistance may produce 'errors': it may be exaggerating, being out of proportion with the real issue; it may report problems without a real basis (phantom problems); prolonged resistance may become entrenched conflict, driven by changes in its own structure of operations;[15] or it may ignore what is happening.

Theories of resistance are one source of error. Schematic information processing is prone to error when it perceives the situation incorrectly, or when it links a situation to an inadequate action programme (see Norman 1980). This likelihood of error is counterbalanced by efficiency gains from schemata when situations are complex but stable, and the project is under time pressure. Schemata allow us to act when the available information is still incomplete.

The possibility of dysfunction must not distract us from our primary focus on functions. It seems more fruitful to clarify functions first, and afterwards to study the conditions of dysfunctions as deviations from these functions. The ultimate criterion for dysfunction is the discontinuation of a social system, in our case of a project of science and technology. This is difficult to observe unless it actually happens. In social systems the analysis of dysfunctions is hypothetical. The risk of the dysfunctions of resistance is a paradoxical part of the communication about resistance. It provides its negative news value that is necessary to attract attention and to exert its function to enhance the future of the project.

With the Bernese framework (Cranach *et al.* 1986; Cranach 1992) I assume that communication primarily serves three functions for social action. It arises in a project when obstacles create a state of emergency, to resolve conflicts between different tendencies, and to transfer knowledge from one section to another; other functions of communication are secondary. Secondarily communication cultivates the symbolic realm, refines the quest for knowledge, and increases its efficiency in the form of communication about communication.

Once resistance is a theme of the communication process in a science and technology project, one can study these three primary functions: is there awareness of an emergency situation, otherwise it would not be worthwhile talking about it; does a conflict call for a solution, and does knowledge transfer from diverse origins prove helpful in solving the conflict? When difficulties threaten the progress of a project, people start thinking and communicating to coordinate their attention temporarily. They focus on issues that need trouble-shooting. Resistance is a form of trouble for the project. Additional information may be necessary to make sense of these irritations. Communication generates and transfers information from one location to another, and makes it available where it was not available before. This may include access to personal memories, or to other storage devices such as documents, research and reports.

Where several action tendencies compete for attention, further communication may facilitate a solution and finally lead to a **decision to close the issue** at least temporarily, so that the project can proceed by accommodating the resistance.

Some implications for empirical research on resistance

This framework of living systems allows us to elaborate the functional analogy between pain and resistance with reference to self-monitoring. Analogies are neither true nor false, but more or less fruitful. Scientific and technological projects are social activities that accommodate challenges from the outer and inner environment in the form of negative feedback signals. Scientific and technological contributions elicit reactions from other social systems; the system faces surprise, challenge and unexpected expectations. Public opinion and resistance are an 'irritating' environment for science and technology.

We observed how innovators and resistants act, think and communicate resistance; and how actions are informed by that communication. Communication directs and legitimates actions. Social systems act on various levels simultaneously. Cognition and communication elaborate symbolically pain, informal resistance and institutionalized resistance movements into events that have implications for immediate and future actions.

Resistance is an alarm signal that things are going wrong. The attention is redirected, the project is evaluated, and alterations occur, together with a temporary decline in the pace of action. The functional analogy to pain also helps us to see how resistance affects the institutional learning process. The communication process mediates the self-monitoring in a flexible and life-enhancing way with selection, schematization, increased self-awareness and multiple processing, and gives rise to the possibility of dysfunction. Resistance that is not communicated has little consequences for action. This framework leads to a number of implications for empirical research on resistance in the process of technological development:

- A characteristic of this way of looking at resistance is that it is self-referential – the theory speaks about its own contribution, it contains its own perspective as a variable. As such it represents a way of thinking that is not afraid of paradox.
- The empirical problem is twofold: how strong is resistance (quantitative content analysis), and how is its meaning symbolically constructed (qualitative analysis) by different actors?

- To elaborate the pain metaphor to an analogy is a theoretical and empirical challenge: what pain does for individual action, resistance does for the social process of technology.
- Because we are observing observers, communication of resistance is the focus of analysis. This symbolic activity is empirically related to significant changes in a technology project.
- The functional analysis shifts the focus from causes to effects. Effects seem to be less complex and more characteristic of resistance than the causes. Resistance is the independent variable, rather than the dependent one.
- For the problem of resistance we turn away from an interest in explaining resistance and in controlling it, to an interest in increasing its effective communication to the long-term benefit of the process of which it is a part.
- An implication of this approach is that we become interested in dynamics, cycles of events, and time-series analysis. The study of resistance to technology is reorientated, for example from sophisticated structural analysis of public opinion toward the analysis of events over time. Longitudinal analysis of public attention to issues is of core empirical interest. This includes time-series analysis of public opinion surveys, content of media coverage and policy developments on new technology.
- Functional analysis suggests the comparison of resistance, its presence or absence over time, and its long-term effects in varied national and technological contexts. For example, one could argue that the suppression of public resistance against nuclear power contributed to the Chernobyl disaster in 1986 by allowing lower safety standards.

In a functional analysis resistance is no longer a deficit, but a resource for the development of technology. This seems much in line with an old logic, often attributed to Edmund Burke, according to which a project may have to change in order to preserve itself. Resistance is the reality principle of any project that indicates the need for alterations, and points to the kind of alterations that are required.

Notes

1 Acknowledgements: I should like to thank to Jane Gregory, Alan Morton and Dan Wright for helpful readings of earlier drafts; however, the responsibility for the result is mine alone.
2 Performance indicators such as patents, funding, publications, staff/student ratios, and teaching efforts are attributed to particular institutional structures.
3 This state of affairs is excellently illustrated in Woody Allan's film 'Zelig' on the level of personality.
4 Similarly with information technology, Weizenbaum (1976), and Chargaff's book on genetic engineering (1978) provided a stimulus for social action. However these authors remained relatively isolated individuals within science and technology.

5 This similarity between pain and resistance does not imply that people resist because they are in pain. This may be the case, but it is a special case, which constitutes a homology rather than an analogy (Lorenz 1974; Cranach 1992).

6 This seems to me a more useful analogy than the one implied by common talk about the 'pains of change' which has different pragmatics. In the latter we assimilate events metaphorically to giving birth. This implies that pain is unavoidable and has to be endured to bring forward the new.

7 A striking phenomenon which urges a shift of analysis of pain from its cause to its effects is 'phantom limb pain'. People who have had a limb amputated may continue to **sense, localize** pain, and **identify** kinds of pains in the absent limb long after the healing of injured tissues. Such sensations may be related to touch on the remaining body part. The production of that perception reflects the internals of the pain system rather than a painful irritation (Howe 1983; Melzack & Wall 1988, pp. 65ff).

8 Resistance as a warning signal – the red flag, like pain – is an idea that, like most ideas, is not entirely new. I trace this idea back to Lawrence (1954) and to Klein (1966); however, they did not develop the metaphor idea any further to the level of analogy and hypothesis.

9 Schmidt (1985) explores historically what he called the 'resistance argument' in epistemology. It is an old argument put forward by realists against their opponents according to which reality shows itself compellingly in the material resistance against our wilful activity.

10 When analysing functions we have to keep in mind (a) that other functions are possible, (b) that the same structure may serve different functions on different levels, and (c) that functions can decay into dysfunctions.

11 Sometimes it is necessary to 'listen inside' in order to gain clarity of how 'outside' things really are, and to gain clarity of where to go; this may require the bracketing of preconceptions and traditions, and the refocusing of attention. The meditative traditions of all cultures seem to converge on this wisdom to solve problems. Pain and resistance stimulate and focus reflexive attention.

12 For an attempt to combine both analyses in one study, see Bauer (1993) on resistance to the implementation of a computer project in a Swiss bank.

13 I owe this idea to Harry Collins. He reflected on the mixed reception of his and Trevor Pinch's book *The Golem* at a seminar at the Science Museum in 1994.

14 This is the analogue of our capacity to control pain by adequate cognitions – a capacity that can be learnt, and is used in pain therapy (Melzack & Wall 1988).

15 This is an analogy to the gradual transition from acute to debilitating chronic pain under inadequate handling of pain.

References

BALFOUR, M. (1988).   *Withstanding Hitler in Germany 1933–45.* London: Routledge.

BATESON, G. (1972).   *Towards an ecology of mind.* San Francisco: Chandler.

BAUER, M. (1991).   Resistance to change – a monitor of new technology? *Systems Practice* **4**(3), 181–96.

BAUER, M. (1993).   Resistance to change. A functional analysis of responses to technical change in a Swiss bank. PhD thesis, London School of Economics.

BAUER, M., DURANT, J. & EVANS, G. (1994).   European public perceptions of science. *International Journal of Public Opinion Research* **6**(2), 163–86.

BECK, U. (1993).   *Der Erfindung des Politischen.* Frankfurt: Suhrkamp.

BERKELEY, D. & HUMPHREYS, P. (1982). Structuring decision problems and the 'bias' heuristic. *Acta Psychologica* **50**, 201–52.

BOULDING, K. (1956). *The image.* Ann Arbor: University of Michigan Press.

CHARGAFF, E. (1978). *Heraclitean fire: sketches from a life before nature.* New York: Rockefeller University Press.

COLLINS, H. M. & PINCH, T. (1993). *The Golem: what everyone should know about science.* Cambridge: Cambridge University Press.

CRANACH, M. von (1992). The multi-level organisation of knowledge and action – an integration of complexity. In *Social representations and the social bases of knowledge,* ed. M. von Cranach, W. Doise, and G. Mugny, pp. 10–22. Lewiston: Hogrefe and Huber.

CRANACH, M. von & OCHSENBEIN, G. (1985). Selbstüberwachungssysteme und ihre Funktion in der menschlichen Informationsverarbeitung. *Schweizerische Zeitschrift für Psychologie* **44**, 221–35.

CRANACH, M. von, OCHSENBEIN, G. & VALACH, V. (1986). The group as self-active system. *European Journal of Social Psychology* **16**, 193–229.

DOWNS, A. (1972). Up and down with ecology – the 'issue–attention cycle'. *Public Interest* **28**, 38–50.

EVERS, A. & NOWOTNY, H. (1987). *Über den Umgang mit Unsicherheit; die Entdeckung der Gestaltbarkeit der Gesellschaft.* Frankfurt: Suhrkamp.

FARR, R. M. & MOSCOVICI, S. (ed.) (1984). *Social representations.* Cambridge: Cambridge University Press.

FISCHHOFF, B. (1982). Debiasing. In *Judgement under uncertainty: heuristics and biases,* ed. D. Kahnemann, P. Slovic and A. Tversky, pp. 422–44. Cambridge: Cambridge University Press.

GAMSON, W. A. & MODIGLIANI, A. (1989). Media discourse and public opinion on nuclear power: a constructivist approach. *American Journal of Sociology* **95**, 1–37.

GIERYN, T. F. (1983). Boundary-work and the demarcation of science from non-science: strains and interests in professional ideologies of scientists. *American Sociological Review* **48**, 781–95.

HALTINGER, K. W. (1986). Widerstandsmotivation in der Schweizer Bevölkerung. *Allgemeine Schweizerische Militärzeitschrift* **7/8**, 403–9.

HEIDENSCHEDER, M. (1992). Zurechnung als soziologische Kategorie. Zu Luhmanns Verständnis von Handlung als Systemleistung. *Zeitschrift für Soziologie* **21**, 440–55.

HOWE, J. F. (1983). Phantom limb pain – a re-afferentation syndrome. *Pain* **15**, 101–7.

KAHNEMANN, D., SLOVIC, P. & TVERSKY, A. (ed.) (1982). *Judgement under uncertainty: heuristics and biases.* Cambridge: Cambridge University Press.

KLEIN, D. (1966). Some notes on the dynamics of resistance to change. The defender role. In *The planning of change,* ed. W. G. Bennis *et al.* New York: Holt, Rinehart, Winston Inc. (3rd edn, 1976.)

KUEPPERS, G. & KROHN, W. (1992). Zur Emergenz systemspezifischer Leistungen. In *Emergenz: Die Entstehung von Ordnung, Organisation und Bedeutung,* ed. W. Krohn and G. Kueppers. Frankfurt: Suhrkamp.

LATOUR, B. (1988). The prince for machines as well as for machinations. In *Technology and social process,* ed. B. Elliot, pp. 20–43. Edinburgh: Edinburgh University Press.

LAWRENCE, P. L. (1954).   How to overcome resistance to change. *Harvard Business Review* **32**(3), 49–57. (Reprint 1969.)

LEYDESDORFF, L. (1993).   'Structure'/'action' contingencies and the model of parallel distributed processing. *Journal for the Theory of Social Behaviour* **23**, 47–77.

LORENZ, K. (1974).   Analogy as a source of knowledge. *Science* **185**, 229–34.

LUHMANN, N. (1984).   *Soziale Systeme. Grundriss einer allgemeinen Theorie*. Frankfurt: Suhrkamp.

LUHMANN, N. (1992).   *Wissenschaft der Gesellschaft*. Frankfurt: Suhrkamp.

MAHAJAN, V. & PETERSON, R. A. (1985).   *Models for innovation diffusion*. Series 'Quantitative applications in the social sciences', No. 48. Beverley Hills, Calif: Sage.

MAZUR, A. (1975).   Opposition to technological innovations. *Minerva* **13**, 58–81.

MAZUR, A. (1984).   Media influences on public attitudes toward nuclear power. In *Public reactions to nuclear power: are there critical masses?* ed. W. R. Freudenberg and E. A. Rosa, pp. 97–114. Boulder, Colo: Westview Press.

MELZACK, R. & WALL, P. (1988).   *The challenge of pain*. Revised edition, Harmondsworth, UK: Penguin.

MILLER, J. G. (1986).   Can systems theory generate testable hypotheses? From Talcott Parsons to living systems theory. *Systems Research* **3**(2), 73–84.

MOSCOVICI, S. & HEWSTONE, M. (1983).   Social representations and social explanations: from naive to the amateur scientist. In *Attribution theory: social and functional extensions*, ed. M. Hewstone. Oxford: Blackwell.

NEIDHARDT, F. (1993).   The public as a communication system. *Public Understanding of Science* **2**, 339–50.

NORMAN, D. A. (1980).   Categorisation of action slips. *Psychological Review* **88**, 1–15.

OTTAWAY, R. N. (1983).   The change agent: a taxonomy in relation to the change process. *Human Relations* **36**, 361–92.

RASCHKE, J. (1988).   Soziale Bewegungen. Ein historisch–systematischer Grundriss. Frankfurt: Campus.

RHINOW, A. R. (1985).   *Widerstandsrecht im Rechtsstaat?* Staat und Politik, 30. Bern: Haupt.

ROGERS, E. M. (1983).   *Diffusion of innovations*, 3rd edn. New York: Free Press.

RÜDIG, W. (1990).   *Anti-nuclear movements: a world survey of opposition to nuclear energy*. London: Longmans.

SANER, H. (1988).   *Identität und Widerstand. Fragen in einer verfallenden Demokratie*. Basel: Lenos.

SANER, H. (1994).   Im Vorschein der Apokalypse. Grossrisiken und die Herausforderung an die Demokratie. *NZZ, Beilage 'Technologie und Gesellschaft'*, 26 January, 36 (Fernausgabe No. 20).

SCHMIDT, B. (1985).   *Das Widerstandsargument in der Erkenntnistheorie. Ein Angriff auf die Automatisierung des Wissens*. Frankfurt: Suhrkamp.

SCHMÄDEKE, J. & STEINBACH, P. (ed.) (1985).   *Der Widerstand gegen den Nationalsozialismus*. Munich: Piper.

SCOTT, J. C. (1985).   *Weapons of the weak. Everyday forms of peasant resistance*. New Haven, Conn: Yale University Press.

SHARP, G. (1973).   *The politics of nonviolent action*, three volumes. Boston: Porter Sargent.

SLOVIC, P., FISCHHOFF, B. & LICHTENSTEIN, S. (1984).   Perception and acceptability of risk from energy systems. In *Public reactions to nuclear power: are there critical masses?* ed. W. R. Freudenberg and E. A. Rosa, pp. 115–35. Boulder, Colo: Westview Press.

STRODTHOFF, G. G., HAWKINS, R. P. & SCHOENFELD, A. C. (1985).   Media roles in a social movement: a model of ideology diffusion. *Journal of Communication* **35**, 134–53.

TOURAINE, A. (1992).   *Critique de la modernité*. Paris: Fayard.

WALL, P. D. (1979).   On the relation of injury and pain. *Pain* **6**, 253–64.

WEIZENBAUM, J. (1976).   *Computers, power and human reason*. San Francisco: W. H. Freeman.

WILLKE, H. (1991).   *Systemtheorie*. Stuttgart: Fischer Verlag. UTB 1161.

WIPPERMANN, W. (1983).   *Der Europäische Faschismus im Vergleich 1922–1982*. Frankfurt: Suhrkamp.

WYNNE, B. (1993).   Public uptake of science: a case for institutional reflexivity. *Public Understanding of Science* **2**, 321–38.

Index